# Irish Studies and the Dynamics of Memory

# Reimagining Ireland

Volume 79

Edited by Dr Eamon Maher
Institute of Technology, Tallaght

PETER LANG
Oxford • Bern • Berlin • Bruxelles • Frankfurt am Main • New York • Wien

Marguérite Corporaal, Christopher Cusack
and Ruud van den Beuken (eds)

# Irish Studies and the Dynamics of Memory

## Transitions and Transformations

PETER LANG
Oxford • Bern • Berlin • Bruxelles • Frankfurt am Main • New York • Wien

Bibliographic information published by Die Deutsche Nationalbibliothek.
Die Deutsche Nationalbibliothek lists this publication in the Deutsche Nationalbibliografie; detailed bibliographic data is available on the Internet at http://dnb.d-nb.de.

A catalogue record for this book is available from the British Library.

Library of Congress Control Number: 2016954863

ISSN 1662-9094
ISBN 978-3-0343-2236-2 (print) • ISBN 978-1-78707-224-4 (ePDF)
ISBN 978-1-78707-225-1 (ePub) • ISBN 978-1-78707-226-8 (mobi)

Cover image: Fourteenth-century east window of the Dominican Priory in Athenry, County Galway, Ireland. Andreas F. Borchert.


Hochfeldstrasse 32, CH-3012 Bern, Switzerland
info@peterlang.com, www.peterlang.com, www.peterlang.net



This publication has been peer reviewed.

Printed in Germany

# Contents

# Figures

# Acknowledgements

This volume is a product of the *Irish Studies and the Dynamics of Memory: Transitions and Transformations* conference, which was held at Radboud University, Nijmegen, between 31 March and 2 April 2015, as part of the ERC-funded research project *Relocated Remembrance: The Great Famine in Irish (Diaspora) Fiction, 1847–1921* (grant agreement no. 262898-FAMINE). Most of the chapters in this volume were first presented as papers at this conference.

We would like to acknowledge the generous financial support offered by several partners, which was vital for the realization of the conference and this volume: the European Research Council, the Embassy of Ireland in the Netherlands, Radboud University's Department of English and the university's Institute for Historical, Literary and Cultural Studies.

We would like to thank all contributors to this volume for their stimulating scholarship. We are grateful to Jeanne Lenders for assisting us in editing the chapters. Finally, we would like to express sincere thanks to Eamon Maher, series editor for the *Reimagining Ireland* series, and Christabel Scaife, our editor at Peter Lang, for their support of this publication, their excellent guidance throughout its development and the pleasant collaboration.

Marguérite Corporaal
Christopher Cusack
Ruud van den Beuken

September 2016

MARGUÉRITE CORPORAAL, CHRISTOPHER CUSACK AND RUUD VAN DEN BEUKEN

# Introduction: Transitions and Transformations

In 'Home Sickness', a story from George Moore's collection *The Untilled Field* (1905), the old emigrant James Bryden reminisces about his childhood village in the West of Ireland: the bar-room in New York's bowery is 'forgotten [...] and the things he saw most clearly were the green hillside, and the bog lake and the rushes about it, and the greater lake in the distance, and behind it the blue lines of wandering hills'.[1] Bryden's nostalgia for his native community demonstrates the fact that emigrants' identities are often hyphenated, that is, divided between homeland and host society.[2] Simultaneously, the local-colour tale illustrates the significant role that memory plays in Irish and Irish diaspora cultures and societies – a phenomenon that Emilie Pine has described in *The Politics of Irish Memory* as the 'Irish cultural obsession with the past'.[3]

Over the past few years, the centrality of remembrance to Ireland's political and cultural landscape as well as to Irish communities worldwide could not be overlooked. The launch of an annual international Famine commemoration day in 2009, which takes place on both sides of the Atlantic, illustrates the urge to remember even the most painful aspects of what Jan Assmann terms the 'fateful events of the past'.[4] Moreover, the fact that, upon the successful end of the bailout in December 2013,

1 George Moore, 'Home Sickness', in *The Untilled Field* (London: T. F. Unwin, 1903), 48–9.
2 Angelika Bammer, 'Introduction', in Angelika Bammer, ed., *Displacements: Cultural Identities in Question* (Bloomington, IN: Indiana University Press, 1994), xii.
3 Emilie Pine, *The Politics of Irish Memory: Performing Remembrances in Contemporary Irish Culture* (Basingstoke: Palgrave Macmillan, 2011), 16.
4 Jan Assmann, 'Collective Memory and Cultural Identity', *New German Critique*, 65 (1995), 129.

Finance Minister Michael Noonan stated that Ireland's financial crash was 'the greatest crisis that this country has experienced since the famine',[5] bears witness to the ways in which present conditions in Ireland are often translated into what Oona Frawley has called the narrative 'scripts' and tropes of past forms of remembering.[6]

Additionally, the present 'decade of centenaries' underscores the urge to re-enact and politicize the past, fostering grand narratives about the emergence of Ireland as a postcolonial nation. The 1916 centenary commemoration saw a wide range of activities that sought to transmit the memory of the past event to present generations, including a nationwide Proclamation Day at schools on 15 March during which flags were hoisted and the Proclamation was read, as well as three exhibitions in Dublin at the National Museum, Collins Barracks, the General Post Office, and the Rotunda Hospital respectively. Such scheduled events often evoke the past with a view to creating a heightened awareness of national identities: for example, Comhrá 2016, a series of seminars organized by Conradh na Gaeilge, commemorated the importance of the Irish language and the role of the Gaelic League as a source of inspiration for the Easter Rising.[7]

Serving to enhance a sense of national identity, the Easter Rising is also being commodified: visitors to one of the three Dublin exhibitions can purchase memorabilia such as mugs, badges, coins, as well as statuettes; T-shirts featuring the Proclamation along with images of the ravaged GPO can be ordered; and even chocolate bars in wrappers depicting the 1916 martyrs are widely available. Such manifestations of memory on the commercial market illustrate what Bill Brown in *A Sense of Things* (2008) calls the material objectification of the cultural past,[8] that often embodies a

5 'Ireland Prepares to Exit Bailout after "biggest crisis since the Famine" – As It Happened', *The Guardian* (13 December 2013).

6 Oona Frawley, 'Toward a Theory of Cultural Memory in an Irish Postcolonial Context', in *Memory Ireland, Volume 1: History and Modernity*, ed. Oona Frawley (Syracuse, NY: Syracuse University Press, 2010), 30.

7 'Easter Rising Commemorations: 50 Events for 2016', *Irish Times* (13 January 2016) <http://www.irishtimes.com/culture/heritage/easter-rising-commemorations-50-events-for-2016-1.2492549> accessed 20 September 2016.

8 Bill Brown, *A Sense of Things: The Object Matter of American Literature* (Chicago, IL: University of Chicago Press, 2003), 184.

national consciousness, and that can function as a strategy through which painful as well as celebratory 'emotions [...] can be objectified'.[9]

This commercialization of 1916 memory and the souvenir industry that thus sells 'Irishness' has elicited a variety of responses. The commodification of the Rising has been satirized by Ulster artist Rita Duffy in her project *The Souvenir Shop*, which exhibited Mexican grave candles that venerate Patrick Pearse, and which gave visitors the opportunity 'to buy a vintage-style paper cutout "make your own Markievicz" doll'.[10] At the same time, the legacies of the Easter Rising have been appropriated by Sinn Féin in its launch of the party's own programme of events to mark the centenary, including a re-enactment of Jeremiah O'Donovan Rossa's funeral at Glasnevin cemetery on 1 August. The promotional film they aired to announce this commemorative programme ends with the slogan 'Join us in building their republic – your republic', a clear politicization of 1916 memories to rally further support for the party's republican ideals.[11] This reveals the dynamics of recollection according to which '[m]emory is used strategically: not merely to explain the group past but also to transform it into a reliable identity source for the group present.'[12]

## Irish Memory Studies: Trends and Topics

In recent years, scholarly interest in Irish memory has continued to grow exponentially. This was partially motivated by the 'memory fever' in the humanities,[13] but also by the legacies of the Troubles, institutional abuse

9 Daniel Miller, *The Comfort of Things* (London: Polity Press, 2008), 26.

10 Vicky Cosstick, 'Laundered Diesel and Black and Tan Boot Polish: The Rising Gets an Artful Injection of Mischief', *Irish Times* (5 April 2016). We are grateful to Peter Gray for directing us to this article.

11 <https://www.youtube.com/watch?v=ojw8KQmvE1U>.

12 Barbara Misztal, 'Memory and History', in Oona Frawley, ed., *Memory Ireland, Volume 1: History and Modernity* (Syracuse, NY: Syracuse University Press, 2010), 3.

13 Andreas Huyssen, *Twilight Memories: Marking Time in a Culture of Amnesia* (London: Routledge, 1995), 7.

and the aforementioned 'decade of centenaries'. Oona Frawley's four-volume *Memory Ireland* series (2010–2014) and the Irish Memory Studies Research Network, established at University College Dublin, are just a few examples illustrating the mnemonic trend in research on Irish and Irish diaspora societies. While this research on Irish memory has addressed significant 'figures of memory',[14] or 'cruxes'[15] in Ireland's past, such as the 1798 rebellion, the Great Famine, the Easter Rising and the Troubles, it has also directed attention to three no less relevant issues: popular memory, transgenerational memory, and contested and/or traumatic memory.

Memory studies experts, including Jefferson A. Singer and Martin A. Conway, have made a distinction between 'availability'and 'accessiblity' of memory, showing that not all recollections are institutionalized and preserved,[16] and other memory scholars such as Aleida Assmann have pointed to the crucial role of canonization in the preservation, dissemination, and public recognition of memories.[17] Popular forms of memory – in oral traditions and folklore – are often not among those canonized recollections of the past, but in the case of Ireland, these alternative mnemonic registers have played a significant role in shaping people's identities.[18]

Several recent seminal studies have explored the reverberations of the past in Irish folklore. Guy Beiner's award-winning *Remembering the Year of the French: Irish Folk History and Social Memory* (2007), for example, examines the memory of the 1798 United Irishmen rebellion in vernacular or 'folk history', taking as its starting point Maurice Halbswachs's

14 Assmann, 'Collective Memory and Cultural Identity', 129.

15 Oona Frawley, 'Introduction: Cruxes in Irish Cultural Memory: The Famine and the Troubles', in Oona Frawley, ed., *Memory Ireland, Volume 3: The Famine and the Troubles* (Syracuse, NY: Syracuse University Press, 2014), 1–17.

16 Jefferson A. Singer and Martin A. Conway, 'Should We Forget Forgetting?', *Memory Studies*, 1(3) (2008), 282.

17 Aleida Assmann, 'Canon and Archive', in Astrid Erll and Ansgar Nünning, eds, *Cultural Memory Studies: An International and Interdisciplinary Handbook* (Berlin; New York: De Gruyter, 2008), 97–109.

18 See Diarmuid Ó Giolláin, *Locating Irish Folklore: Tradition, Modernity, Identity* (Cork: Cork University Press, 2000), 63–94.

well-known concept of 'social memory'.[19] Ray Cashman's *Storytelling on the Northern Irish Border: Characters and Community* (2008) also elaborates on Halbswachs's influential work, investigating the role of popular memory in the stories that people tell each other in Aghyaran, a mixed Catholic-Protestant border community in Northern Ireland, during wakes and ceilis. As Cashman shows, these remembered tales about local characters play a role in processes of identity and community formation that transcend religious sectarianism.[20]

Research on transgenerational mediations of memory has received new impetus from the introduction of terminology that helps explain processes of memory transfer across time. Marianne Hirsch's 'postmemory', a term that describes the ways in which a past is reconfigured by the children of those who directly experienced an event,[21] and Alison Landsberg's term 'prosthetic memory', which extends Hirsch's concept to anyone who has indirect memories of an event,[22] have been adopted by various Irish studies scholars. For instance, Richard S. Grayson and Fearghal McGarry's recent collection of essays, *Remembering 1916: The Easter Rising, the Somme and the Politics of Memory in Ireland* (2016) contains contributions by Guy Beiner and Fearghal McGarry that deal with the 'postmemory' of these 1916 events.[23] Gearóid Ó hAllmhuráin's *Flowing Tides: History and Memory in*

19 Guy Beiner, *Remembering the Year of the French: Irish Folk History and Social Memory* (Madison, WI: University of Wisconsin Press, 2007), 9, 30.

20 Ray Cashman, *Storytelling on the Northern Irish Border: Characters and Community* (Bloomington, IN: Indiana University Press, 2008).

21 See *Family Frames: Photography, Narrative and Postmemory* (Cambridge, MA: Harvard University Press, 1997), 22. For a further discussion of this concept, see Marianne Hirsch and Leo Spitzer, '"We Would Not Have Come Without You": Generations of Nostalgia', in Katharine Hodgkin and Susannah Radstone, eds, *Memory, History, Nation: Contested Pasts* (New Brunswick, NJ: Transaction Publishers, 2006), 79–97.

22 See Alison Landsberg, *Prosthetic Memory: The Transformation of American Remembrance in the Age of Mass Culture* (New York: Columbia University Press, 2004), 2.

23 Guy Beiner, 'Making Sense of Memory: Coming to Terms with Conceptualizations of Historical Remembrance', 21–3. See also Fearghal McGarry, 'Hard Service: Remembering the Abbey Theatre's Rebels', 86–112. Both in Richard S. Grayson and

*an Irish Soundscape* (2016) investigates the cultural memory of Irish traditional music, focusing especially on Co. Clare as a site of musical memory and employing the notion of 'prosthetic memory' to describe how musicians worldwide relate to these musical legacies.[24]

A strong engagement with theoretical frameworks on memory can also be found in research on problematic episodes from the Irish past. Emilie Pine's aforementioned *The Politics of Irish Memory* (2011) discusses the cultural memory of institutional abuse in film and drama against the background of trauma theory. Her book also contains a chapter on the controversial cultural legacies of Bobby Sands and the other IRA hunger strikers, and much research in Irish studies that deals with concepts of traumatic pasts concerns the Northern Irish Troubles. Graham Dawson's *Making Peace with the Past?: Memory, Trauma and the Irish Troubles* (2007) examines the function of memories of political violence in Northern Ireland before, during, and after the Good Friday Agreement. His study provides examples of how competing narratives of the past, whether in the form of personal or cultural memory, operate in contexts of 'political transition'.[25] Similarly, Cillian McGrattan's *Memory, Politics and Identity: Haunted by History* (2013) traces the role of memory related to violent, disruptive events (such as Bloody Friday and Bloody Sunday in Northern Ireland) after the peace process, while also addressing the complex issue of truth recovery.[26]

The existence of competing memories in relation to troubled pasts was also central to the exhibition *Art of the Troubles* (2014), at the Ulster museum in Belfast. Displaying works of art created during and after the Troubles, which reinterpret the events in retrospect, the exhibition included

---

Fearghal McGarry, eds, *Remembering 1916: The Easter Rising, the Somme and the Politics of Memory in Ireland* (Cambridge: Cambridge University Press, 2016).

24 Gearóid Ó hAllmhuráin, *Flowing Tides: History and Memory in an Irish Soundscape* (Oxford: Oxford University Press, 2016), 235–48.

25 Graham Dawson, *Making Peace with the Past?: Memory, Trauma and the Irish Troubles* (Manchester: Manchester University Press, 2007), 4.

26 Cillian McGrattan, *Memory, Politics and Identity: Haunted by History* (Basingstoke: Palgrave Macmillan, 2013), 123–45.

contributions by artists from both loyalist and republican, Protestant and Catholic backgrounds, such as Philip Napier, John Keane, and Rita Duffy. Significantly, the exhibition was co-curated by staff from the Institute of Irish-British Studies at University College Dublin, and complemented by an academic conference on the subject that was hosted by the museum. *Art of the Troubles* is by no means the only project aimed at translating and disseminating the past and making visible processes of memory in which academics specializing in Irish history, literature, or culture played a pivotal role. In fact, one of the current trends in Irish memory studies is a progressive integration of academic discourses on memory studies and cultural practices through collaborations between social and cultural institutions and scholars. Examples are, *inter alia*, the present partnership between the National Museums of Northern Ireland, Queen's University Belfast, and the University of Ulster in the establishment of the First World War Coordinating Centre, *Living Legacies*; the contributions by historians Mary Daly and Emmett O'Connor to the six-part documentary series *Citizens' Lockout, 1913–2013* that was broadcast on RTÉ Radio 1; and the participation of scholars such as Luke Gibbons and Fearghal McGarry in documentaries shown as part of the *1916 Rising* exhibition at the GPO Witness History Visitor Centre in Dublin.

Central to many such explorations of the Irish past is the language of trauma. Indeed, since the Great Famine's sesquicentenary in the 1990s, the notion of trauma has become a dominant paradigm in the study of Irish cultural memory. As numerous scholars of collective trauma have shown, the notion that a community or society has undergone a trauma contributes to social cohesion and the construction of cultural identities. In the words of Jeffrey C. Alexander, '[p]rojected as ideologies that create new ideal interests, trauma narratives can trigger significant repairs in the civil fabric'.[27] In the Irish (and Irish-diasporic) context, the Famine in particular has been imagined as what Dominick LaCapra calls a 'founding trauma', 'the valorized or intensely cathected basis of identity for an

27 Jeffrey C. Alexander, *Trauma: A Social Theory* (Cambridge: Polity Press, 2012), 2.

individual or a group'.[28] Nevertheless, the use of trauma as an interpretive category for the study of the Irish past can be problematic. While the notion of cultural trauma has been a useful instrument both in shaping and analysing narratives about influential events such as the Famine and the Troubles, many commentators do not differentiate between individual psychological and collective cultural trauma, and uncritically appropriate the (largely Freudian) language of psychological, individual trauma, with its focus on psychological repression and 'working through', using what Joseph Valente describes as 'a loosely psychoanalytical model of repressed memory syndrome' to construe how certain events have impacted on Irish society and culture.[29] While it is beyond the scope of this introduction to address this topic in greater detail, scholars working on Irish memory must be aware of the problematic nature of the discourse of trauma in relation to Irish culture. Critics such as David Lloyd, Margaret Kelleher, Emily Mark-FitzGerald and Oona Frawley have interrogated and complicated the use of the rhetoric of trauma in the Irish context, and their work demonstrates how important it is to remain critical about the concepts and terminology we use to make sense of the past.[30]

## Future Directions

One of the major developments in memory research is a focus on transnational or transcultural memory. Rooted in the premise that 'migrants carry their heritage, memories and traumas with them', which 'are transferred and

28 Dominick LaCapra, *Writing History, Writing Trauma* (Baltimore, MD: Johns Hopkins University Press, 2001), 23.

29 Joseph Valente, 'Ethnostalgia: *Irish Hunger* and Traumatic Memory', in Oona Frawley, ed., *Memory Ireland, Volume 3: The Famine and the Troubles* (Syracuse, NY: Syracuse University Press, 2014), 174–5.

30 For a good recent analysis of the role of trauma in the study of the Famine, see Frawley, 'Introduction: Cruxes in Irish Cultural Memory', 4–11.

brought into new social constellations and political contexts',[31] scholars are increasingly interested in the emergence of shared, transnational memories between countries of birth and settlement. More recently, however, scholars have interpreted transcultural memory beyond the context of emigration, to include any form of shared recollections that transcend communal borders or acts of memory transfer between societies. In her contribution to the collection *Transcultural Memory* (2014), Astrid Erll, for example, argues against any approach which ties memory to 'clear-cut territories and social formations', since it 'circulates *across* [...] and also *beyond* cultures',[32] and Lucy Bond and Jessica Rapson have defined transcultural memory as 'the travelling of memory *within* and *between* national, ethnic and religious collectives'.[33] Furthermore, scholars such as Michael Rothberg have particularly looked at the interactions and intersections between different memory cultures, arguing in favour of a perception of the public sphere as a fluid space in which memory is essentially 'multidirectional'.[34]

The transnational turn in memory studies in general has also left its imprint on Irish memory studies, and the past few years have seen the publication of some important studies that broaden the issue of memory to include diaspora communities. Emily Mark-FitzGerald's *Commemorating the Famine: Memory and the Monument* (2013) examines Famine memorials in the Republic of Ireland, Northern Ireland, Canada, the United States and Australia from a comparative perspective, thereby laying bare important transnational mnemonic registers.[35] Cian T. McMahon's study

31 Aleida Assmann and Sebastian Conrad, 'Introduction', in Aleida Assmann and Sebastian Conrad, eds, *Memory in a Global Age: Discourses, Practices and Trajectories* (Basingstoke: Palgrave Macmillan, 2010), 2.

32 Astrid Erll, 'Travelling Memory', in Rick Crownshaw, ed., *Transcultural Memory* (London: Routledge, 2014), 20–4. Italics in original.

33 Lucy Bond and Jessica Rapson, 'Introduction', in Lucy Bond and Jessica Rapson, eds, *The Transcultural Turn: Interrogating Memory Between and Beyond Borders* (Berlin: De Gruyter, 2014), 19.

34 Michael Rothberg, *Multidirectional Memory: Remembering the Holocaust in the Age of Decolonization* (Stanford, CA: Stanford University Press, 2009), 5, 11.

35 Emily Mark-FitzGerald, *Commemorating the Famine: Memory and the Monument* (Liverpool: Liverpool University Press, 2013).

*The Global Dimensions of Irish Identity: Race, Nation, and the Popular Press* (2015) also employs a transnational scope by showing how memories shared by communities of Irish in Ireland, Australia, and North America were revived in the popular press to create a transcultural sense of identity.[36]

While such studies make important strides in looking beyond Ireland in their investigation of recollection and commemoration, the general focus in scholarship and cultural initiatives remains primarily tied to the idea of a rather monolithic shared Irish culture and identity. Conversely, the ways in which Ireland's pasts have been reconfigured by contact with the recollections of other cultural communities – through the settlement of Irish people in other parts of the world, through the arrival of immigrants in Ireland, and through the present-day media – could receive further attention in Irish memory studies. Overcoming the sense of insularity that emanates from many recent projects, such as the newly established diaspora museum Epic Ireland in Dublin (2016) which mainly focuses on Irish cultural memory as an export product and fails to address the contributions by immigrants to Irish reimaginations of the past, might yield refreshing insights on the complex dynamics of memory and encourage Irish studies in general to move beyond the postcolonial and national paradigms that have dominated the field.

## The Outline of this Volume

This volume is divided into four parts. The first part, 'Commemorative Practices', comprises three essays which analyse the ways particular figures and events in Irish history have been commemorated – or reappropriated – by means of a variety of cultural expressions. Ruud van den Beuken's essay explores the representation of Jonathan Swift and Daniel O'Connell in two plays by Edward Longford that were performed at the Dublin Gate

36 Cian T. McMahon, *The Global Dimensions of Irish Identity: Race, Nation, and the Popular Press* (Chapel Hill, NC: University of North Carolina Press, 2015).

Theatre in the 1930s. Van den Beuken argues that Longford's historical plays use complex memorial strategies to reflect on the sectarianism of Ireland in the contemporary Irish Free State. Tracy Fahey's article reports on a research project which studied the folkloric memory of the burning of Wildgoose Lodge in 1816. Considering a number of varying oral accounts of this agrarian crime that have been transmitted from generation to generation, Fahey casts light on the dynamic between 'official' historical narratives and local folkloric memories, and underscores the usefulness of such social memory. In her article on the role of photography in the commemoration of the leaders of the Easter Rising, Gail Baylis explores how this quintessentially modern technology contributed to the mediation of the memory of the insurrection and the martyrization of its commanders. In her essay, she thus combines a focus on the materiality of memory with its politicization.

The essays in the second part, 'Contested Politics', consider the role of politics in the commemoration of certain historical events, looking at a variety of modes of recollection. Eve Morrison's article addresses the interfaces between various types of representations of the War of Independence and the subsequent Civil War, including personal testimonies and folklore. Paying particular attention to images of haunting in her source material, Morrison shows the value of approaching such sources from a variety of angles. In his essay on the work of Francis Stuart and John McGahern, Eamon Maher explores the dialectical relationship between autobiography and fiction in their works. Maher is particularly interested in the ways memories of the War of Independence and the Civil War impinge on the oeuvres of these writers, and how the problems inherent to recollection influence their works. Located at the intersection between political science and memory studies, Sara Dybris McQuaid's article uses the Northern Irish Parades Commission as a case study for an integrated methodological framework which combines these two disciplines. Analysing how the policies of the Parades Commission interact with (sectarian) cultural memory, McQuaid concludes that such dynamics can only be fully brought to light by means of a synergetic approach. Concluding this section, Stephen Hopkins analyses the tension between the 'official' republican memory that was formed and disseminated following the Good Friday

Agreement by Northern Irish republican leaders, and 'dissident' memories which counter this formalization of accounts of the past. Hopkins suggests that the conflict between official and dissenting accounts of the past will not abate as long as the republican movement's traditional goal has not been realized.

The third part, 'Memory and Trauma', looks at the inscription of memory and trauma through fiction and theatre. Niamh NicGhabhann's article uses Edna O'Brien's short fiction to explore the ways the representation of public space interacts with the construction and dissemination of memory. Focusing on the depiction of the asylum in O'Brien's work, NicGhabhann argues that such public spaces both produce and are invested with meaning in a variety of ways. Emilie Pine's essay demonstrates how documentary theatre can play an important role in the negotiation of traumatic pasts, such as the child abuse in Irish institutions run by the Catholic Church as exposed by the 2009 Ryan Report. Theorizing the notion of 'theatrical witnessing', Pine argues that plays such as Mary Raftery's *No Escape* (2010) contribute to the development of cultural memories and public awareness about important social issues. In his essay on two plays by Enda Walsh, Nelson Barre reflects on the connection between memorialization and performance, arguing that Walsh's plays problematize the notion of a stable historical truth. As Barre shows, repetition and performance function to create personal realities which can be highly variable and depend on the continual revision of memory and the past.

The final part, 'Theoretical Developments', features contributions which suggest directions that Irish studies scholars might explore in the future through various critical engagements with memory theory. Concentrating on Irish North-American fiction, the article by Marguérite Corporaal, Christopher Cusack and Lindsay Janssen demonstrates how the cultural memory of the Great Famine in fiction has developed over time and across space. Analysing how memory is reconfigured transnationally and transgenerationally, their essay demonstrates how a dynamic approach to memory studies and theories about (diasporic) cultural identity might benefit the study of the Irish past. Graham Dawson's expansive contribution surveys the temporalities of memory in post-conflict Northern Ireland, highlighting the complex interplay – and often conflict – between past events, memories of these events and future aims which complicate Northern Irish efforts to come to terms with the legacies of the Troubles. In so doing, his article also considers the ways affect and emotion influence the move towards social reconciliation. Concluding this volume, Guy Beiner's essay theorizes the notion of disremembering and suggests a number of ways it might be applied in an Irish studies context. Beiner introduces the notion of 'social forgetting' to conceptualize the tension between private remembrance and public silence, and, by extension, the dialectic between memory and amnesia, which marks not just Irish engagements with the past but is in fact a core element of any memorial culture.

Taken together, these essays showcase the fecundity of Irish memory studies. Moreover, in the ways they engage with recent developments in memory theory and critically interrogate the manifestations and functions of memory in Irish and Irish-diasporic culture and society, they also suggest many avenues for further research. While this volume thus shows that Irish memory studies is thriving, it also intends to underline the editors' belief that Ireland's history and contemporary society, no less than its multifaceted culture and its transnational legacies, will continue to lend themselves to innovative critical engagements. After all, the old hairy chestnut has it that 'the Irish never forget'. And yet, while sometimes the Irish actually *do* forget, in Ireland, too, Mnemosyne is mother to all Muses.

## Bibliography

Alexander, Jeffrey C., *Trauma: A Social Theory* (Cambridge: Polity Press, 2012).

Assmann, Aleida, 'Canon and Archive', in Astrid Erll and Ansgar Nünning, eds, *Cultural Memory Studies: An International and Interdisciplinary Handbook* (Berlin and New York: De Gruyter, 2008), 97–109.

——, and Sebastian Conrad, 'Introduction', in Aleida Assmann and Sebastian Conrad, eds, *Memory in a Global Age: Discourses, Practices and Trajectories* (Basingstoke: Palgrave Macmillan, 2010), 1–16.

Assmann, Jan, 'Collective Memory and Cultural Identity', *New German Critique*, 65 (1995), 125–33.

Bammer, Angelika, 'Introduction', in Angelika Bammer, ed., *Displacements: Cultural Identities in Question* (Bloomington, IN: Indiana University Press, 1994), xi–xx.
Beiner, Guy, 'Making Sense of Memory: Coming to Terms with Conceptualizations of Historical Remembrance', in Richard S. Grayson and Fearghal McGarry, eds, *Remembering 1916: The Easter Rising, the Somme and the Politics of Memory in Ireland* (Cambridge: Cambridge University Press, 2016), 13–23.
——, *Remembering the Year of the French: Irish Folk History and Social Memory* (Madison, WI: University of Wisconsin Press, 2007).
Bond, Lucy, and Jessica Rapson, 'Introduction', in Lucy Bond and Jessica Rapson, eds, *The Transcultural Turn: Interrogating Memory Between and Beyond Borders* (Berlin: De Gruyter, 2014), 1–26.
Brown, Bill, *A Sense of Things: The Object Matter of American Literature* (Chicago, IL: University of Chicago Press, 2003).
Cashman, Ray, *Storytelling on the Northern Irish Border: Characters and Community* (Bloomington, IN: Indiana University Press, 2008).
Cosstick, Vicky, 'Laundered Diesel and Black and Tan Boot Polish: The Rising Gets an Artful Injection of Mischief', *Irish Times* (5 April 2016).
Dawson, Graham, *Making Peace with the Past?: Memory, Trauma and the Irish Troubles* (Manchester: Manchester University Press, 2007).
'Easter Rising Commemorations: 50 Events for 2016', *The Irish Times* (13 January 2016) <http://www.irishtimes.com/culture/heritage/easter-rising-commemorations-50-events-for-2016-1.2492549> accessed 20 September 2016.
Erll, Astrid, 'Travelling Memory', in Rick Crownshaw, ed., *Transcultural Memory* (London: Routledge, 2014), 9–23.
Frawley, Oona, 'Introduction: Cruxes in Irish Cultural Memory: The Famine and the Troubles', in Oona Frawley, ed., *Memory Ireland, Volume 3: The Famine and the Troubles* (Syracuse, NY: Syracuse University Press, 2014), 1–17.
——, 'Toward a Theory of Cultural Memory in an Irish Postcolonial Context', in *Memory Ireland, Volume 1: History and Modernity*, ed. Oona Frawley (Syracuse, NY: Syracuse University Press, 2010), 18–34.
Hirsch, Marianne, *Family Frames: Photography, Narrative and Postmemory* (Cambridge, MA: Harvard University Press, 1997).
——, and Leo Spitzer, '"We Would Not Have Come Without You": Generations of Nostalgia', in Katharine Hodgkin and Susannah Radstone, eds, *Memory, History, Nation: Contested Pasts* (New Brunswick, NJ: Transaction Publishers, 2006), 79–97.
Huyssen, Andreas, *Twilight Memories: Marking Time in a Culture of Amnesia* (London: Routledge, 1995).

'Ireland Prepares to Exit Bailout after "biggest crisis since the Famine" – As It Happened', *The Guardian* (13 December 2013).

LaCapra, Dominick, *Writing History, Writing Trauma* (Baltimore, MD: Johns Hopkins University Press, 2001).

Landsberg, Alison, *Prosthetic Memory: The Transformation of American Remembrance in the Age of Mass Culture* (New York: Columbia University Press, 2004).

McGarry, Fearghal, 'Hard Service: Remembering the Abbey Theatre's Rebels', in Richard S. Grayson and Fearghal McGarry, eds, *Remembering 1916: The Easter Rising, the Somme and the Politics of Memory in Ireland* (Cambridge: Cambridge University Press, 2016), 86–112.

McGrattan, Cillian, *Memory, Politics and Identity: Haunted by History* (Basingstoke: Palgrave Macmillan, 2013).

McMahon, Cian T., *The Global Dimensions of Irish Identity: Race, Nation, and the Popular Press* (Chapel Hill, NC: University of North Carolina Press, 2015).

Mark-FitzGerald, Emily, *Commemorating the Famine: Memory and the Monument* (Liverpool: Liverpool University Press, 2013).

Miller, Daniel, *The Comfort of Things* (London: Polity Press, 2008).

Misztal, Barbara, 'Memory and History', in Oona Frawley, ed., *Memory Ireland, Volume 1: History and Modernity* (Syracuse, NY: Syracuse University Press, 2010), 3–17.

Moore, George, 'Home Sickness', in *The Untilled Field* (London: T. F. Unwin, 1903).

Ó Giolláin, Diarmuid, *Locating Irish Folklore: Tradition, Modernity, Identity* (Cork: Cork University Press, 2000).

Ó hAllmhuráin, Gearóid, *Flowing Tides: History and Memory in an Irish Soundscape* (Oxford: Oxford University Press, 2016).

Pine, Emilie, *The Politics of Irish Memory: Performing Remembrances in Contemporary Irish Culture* (Basingstoke: Palgrave Macmillan, 2011).

Rothberg, Michael, *Multidirectional Memory: Remembering the Holocaust in the Age of Decolonization* (Stanford, CA: Stanford University Press, 2009).

Singer, Jefferson A., and Martin A. Conway, 'Should We Forget Forgetting?', *Memory Studies*, 1(3) (2008), 279–85.

Valente, Joseph, 'Ethnostalgia: *Irish Hunger* and Traumatic Memory', in Oona Frawley, ed., *Memory Ireland, Volume 3: The Famine and the Troubles* (Syracuse, NY: Syracuse University Press, 2014), 174–92.

PART I

# Commemorative Practices

RUUD VAN DEN BEUKEN

# 1 Remembering the Drapier and King Dan: The Sectarian Legacies of Swift and O'Connell in Edward Longford's *Yahoo* (1933) and *Ascendancy* (1935)

During the Earl of Longford's six-year tenure as chairman of the Dublin Gate Theatre's board of directors, Hilton Edwards and Micheál MacLiammóir not only benefited from his financial generosity, which had warded off the Gate's impending bankruptcy in 1930 and kept the budding avant-garde playhouse afloat during the subsequent decade, but also produced several plays that the Earl had written, translated or adapted to the stage. The latter category includes his rendition of Sheridan Le Fanu's vampire thriller *Carmilla* (1932), with a cast featuring Cyril Cusack and Coralie Carmichael, while Edwards and MacLiammóir themselves played the roles of Aigisthos and Orestes respectively in Longford's translation of Aeschylus's *Agamemnon* (1933). No less importantly, the Gate's founders also produced three original plays by the Earl of Longford, two of which engage with the contested legacies of self-professed Irish champions of liberty: *Yahoo* (1933), a historical play about Jonathan Swift that addresses the Dean's attempts to transcend religious divides in his vision of Irish nationhood, and *Ascendancy* (1935), which depicts the religious tensions that intensified in the wake of Daniel O'Connell's campaign for Catholic Emancipation.

Although the historical settings of these plays might seem incongruous, ranging from the early eighteenth to the early nineteenth century, with yet another century separating Longford's contemporary audiences from the events depicted in *Ascendancy*, their pertinence to Irish theatregoers in 1930s cannot be understated. By questioning the discriminatory discourses that imbued various conflicting representations of Swift's and O'Connell's historical legacies, both plays addressed the ways in which contemporary postcolonial constructions of Irish nationhood were likewise compromised by unremitting sectarian strife. In this way, Longford's historical dramas might be argued to have employed potent memory strategies to articulate a more inclusive conceptualization of Irish identities that served as an alternative to the bitter political quarrels that continued to divide the newly independent nation.

## 'Leave the Dean in the obscurity he deserves': Jonathan Swift Contested in *Yahoo* (1933)

In 1724 and 1725, Jonathan Swift pseudonymously published the Drapier's Letters, a polemic series of pamphlets that decried the patent that King George I had granted to William Wood, an ironmonger, to mint base copper coins that threatened to destabilize the Irish economy.[1] Although Dublin Castle issued a reward to be paid to the man who would reveal the true identity of 'M. B. Drapier', Swift's authorship only became publicly known after Wood's patent had been annulled.[2] Hailed by contemporaries

1 Jonathan Swift, 'Drapier's Letter I': 'Mr WOOD made his HALFPENCE of such *base* metal, and so much smaller than the English ones, that the *brazier* would not give you above a *penny* of good money for a *shilling* of his; so that the sum of *four-score* and *ten thousand pounds* in good gold and silver, must be given for TRASH that will not be worth above *eight* or *nine thousand pound* real value' (423–4). For an explanation of this controversy and its historical roots, see Carole Fabricant, 'Swift the Irishman', 56–7; and the notes to the Drapier's Letters in Jonathan Swift, in Angus Ross and David Woolley, eds, *Major Works*, 669–71.

2 Despite this danger, Swift not only wrote a personal letter to Lord Carteret, the 2nd Earl Granville and Lord Lieutenant of Ireland, to apprise the viceroy of his opposition to Wood's patent, but also included a copy of his first Drapier Letter, cunningly claiming that it 'is entitled to a weaver, and suited to the vulgar, but thought to be the work of a better hand'. See 'Swift to Lord Carteret' (28 April 1724), in *Jonathan Swift: The Major Works*, 431.

as an act of national heroism, Swift's polemic has come to be interpreted as a turning point in Irish history; according to Joseph McMinn, the Dean of St Patrick's Cathedral had granted 'English-speaking Ireland a new sense of a separate identity, one which would enter into the mainstream of later Irish nationalism'.[3] Likewise, Carole Fabricant has commented on the remarkable comprehensiveness of the notion of Irishness in the Drapier's Letters, arguing that 'Swift's outlook extends beyond the boundaries of a narrowly defined 'colonial nationalism' to embrace a more expansive vision, one that undoubtedly assumed the continued hegemony of the Anglo-Irish elite but that also makes room for a range of other groups in Irish society'.[4]

In 1933, Hilton Edwards portrayed Jonathan Swift in the Earl of Longford's third Gate play, *Yahoo*, which is set in 1724 and depicts the Dean's decision to challenge Wood's patent by writing the Drapier's Letters.[5]

3 Joseph McMinn, 'Swift's Life', in *The Cambridge Companion to Jonathan Swift* (2003), 25.

4 Carole Fabricant, 'Swift the Irishman', in *The Cambridge Companion to Jonathan Swift* (2003), 57.

5 *The Irish Press* commented favourably on the Gate's promotion of original Irish plays: 'Was there ever a theatre more bravely adventurous than the Dublin "Gate"? After the piratical romanticism of *Grania of the Ships* they present another new play, and again it is a play by an Irish author and with, for subject, an Irish personality'. Although the reviewer found some of the actors' performances to be uneven, he considered *Yahoo* to be a success: 'Lord Longford has brought to the theme what it needed – an almost impassioned sympathy and insight into Swift's tortured mind'. With regards to Longford's conclusion, the critic observed that '[t]he third and last Act is daringly experimental in conception, and will provide nights of controversy in Dublin, both as to its estimate of Swift's position and as to its theatre technique' (D. M., '*Yahoo* at the Gate Theatre: Lord Longford's Play on Dean Swift', *The Irish Press*, 20 September 1933). Writing for *The Irish Independent*, David Sears considered Longford's characterization of Swift as having been 'drawn for us with a faithfulness that carries absolute conviction', and argued that 'expressionism is used with striking success' during the play's finale. More importantly, Sears admitted that 'some of the reminders of our forgetfulness and ingratitude to Swift came like a slap in the face. The final curtain leaves us ashamed of ourselves, which is probably how the author meant us to feel' ('A Portrait of Dean Swift: Lord Longford's Play', *The Irish Independent*, 20 September 1933).

In a sense, the performance of Longford's play was not an isolated occurrence, for the immediate post-independence decades saw something of a flurry of plays depicting Swift's tumultuous life: *Yahoo* had been preceded at the Abbey in 1930 by W. B. Yeats's *The Words upon the Window-Pane*, while Edwards and MacLiammóir would go on to produce Denis Johnston's *The Dreaming Dust* during their first season at the Gaiety in 1940.[6] Curtis Canfield referred to this 'newly-awakened enthusiasm for Swift' in rather overstated terms: although it is understandable that he believed that this departure from the usual fare of mythological dramas was desirable, his observation that 'with Yeats as with the new Ireland the break with the past is complete' is more tenuous.[7] While it is true that Yeats's and Johnston's plays mostly explore the relatively frivolous matter of the Dean's romantic affiliations, Lord Longford's *Yahoo* ultimately transcends this mundane topic by also addressing Swift's ambiguous attitude towards Irish nationalism and his contested place in Ireland's history. In doing so, Longford provides a strong critique of the embedded anti-Irish colonial discourse that persisted even after independence had been achieved and employs prospective memory strategies to accuse his postcolonial audience of espousing an all-too-narrow conceptualization of Irishness.

6 Yeats's play depicts a séance at which Swift's spirit is invoked and suggests that the Dean was unwilling to marry because he did not want to father children: he feared that they would inherit his mental illness and despaired at the state of the world into which they would be born. In Johnston's metatheatrical play, the characters refer to each other as the seven deadly sins and perform various scenes from the Dean's life in an attempt to understand his motives in spurning his lovers. They come to the conclusion that Swift and Stella were unable to marry because they were related by blood, since both of them were supposedly born out of wedlock. Swift's subsequent attempt to marry Vanessa is thwarted by Stella, who demands that he reveal their shared illegitimacy before consenting to the match. For a discussion of Johnston's play, see Christopher Morash, 'Denis Johnston's Swift Project: "There Must Be Something Wrong with the Information"', *Canadian Journal of Irish Studies*, 33(2) (2007), 56–9; and Virginie Girel-Pietka, '"Winds of Change" in *The Moon in the Yellow River* and *The Dreaming Dust* by Denis Johnston: Staging Identity in a Crisis', *Journal of Franco-Irish Studies* 3(1) (2013), 84–99.

7 Curtis Canfield, 'Note on *The Words upon the Window-Pane*', in *Plays of Changing Ireland* (1936), 7.

The play opens with Swift reading the final passage of his *Gulliver's Travels* (1726) manuscript to his housekeeper, Mrs Dingley, and Esther Johnson (Stella), whose rivalry with Esther Vanhomrigh (Vanessa) for the Dean's affections remains one of the most debated aspects of Swift's biography.[8] Although the Dean has just completed his masterpiece, he is dejected by his recent political downfall and subsequent exile to Dublin, which he describes in blatantly imperialist terms as 'this foul city, this nest of malice and slander, this country of bogs and fogs and savages'.[9] Stella tries to rebuke Swift for his slander by referring to his vocal support for local Irish industries, but the Dean is adamant: 'I might as well throw my pamphlets into the stinking, black waters of the Liffey and myself after them as expect the people of the place to have the money to buy them, the wit to read them, or the patriotism to obey them'.[10] Moving beyond the stereotypical dichotomy between colonizer and colonized, however, Swift equates both the Irish as well as the English with the despicable Yahoos that feature in *Gulliver's Travels*, and consequently refuses to heed Stella's pleas to '[c]ry aloud and wake' the Irish people and incite them to win their freedom: 'I will not fight for a tribe of these loathsome vermin against the depredations of another filthy tribe, because the one is weak and slavish, the other strong and rapacious'.[11] While Swift's characterizations betray at least a partial complicity with colonial discourse, the

8 For a discussion of Swift's romantic relationships, see Margaret Ann Doody, 'Swift and Women', in *The Cambridge Companion to Jonathan Swift* (2003), 87–111.

9 Lord Edward Longford, *Yahoo* (1933), 156. For a description of the political upheavals (and the subsequent downfall of the Tories) that had forced Swift to retire to Dublin, see Joseph McMinn, 'Swift's Life', 21–3. It should be noted, however, that Caroline Fabricant has argued that 'the popular view of Swift living out his years in Ireland as a disgruntled exile, filled with bitterness at his entrapment in a hateful land and constantly obsessing about his "glory days" in England, requires drastic modification'; see 'Swift the Irishman', 51.

10 Longford, *Yahoo* (1933), 156. Stella's remark undoubtedly refers to Swift's *Proposal for the Universal Use of Irish Manufacture* (1720).

11 Longford, *Yahoo* (1933), 157. Likewise, when Stella asks Swift to 'stand forth as Ireland's defender against tyrants', Swift states that '[t]hey are all Yahoos, vermin, ingrates, tyrant and slave alike' (158).

roots of his denunciations are universal rather than particular: the Dean's ostensible anti-Irish racism ultimately derives from his general distaste of humankind. As a result, Swift remains uninterested when William Wood's controversial patent is first brought to his attention, and the larger part of the play is dedicated to quarrelling with his two romantic rivals while George Berkeley (the future Bishop Berkeley) attempts to mediate – like many of Swift's commentators, the Earl of Longford underwrites the popular theory that the Dean secretly married Stella to undermine Vanessa's attempts at blackmailing him.[12]

Over the course of the play, however, Swift slowly warms to the idea of becoming 'an Irish Leonidas [who] might stand at a new Thermopylæ' and speculates that if Catholics and Protestants 'could [...] but be combined into one whole, were human nature not so utterly vile that man and his neighbour will not agree together even in their common misery, could these slaves but understand the horror of servitude and the splendour of freedom, Ireland might yet be the admiration of the world'.[13] Longford even construes this remarkably inclusive notion of Irishness as eventually having been Swift's principal concern, for during a heated argument with Vanessa about his perceived faithlessness, he suddenly reveals that his 'private griefs, [which are] such as no man hath ever borne', are the conditions in which the Irish people live:

> A whole nation is slave to a tyrant and does not know that it is enslaved. The people of Ireland are miserable and oppressed, they are filthy and ragged, and hungry and diseased, and how am I to rouse them to a sense of their miseries? A king's mistress and a cheating hardware man take a toll on the halfpenny in the pocket of an Irish beggar. Is not that enough to rend my heart with grief and fury?[14]

12 As Louise Barnett has observed in *Jonathan Swift in the Company of Women* (2007), there is no historical evidence that supports this hypothesis; rather, '[t]he putative secret marriage between Swift and Stella begged to be invented to explain what otherwise appears to the ordinary world to be inexplicable behaviour: that is, a significant and long-lasting yet nonsexual relationship between two people who appear to have had no impediment to marriage' (40).

13 Longford, *Yahoo* (1933), 158, 162.

14 Ibid., 175.

Although Vanessa refuses to be drawn into colonial politics and expresses her superiority by coldly dismissing his concerns – 'I am none of your Irishwomen' – this explication of Swift's nationalism reveals Longford's unequivocal assertion of the Dean's attempts to transcend religious divides and articulate a unified Irish identity.[15]

While this emotionally charged depiction of Swift's Irishness provides an interesting counterpoint to his earlier celebrations of the literary and political triumphs that characterized his London life, the play's most important statement about Irish history occurs in its concluding act, which Canfield described as 'a *mélange* of impressions all bearing either seriously or satirically, as the case may be, on Swift's place in the modern consciousness'.[16] This final part of the play, which disparages the audience's conceptualization of Irish nationalism, is set a few months after Swift's marriage to Stella and his denunciation of Vanessa, and the Dean has grown increasingly reclusive; some of his associates even believe that he is losing his mind.[17] Stella, who has not seen her husband in months either, gathers her resolve and decides to visit Swift so she can provide him with moral support, since

15 Ibid. In his fourth Drapier's Letter, Swift stated that 'people of all ranks, parties and denominations are convinced to a man, that the utter undoing of themselves and their posterity for ever, will be dated from the admission of that execrable coin' (440), and, notwithstanding the Irish people's undying loyalty to King George I, even hinted at the possibility of establishing their autonomy by claiming that 'the remedy is wholly in your own hands, [… for] by the laws of GOD, of NATURE, of NATIONS, and of your own Country, you ARE and OUGHT to be as FREE a people as your brethren in England' (442).

16 Canfield, 'Note on *Yahoo*', in *Plays of Changing Ireland* (1936), 150.

17 Longford, *Yahoo* (1933): Swift's servant remarks that '[h]e is so changed does naught now but write and write, and write, like a madman in Bedlam' (179). It should be noted that Longford has adapted the chronology of Swift's life for his own purposes: while the play covers a period of only a few months, Swift had already returned to Ireland by 1714; he published the Drapier's Letters in 1724, lost his sanity in 1742 and died in 1744. Also, as Louisa Barnett observes (in *Jonathan Swift in the Company of Women*, 40), Swift's alleged marriage to Stella is conventionally said to have taken place in 1716, with St George Ashe presiding over the ceremony rather than George Berkeley, while Vanessa died in 1723 – both events, then, actually occurred *before* the Drapier Letters controversy.

she fears that he has come under scrutiny as the pseudonymous author of the Drapier's Letters.

As Stella reveals that she is succumbing to a fatal illness and pleads with him to finally consummate their marriage, a crowd gathers outside to celebrate the outcome of Swift's polemics, which have achieved the abolishment of the counterfeit currency.[18] The people even rush in to congratulate Swift, whose identity had been well-guarded by the general population in defiance of Dublin Castle. However, while the Dean is giving a speech to assert that 'England has no right to rule Ireland' and praise the fact that 'you have shown to all that you will not permit your country to be subordinated to English corruption and tyranny', he suddenly collapses.[19] During his stupor, the ghost of Vanessa, who did not survive Swift's rejection, comes to torment him on what might be described as a literal version of Marvin Carlson's 'haunted stage', to which Emilie Pine, in *The Politics of Irish Memory* (2011), attributes the status of being 'representative of the larger haunting of Irish culture by the past'.[20] Arguing that '[g]hosts are a sign of what Ricoeur calls the 'excesses' of memory, a manifestation of the excessive grip of the past on the imagination of culture in the present', Pine observes that that this pathological obsession with the past often fails to properly address its 'implications for the future'.[21] However, in the subsequent scenes of *Yahoo*, Longford might be said to perform the exorcism that Pine explicitly calls for, as the Dean experiences nightmarish visions of expurgated editions of *Gulliver's Travels* being sold as children's books, of people arguing against naming a road after him, of his possessions being sold at an auction, of a psycho-analytical lecture that denounces his works, and of a mental hospital being founded in his name.[22] While Swift is being

18 Longford, *Yahoo* (1933), 183–5.

19 Ibid., 185.

20 Pine, *The Politics of Irish Memory* (2011), 16.

21 Ibid., 16–17.

22 By 1933, bowdlerized versions of *Gulliver's Travels* had indeed become a staple of children's literature. In his will, Swift decreed the establishment of 'an hospital large enough for the reception of [...] idiots and lunatics' ('Dr Swift's Will, With the Codicil Annexed', 519). This mental asylum, St Patrick's Hospital, still exists today.

paraded before a crowd that repeatedly shouts that '[h]e's mad!', a Distant Voice describes him as 'the Mad Dean' and states that 'Swift expires, a driveller and a show'.[23] Before the curtain drops, however, Swift faces God and confesses his sins as an angelic chorus sings him to his rest.

This expressionistic hallucination, which is strongly reminiscent of the innovative techniques that Denis Johnston had used in *The Old Lady Says 'No!'* (1929), is imbued with mnemonic strategies that problematize Swift's position in the Irish Free State and implicitly criticize the play's postcolonial audience. Even before he descends into madness, the Dean expresses his doubts about the longevity of Irish memory: 'if ever a time came when the slave had thrown off the last of his fetters, would he remember in his liberty who it was stirred him to strike the first blow?'.[24] Swift even composes a satirical poem to illustrate his point:

> Can we the Drapier then forget?
> Is not our nation in his debt?
> 'Twas he that wrote the Drapier's letters;
> He should have left them to his betters.
> We had a hundred abler men,
> Nor need depend upon his pen.
> Say what you will about his reading,
> You never can defend his breeding.[25]

Such mnemonic pessimism is likewise evinced by Swift's conviction that 'in the Annals of Ireland you will scarce find a grudging reference to his name', which, by describing a future that is actually the audience's *present*, constitutes a sarcastic application of a prospective memory structure that serves as an unambiguous indictment of the modern Irish nation. Indeed, when Berkeley contends that '[t]he people of this kingdom will ever love and venerate their champion! Will he be forgotten while an Irishman lives? Will his memory fade?', Swift answers that this effacement will occur '[a]s fast as friendship. Time, prejudice, calumny and the jealous pretensions

23 Longford, *Yahoo* (1933), 190.
24 Ibid., 181.
25 Ibid., 182.

of men less honest in their hate and scorn, yet no more sincere in their love, will thrust into oblivion the greatest benefactor'.[26] By prophesying the bleakness of Swift's future, then, Longford berates his contemporary audience as ungrateful, forgetful, and, more generally, faithless to the cause of Irish nationalism.

Although the crowd that celebrates Swift's victory repeats the first two lines of his scathing poem in a much more assertive manner and goes on to alter its conclusion in a way that emphasizes their loyalty, the aforementioned degradations of the Dean during his dark night of the soul belie their jubilant devotion.[27] While there are unseen voices that explicitly endow him with a primary position in an enumeration of Irish nationalists to come – 'Swift, Flood, Grattan, Tone, Emmet, David, Mitchel, Parnell, Griffith, Pearse' – the creation of this teleology is problematized by the continuous effacement of his legacy that occurs throughout his otherworldly ordeal.[28] The Man in a Bowler Hat, for example, speaks against the naming of a 'Dean Swift Road': he describes Swift as 'a man of anti-Irish outlook and degraded morals, and a writer of dirty reading matter, not in any way suited for use in our schools and colleges' and even urges the inhabitants of the district to 'leave the Dean in the obscurity he deserves'.[29] Likewise, during the psycho-analytical lecture, which asserts that Swift had a pathological anal fixation, the Dean is described as a 'great Englishman', to which the on-stage audience applauds, thereby erasing his Irish identity and validating Britain's neo-imperialistic hegemony.[30] In foregrounding these processes, *Yahoo* reveals not only the artificiality of both personal as well as national

26 Longford, *Yahoo* (1933), 182.

27 The crowd's version stresses their steadfast devotion to the Dean: 'Can we the Drapier then forget?/Is not our nation in his debt?/Two kingdoms by a faction led,/Had set a price upon his head./But not a traitor could be found/To sell him for six hundred pound!' (ibid., 185).

28 Ibid., 188.

29 Ibid. Ironically, he is unable to pronounce the alternative road name that he tries to propose: 'I can't read this, it's in Irish' (188).

30 Ibid., 189.

identities, but also the necessity to maintain viable mnemonic strategies to reinforce such identities despite their tenuous nature.[31]

## 'My blessing on the pistol and the powder and the ball!': Daniel O'Connell and Political Violence in *Ascendancy* (1935)

*Ascendancy* (1935), the last play that the Earl of Longford wrote before terminating his partnership with Edwards and MacLiammóir, likewise addresses the construction of national identities through modes of prospective historiography – albeit in a less dramaturgically experimental vein.[32] Even though the play's main theme – the Protestant elite's fear of the Catholic Emancipation movement of the 1820s – might seem to be historically circumscribed, the political questions that *Ascendancy* raises

31 As Mary Trotter has argued in *Modern Irish Theatre* (2008), Longford's background as a Protestant landlord is no less important to his message: 'Writing about Swift allowed Anglo-Irish like Yeats and Longford to consider the Irish patriot who sacrifices his position of imperial power for his national patriotism, but who is never completely accepted by his countrymen because of his ancestral position of imperial privilege' (100).

32 David Sears, writing for *The Irish Independent*, praised the actors' respective performances and described the play as 'a merciless exposé of the weakness of the position of the Irish Protestant ascendancy as early as the first decades of the nineteenth century'. However, Sears also critiqued some elements of Longford's dramaturgy, claiming that 'where the Protestant Ascendancy failed was in lack of intelligent patriotism. In the author's anxiety to show that they were Irish, and regarded themselves as such, he rather gave the impression that the cause of their downfall was drunkenness which, at the period, was a European, not merely an Irish Protestant, vice. Their acceptance of the support of British bayonets […] was the real cause of their downfall' ('New Play at the Gate Theatre', *The Irish Independent*, 30 January 1935). The actors were also praised by the reviewer for *The Irish Press*, who underlined the play's contemporary relevance by observing that it had 'so close a political contact with our minds that one was inclined at first to regard it as politics rather than as a play' ('A New Lord Longford Play', *The Irish Press*, 30 January 1935).

were no less pertinent to its contemporary audience. Indeed, Longford's historical drama might be said to problematize postcolonial attempts at national identity formation by revealing the trenchant nature of sectarian strife and political violence, on the one hand, and articulating a more inclusive (if elusive) notion of Irishness, on the other.

Set roughly a century after the publication of the Drapier's Letters, *Ascendancy* depicts the fictional Earl of Clonave, a somewhat lascivious but generally sympathetic Westmeath landlord. While the Earl is surveying his estate with his two sons, Arthur and Robert, who are overly hedonistic and ascetic respectively, they discuss the swift rise of Daniel O'Connell, whom the Earl believes to be well on his way to becoming the 'ruler of Ireland and of England, too, perhaps'.[33] Indeed, O'Connell, who had founded the Catholic Association in 1823, would spend the better part of that decade campaigning for Catholic Emancipation, which aimed to end the Protestant dominion over Irish public life.[34] O'Connell's efforts succeeded in rallying the disenfranchised Irish population and his plans came to fruition in 1829, when he stood for election in Parliament even though, as a Catholic, he was constitutionally barred from taking his seat. Fearing widespread revolt if O'Connell's election were overturned, the Duke of Wellington, the British Prime Minister, decided to sanction a circumscribed bill in favour of Catholic Emancipation.[35] Although the strictures that Wellington imposed severely limited the Act's effectiveness, it marked an important step towards mitigating the sectarian tensions that had been flaring up in Irish society.

In Longford's play, the Earl of Clonave openly acknowledges the deep religious divides of the 1820s: he would rather die than see 'this country being handed over to Papists to ruin'.[36] In the Earl's mind, this imminent disaster is linked to the dwindled legacy of the Orangemen, who had defended their new Protestant King William III against the deposed

33 Longford, *Ascendancy* (1935), 8.

34 Patrick M. Geoghegan, *King Dan: The Rise of Daniel O'Connell 1775–1829* (2008), 187–247.

35 Geoghegan, *King Dan* (2008), 248–70.

36 Longford, *Ascendancy* (1935), 8.

Catholic monarch James II at the Battle of the Boyne in 1690, but who had grown complacent '[w]ith their glorious, pious and immortal memory, their sashes and their banquets! Eating and drinking damnation to their enemies!'.[37] To prevent further degeneration, the Earl has decided to establish a 'new virile society for the maintenance of Protestant ascendancy', which will defend their interests by force if necessary.[38] Arthur shows little interest in his father's endeavour, however, stating that he will simply move to England if 'O'Connell starts to cut Protestant throats', while Robert claims that the Protestant cause is already lost: 'Men of your type have held power in Ireland for a long time, but now you have not the strength and hardly even the will to hold it any longer'.[39]

Robert substantiates his assessment of the Ascendancy's terminal decline by referring to recent history, reminding his father that '[s]ome twenty years ago you threw away the power to help yourselves' by voting in favour of the Act of Union (1800), which dissolved the Parliament of Ireland and established the United Kingdom of Great Britain and Ireland (1801) in the wake of Theobald Wolfe Tone's failed United Irishmen rebellion (1798).[40] The Earl has begun to doubt the wisdom of that decision, fearing that the British will not rally against O'Connell, but Arthur cynically reminds him that '[o]f course you were not mistaken, sir, your vote won you an Earldom'.[41] Robert, who is indifferent to wealth and power, goes one step further and casts this self-inflicted loss of the right to govern into a prospective statement: 'The gentry are doomed. […] There'll be big changes in Ireland in the next hundred years'.[42] His father chooses to interpret Robert's pronouncement in fatalistically dichotomous terms, for he believes that 'if Protestant ascendancy goes, everything goes. If Catholic

37 Ibid.
38 Ibid.
39 Ibid., 9.
40 Ibid.
41 Ibid., 10.
42 Ibid., 12.

Emancipation becomes law it's the end of us'.[43] For the Earl, then, societal inequity is a prerequisite for the survival of his class.

Although the aristocracy's political consciousness is entrenched in Irish history, the deeper roots of their imminent decline are revealed at the end of the first act, when the Clonave family leaves the scene and Patrick Moynagh, a disgruntled tenant, takes centre stage. Despite his anti-Catholic sentiments, the Earl is generally depicted as a considerate landlord, but Patrick resents his decision to offer a housekeeping job at this mansion to Katie Dwyer, a local girl whom the irate tenant wanted to marry. In Patrick's mind, his personal loss – and the possibility that Katie will fall prey to the Earl's seductions – becomes an emblem of Ireland's colonial subjugation, and in a towering rage that reaches back two centuries rather than twenty years, he delivers his 'curse upon you and upon your master and upon the race of Cromwell one and all', for they 'took away our lands and our goods and our fine houses, our gold and our silver, our abbeys and our churches'.[44] In this way, Robert's reference to the Act of Union and Patrick's diatribe against Cromwell's Irish campaign might be said to effect a double historicization of the play's events: both remarks not only establish a historical backdrop for sectarian conflict in the 1820s, but also serve to set the stage for a more viable conceptualization of Irishness that might be pursued in the 1930s.

This orientation towards the future becomes more pronounced in the second act, which takes place a week later and begins just before the party at which the Earl will announce the formation of the William Society. In a last-ditch effort, Robert pleads with his father to reconsider this belligerent enterprise and thereby avoid the 'bloodshed and war in this country' that will result from the Earl's proposal.[45] In light of Robert's anticipations about Ireland's future, this remark might be interpreted as referring not only to his father's military opposition to O'Connell, but also to the Land Wars of the later nineteenth century and the recent massacres of the Civil War (1922–23). The Earl is adamant, however, even though his other son, Arthur, has grown

43 Ibid., 14.

44 Ibid., 30.

45 Ibid., 34.

weary of his father's strident patriotism, complaining that it is '[a]lways Irish this and Irish that'.[46] His subsequent declaration – Arthur states that he is simply 'utterly tired of Ireland' – would also have rung true to contemporary ears, which had become all too used to chauvinistic rhetoric.[47]

Despite his sons' fatalism, the Earl goes on to deliver his speech to the assembled gentry, stating that 'it is only through the maintenance of this Ascendancy that this country of ours can prosper' – a sentiment which is met with drunken approval.[48] His elation begins to wane, however, when Lady Maxwell approaches him to prevent her husband, Sir Benjamin Maxwell, who is one of the Earl's most important supporters, from rounding up and executing prominent Catholics in a pre-emptive strike.[49] While the Earl is willing to moderate his policies to avoid this impending massacre, he is shocked by her subsequent request to deny Maxwell membership of the William Society altogether to ensure the latter's safety from Catholic reprisals. This plea undermines the Earl's staunch nationalism, for he interprets her dread of 'men lying in their blood, of little children stabbed to death with pikes, of helpless women perishing in their burning houses' as an absolute rejection of not only his patriotism, but of his historical identity: 'You are asking me to betray Ireland, our liberties, our children's future, the hopes of our children's children, even our religion, our God'.[50] By thus combining images of violence that would have been familiar to an audience that had lived through the Civil War with allusions to securing Ireland's future, the play also appears to address the threat of ever-growing paramilitary organizations such as Maurice Twomey's Anti-Treaty IRA and Eoin O'Duffy's fascist ACA (the 'Blueshirts'), and the danger of violence when Fianna Fáil defeated Cumann na nGaedheal in the 1932 election.[51]

46 Ibid.

47 Ibid.

48 Ibid., 39.

49 Ibid., 48. As Geoghegan notes in *King Dan* (2008), O'Connell frequently received death threats from 1824 onwards (205).

50 Longford, *Ascendancy* (1935), 50.

51 For discussions of the role of these paramilitary organizations in 1930s politics and elections, see, for example, Eunan O'Halpin, *Defending Ireland: The Irish State and its Enemies since 1922* (1999), 105–21, as well as 'Politics and the State, 1922–1932',

In his dejection, the Earl turns to Robert for help, proclaiming his sacred responsibility 'never to be an absentee, but to remember that an Irishman's first duty is to his country', yet his younger son remains steadfast in his denial, claiming that the Earl's cause is unjust and that his supporters will abandon him soon enough.[52] By asking a set of rhetorical questions to answer his father's blind zeal, Robert tries to offer an alternative vision of Irishness:

> can this country you profess to love, ever have peace or happiness whilst some four out of every five of her inhabitants are slaves to the fifth, and if slaves, then enemies, secret, revengeful, implacable enemies, ever at war with their masters? Must this country be bound for ever to fester with this endless internal warfare? Can you not imagine as something finer than your Protestant Ascendancy an Ireland where the heart of Catholic and Protestant, native and planter, lord and peasant, will be as one? A country where there will be no more factions and parties, but only Irishmen?[53]

Robert's articulation of his sense of societal justice and his desire for communal harmony can hardly be interpreted as a pragmatic attempt to sway his father's sentiments – indeed, the Earl quickly dismisses his son's ideas as irrelevant.[54] To the play's audience, however, Robert's plea would have had a different resonance: the grandiloquence and infighting of the various political groups and paramilitary organizations that still claimed to be struggling for the nation's survival more than a decade after independence had been achieved underline the fact that Robert's dream of '[a] country where there will be no more factions and parties, but only Irishmen' had not yet been realized in 1935.

Indeed, throughout the play, Longford's characters clash over their different appraisals of true Irishness, and their lack of harmony proves especially debilitating when it manifests itself as the blatant racism that

in *A New History of Ireland, Vol. VII: Ireland, 1921–84* (2003), 124–5; Joseph Lee, *Ireland 1912–1985: Politics and Society* (1989), 178–84; Tim Pat Coogan, *The IRA* (2002), rev. edn, 69–75, and *Ireland in the Twentieth Century* (2003), 197–203.

52 Longford, *Ascendancy* (1935), 54.

53 Ibid., 55.

54 Ibid.

flares up to construe the play's final plot twist. Fanny, an English girl whom Arthur married because of her wealth, is having an affair with Captain Lyndhurst, an English officer who is stationed in Ireland. When she gets into a fight with Louisa, one of her sisters-in-law, Arthur merely laughs at her, and she decides to run off with Lyndhurst. Fanny's distaste for Ireland is explicated time and again; when she first appears on stage, she does not deign to conceal her revulsion at both the Irish weather and its people's manners: 'What a climate! What a country! You never have a chance to dress prettily here. And who would appreciate it if you did?'[55] The disparity that she experiences is absolute and even functions as a confirmation of her own identity, for she feels that '[e]very day I stay here I grow more English. I'll never get used to Ireland'.[56] Her scathing remarks, which reflect an imperial discourse that depicts the Irish as 'white niggers', are not lost on her husband, who knows that '[s]he thinks us all savages, peer as well as peasant', and his brother even agrees with this statement to some extent. When Louisa lauds 'the tradition of the Irish gentleman', Robert confronts her with an enumeration of the failings of the Irish gentry: 'drinking, swearing and blaspheming, bullying, duelling, wenching'.[57] Captain Lyndhurst's distaste for the Irish people is even more explicit – he would 'like to break the bones of any dirty Paddy that laid a finger on [Fanny's] little white body'.[58]

When the couple meet up to run away at the end of the second act, they are confronted by Robert, who questions Lyndhurst's character; in response, Lyndhurst hits Fanny's brother-in-law, calling Robert a 'canting hound', and challenges him to a duel.[59] The brief third act takes place at dawn: Robert fires into the air to signify his pacifism, but Lyndhurst fatally

55 Ibid., 19.

56 Ibid., 45.

57 Ibid., 6, 23.

58 Ibid., 45. Likewise, Lyndhurst states that '[i]t's bad enough to be quartered in this God-forsaken country, but, by gad, if I thought I had to live here the rest of my life, I'd blow my brains out'. He follows this remark by claiming that '[t]here's no understanding these confounded Paddies. But I understand you [i.e. Fanny] and you understand me because we're rational, normal English people'.

59 Ibid., 58.

wounds him. The English officer is overcome with grief and blames Fanny for his deeds, as does her husband, who demands that his wife leave the Clonave estate. The Earl himself, however, forgives his daughter-in-law for causing the death of his younger son, but when he goes inside the house to see Robert's corpse, Patrick, the resentful tenant, suddenly appears with a gun and shoots him, laughing like a maniac while the curtain drops.

Ironically, then, the play's killings initially do not seem to occur as a result of O'Connell's alleged thirst for Protestant blood or the Earl's paramilitary posturing. Rather, they are the result of straightforward personal quarrels: Lyndhurst feels that Robert has offended his honour, and Patrick believes that the Earl has taken away the girl that he wanted to marry. In another sense, however, both deaths are historically emblematic: the most principled and conciliatory Irishman falls victim to British aggression, and the Protestant landlord is targeted as the embodiment of collective exploitation. Robert's and his father's death might thus be said to prefigure the vicious British reprisals that would occur during the Anglo-Irish War (1919–21), on the one hand, and the landlord murders that would take place during the Famine (1845–51), of which the shooting of Major Denis Mahon at Strokestown Park in 1847 is the most well-known example, on the other.[60]

More importantly still, Robert's presentiments, Patrick's misgivings, and the rumours that another one of the Earl's tenants has heard – 'the people of Ireland will rise up and throw down the great people of the land' – can be interpreted as prospective memory strategies that would have resonated strongly in the 1930s.[61] Despite their simultaneously prescriptive nature, however, these mnemonic structures belie assimilation: to a contemporary audience, who are aware of the violent historical events that the play foreshadows but whose present is no less fraught with sectarian and paramilitary aggression, Robert's appeals for unanimity must have seemed elusive.

60 During the party, the First Gentleman foreshadows the Earl of Clonave's death by remarking that he will decline membership of the William Society because has 'no intention of risking a stab in the back from an indignant tenant' (ibid., 43).

61 Ibid., 4.

## Conclusion

In this respect, plays such as *Yahoo* and *Ascendancy* are much more than mere historical dramas: while rooted in the nation's past, they offer postcolonial critiques – if not indictments – of the Free State and the conflicting notions of Irishness that threatened to tear it apart. By drawing implicit and explicit parallels between various contested episodes of colonial and sectarian strife in Irish history, on the one hand, and the nation's contemporary political climate, on the other, both plays employ complex mnemonic strategies that reveal the disconcerting nature of such analogies. Accordingly, the nightmarish spectres that haunt Jonathan Swift and the bloodcurdling laughter that attends Patrick Moynagh's murder of the Earl of Clonave seem to underline Emilie Pine's observation that '[t]he ghost represents the unbiddable, irrepressible, and uncontainable nature of memory, which disrupts linear progress and thus haunts not only the present, but the future also.'[62]

Moreover, Pine's subsequent claim that, '[i]n order to become an ethical form of memory, remembrance culture needs to move on from being haunted by trauma, to learn to reconcile the past with the future, and to find ways to exorcise the ghost of the past', is exemplified by Longford's progressive historicist defiance in the face of this struggle.[63] While many characters in *Yahoo* and *Ascendancy* embody the entrenched political and religious divides that also destabilized the contemporary Irish nation, Longford offsets such sectarianism with much more inclusive and conciliatory conceptualizations of Irishness. Even if such sentiments fall on deaf ears in their respective historical settings, they exploit the fact that the establishment of the Free State, by definition, marked the resolution of a teleological endgame: in an independent Irish nation, ancient factions and antipathies should become irrelevant, so that celebrations of unity and social harmony might finally offer the possibility of mnemonic redemption.

62 Pine, *The Politics of Irish Memory* (2011), 16.
63 Ibid., 17.

## Bibliography


Barnett, Louise, *Jonathan Swift in the Company of Women* (Oxford: Oxford University Press, 2007).

Canfield, Curtis, ed., *Plays of Changing Ireland* (New York: Macmillan, 1936).

Coogan, Tim Pat, *The IRA*, rev. edn (Basingstoke: Palgrave Macmillan, 2002).

——, *Ireland in the Twentieth Century* (London: Arrow Books, 2003).

D. M., '*Yahoo* at the Gate Theatre: Lord Longford's Play on Dean Swift', *The Irish Press* (20 September 1933).

Doody, Margaret Ann, 'Swift and Women', in Christopher Fox, ed., *The Cambridge Companion to Jonathan Swift* (Cambridge: Cambridge University Press, 2003), 87–111.

Fabricant, Carole, 'Swift the Irishman', in Christopher Fox, ed., *The Cambridge Companion to Jonathan Swift* (Cambridge: Cambridge University Press, 2003), 48–72.

Geoghegan, Patrick M., *King Dan: The Rise of Daniel O'Connell 1775–1829* (Dublin: Gill & Macmillan, 2008).

Girel-Pietka, Virginie, '"Winds of Change" in *The Moon in the Yellow River* and *The Dreaming Dust* by Denis Johnston: Staging Identity in a Crisis', *Journal of Franco-Irish Studies*, 3(1) (2013), 84–99.

Lee, Joseph, *Ireland 1912–1985: Politics and Society* (Cambridge: Cambridge University Press, 1989).

Longford, Lord Edward, *Ascendancy* (Dublin: Hodges Figgis, 1935).

——, *Yahoo* (1933), in Curtis Canfield, ed., *Plays of Changing Ireland* (New York: Macmillan, 1936).

McMinn, Joseph, 'Swift's Life', in Christopher Fox, ed., *The Cambridge Companion to Jonathan Swift* (Cambridge: Cambridge University Press, 2003), 14–30.

Morash, Christopher, 'Denis Johnston's Swift Project: "There Must Be Something Wrong with the Information"', *Canadian Journal of Irish Studies*, 33(2) (2007), 56–9.

'A New Lord Longford Play', *The Irish Press* (30 January 1935).

O'Halpin, Eunan, *Defending Ireland: The Irish State and its Enemies since 1922* (Oxford: Oxford University Press, 1999).

——, 'Politics and the State, 1922–1932', in J. R. Hill, ed., *A New History of Ireland, Vol. VII: Ireland, 1921–84* (Oxford: Oxford University Press, 2003), 86–126.

Pine, Emilie, *The Politics of Irish Memory: Performing Remembrance in Contemporary Irish Culture* (Basingstoke: Palgrave Macmillan, 2011).

Sears, David, 'New Play at the Gate Theatre', *The Irish Independent* (30 January 1935).

——, 'A Portrait of Dean Swift: Lord Longford's Play', *The Irish Independent* (20 September 1933).

Swift, Jonathan, 'Dr Swift's Will, With the Codicil Annexed', in Sir Walter Scott, ed., *The Works of Jonathan Swift*, Vol. I, 2nd edn (Edinburgh: Archibald Constable and Co., 1824), 516–27.
——, 'Drapier's Letter I' (1724), in Angus Ross and David Woolley, eds, *Jonathan Swift: The Major Works*, rev. edn (Oxford: Oxford University Press, 2003), 422–31.
——, 'Drapier's Letter IV' (1724), in Angus Ross and David Woolley, eds, *Jonathan Swift: The Major Works*, rev. edn (Oxford: Oxford University Press, 2003), 134–47.
——, 'Swift to Lord Carteret' (28 April 1724), Angus Ross and David Woolley, eds, *Jonathan Swift: The Major Works*, rev. edn (Oxford: Oxford University Press, 2003), 431–2.
Trotter, Mary, *Modern Irish Theatre* (Cambridge: Polity Press, 2008).

TRACY FAHEY

# 2 Remembering Wildgoose Lodge: Gothic Stories Recalled and Retold

The 1816 burning of Wildgoose Lodge in Reaghstown, Co. Louth is a well-known yet complex atrocity story of Irish agrarian conflict. This traumatic incident, which triggered a wave of murder, retaliation, informing and executions, is well documented by a fairly contemporaneous account by James Anton of 1846; later journal articles by Paterson and Casey; a newspaper feature by Kiely and historical monographs on the subject by Murray and Dooley.[1] The shocking nature of the burning and its aftermath resonated powerfully in nineteenth-century Reaghstown. Between 1816 and 1819, eight people were burned alive in the conflagration in the Lodge: Edward Lynch, his daughter Bridget, her husband Thomas Rooney, their five-month-old son Peter, and three young servants, Biddy Richards, James Rispin and Ann Cassidy. Eighteen men were executed, of whom twelve were gibbeted in roadside trees in the locality until 1819; and several alleged informers were subsequently murdered in the neighbourhood. This event and its repercussions are famously retold by William Carleton in his short

1 James Anton, *Retrospect of a Military Life* (Edinburgh: Lazars,1846); Thomas George Farquhar Paterson, 'The burning of Wildgoose Lodge', *County Louth Archaeological Journal*, 12(2) (1950), 159–80; Daniel J. Casey, 'Wildgoose Lodge, the Evidence and the Lore', Journal of the County Louth Archaeological and Historical Society, 18(3) (1975), 211–31; Benedict Kiely, 'Return to Wildgoose Lodge –: Leapfrog and a Magic Ash Tree', *The Irish Times* (28 July 1972); Raymond Murray, *The Burning of Wildgoose Lodge: Ribbonism in Louth, Murder and the Gallows* (Armagh: Armagh Diocesan Historical Society, 2005); Terence Dooley, *The Murders at Wildgoose Lodge: Agrarian Crime and Punishment in Pre-Famine Ireland* (London: Four Courts Press, 2007).

story of 1830, 'Confessions of a Reformed Ribbonman', later retitled and republished in 1933 as 'Wildgoose Lodge'.[2]

The story of Wildgoose Lodge occupies a paradoxical position in terms of national and local memory. It is a story that looms large in regional cultural memory through stories told in family homes, yet these stories are not shared publicly. Although the story is nationally known, principally through Carleton's tale and the two monographs on the subject by Dooley and Murray, native versions of the story have traditionally been only discussed within families in the area of Reaghstown. Cronin discusses the unwillingness of subjects to engage in oral memory projects about incidents of historical trauma; she concludes that 'such "porosity of memory" may be due as much to people's unwillingness to remember as to their inability'.[3] In Dooley's *The Murders at Wildgoose Lodge*, he notes the reluctance of local people to talk about the incident, until well into the 1950s:

> As T. G. F. Paterson pointed out, the atrocity at Wildgoose Lodge was still spoken of 'with horror' in the neighbourhood at Reaghstown in the 1950s. Decades later it was still difficult to find anyone who would talk openly of the incident ... There were too many raw nerve ends.[4]

This rawness described by Paterson and Dooley has many possible sources; the horror that still lingers in the locality at the burning; the punishment meted out to the alleged perpetrators, especially the gibbetings of these convicted men which displayed the rotting bodies of friends and neighbours on local roads for several years as a terrible *memento mori*; and, most potent of all, the lingering distrust caused by the neighbour-on-neighbour violence and murder in the narrative of Wildgoose Lodge. The poet Patrick

2 William Carleton, 'Confessions of a Reformed Ribbonman. (An Owre True Tale)', *The Dublin Literary Gazette*, 4 (1830), 49–51; William Carleton, *Traits and Stories of the Irish Peasantry*, Vol. 4. (London: George Routledge and Sons, 1933).

3 Maura Cronin, 'Oral History, Oral Tradition and the Great Famine', in Christian Noack, Lindsay Janssen and Vincent Comerford, eds, *Holodomor and Gorta Mór: Histories, Memories and Representations of Famine in Ukraine and Ireland* (London: Anthem Press, 2012), 235.

4 Terence Dooley, *The Murders at Wildgoose Lodge: Agrarian Crime and Punishment in Pre-Famine Ireland* (London: Four Courts Press, 2007), 57.

Kavanagh, a local man, also wrote of the secrecy that surrounded Wildgoose Lodge in the introduction to Carleton's *Autobiography*:

> The name of Wildgoose Lodge was whispered in my native parish of Inniskeen right up to my own time … I remember one October evening, when I was gathering potatoes as a boy for a neighbour, asking him to tell me something about the business. He told me to ask my father.[5]

Kavanagh's final sentence 'He told me to ask my father' is telling. The story of Wildgoose Lodge was not a tale for public discussion or consumption. It was a story told within a family, around the hearth, under conditions of secrecy. In terms of balancing the dynamics of memory, the official versions with the local lore, there was a clear need for a memory project to be conducted in the community of origin. Discovering and recording these family narratives became the focus of *Remembering Wildgoose Lodge*, a three-year socially engaged memory project conducted by fine-art collective Gothicise[6] in the locality of the Wildgoose Lodge incident. This chapter offers an overview of this project, its fieldwork, the narratives collected, an analysis of these stories and some conclusions as to the significance of this project as a memory project.

## *Remembering Wildgoose Lodge:* The Project

*Remembering Wildgoose Lodge*, a project sparked off by the author's own autoethnographic knowledge of these stories, was carried out between 2013 and 2016. The intention behind this project was simple: to engage with the community who live in the locality of this event, to interview the descendants of the local people involved in the incident, and to record their own particular and distinct family versions of the story. The aim of this exercise was to

5 Patrick Kavanagh, 'Introduction', in *The Autobiography of William Carleton* (London: MacGibbon & Kee, 1968), 10.

6 <http://www.gothicise.weebly.com> accessed 31 March 2016.

recognize the importance of family and regional memories of the event, and to consider the new knowledge emanating from these collected variants of the story of Wildgoose Lodge. This chapter documents the initial stages of the project from 2013 to 2015, which involved working with the older community of Reaghstown, Co. Louth and the surrounding area, in order to collect five different accounts of the tragedy, emanating from five different family stories told by Jim McArt, Pat Drumgoole, Ambrose Finn, Enda Matthews and Gretta Fahey. This research has revealed a body of previously unrecorded folktales and variants still told today in the locality of the original incident. The project sought to analyse the variants that emerged from this fieldwork. This essay outlines the process of mapping surviving variations of this story before focusing on an analysis of these narratives, and outlining the new knowledge generated by this project in terms of the curses and ghost stories that emerge from these variants. It recognizes these folk memories as part of a valuable body of knowledge held in community memory, and explores how best to represent these memories from oral culture as part of this process of recognition. This essay looks at how a socially engaged art project can help to preserve folk narratives, and share different versions of the story in their community of origin. This is something Lippard considers very important: 'The reconstructive potential of an art practice that restores or reveals the meaning of a place to those who live within it cannot be overestimated'.[7]

The *Remembering Wildgoose Lodge* project therefore examines how community stories are constituted and reconstituted in memory. By adopting Fentress and Wickham's anthropological stance on social memory, this project underscores their idea that all memory is inherently social, and never devoid of its immediate, local and collective context:

> [...] even individual memory is not simply personal: the memories which constitute our identity and provide the context for every thought and action are not only our own, but are learned, borrowed and inherited – in part, and part of, a common stock, constructed, sustained, and transmitted by the families, communities, and cultures to whom we belong.[8]

7 Lucy Lippard, 'Notes from a Recent Arrival', in Clare Doherty, ed., *Situation* (London: Whitechapel Gallery, 2009), 155.

8 James Fentress and Christopher Wickham, *Social Memory*, from New Perspectives on the Past series (London: Wiley Blackwell, 1992), viii.

*Remembering Wildgoose Lodge* was conducted through fine-art collective Gothicise, a collaborative established in 2010. This collective has a floating membership, depending on the nature of the project. For this project, the collaborators included primarily local community members with the present author as principal fieldworker. Gothicise projects have always been concerned with memory and the interplay between past and present, using methods of social engagement to create interactive, participative experiences. As a collective, it works with sites and stories, investigates how they intertwine, and what arises from these intersections. The name Gothicise obviously derives from the Gothic – in particular the nostalgic and psychoanalytic implications of the Gothic. The Gothic as a field of academic study is defined by its leading exponents such as David Punter, Fred Botting and Catherine Spooner as a study of in-betweenness, of *otherness*; concerned with notions of the liminal, the interstitial, the marginal, the dispossessed. The collective therefore primarily deals with situations where narratives are reimagined, revisited, reinterpreted and revoiced. In this, *Remembering Wildgoose Lodge* is emblematic of the genre of *postcolonial Gothic* which focuses on the return of repressed histories, as described by Khair in his *The Gothic, Postcolonialism and Otherness: Ghosts from Elsewhere* (2009). The project *Remembering Wildgoose Lodge* also examines how memories become Gothic stories, and how the repercussions of historical events become embodied in the oral folklore and social memory of a specific community.

## Project Origins: Discovering the Variants

This project had its origins in the autoethnographic project of 2012 that set out to document family folktales through a series of photo-texts that mapped strange stories onto real sites. Through this process I discovered that the story of Wildgoose Lodge, related by my grandmother, had come with a unique post-script: the narrative of 'The Curse of the Four F's'. This was a story linked to the fate of four local Big Houses in the neighbourhood of the Lodge that culminated in the cursing of four local landlords with the lack of an heir. After research conducted through primary sources in the Wildgoose Lodge archives in Dundalk library, through secondary sources, and through informal research through conversations conducted with other members of the Reaghstown community, I failed to find another person who had heard this story. I then realized that my grandmother's version of the story was unique within existing historical and fictional narratives. In 2013 the *Remembering Wildgoose Lodge* project began in earnest, as I started to conduct preliminary fieldwork in order to investigate if other dissident versions of this story existed. This initial research, conducted through discussions with locals, revealed that other accounts of the story also had unique narratives tagged on to the story of Wildgoose Lodge, and that in this case, social memory of the event encompasses divergent family narratives.[9]

However, there was still the problem of how to formally conduct fieldwork to capture these stories about family memories. We have already discussed the silence that surrounds the historical events of Wildgoose Lodge in Reaghstown, and the possible reasons for this. One project participant, Jim McArt talked about this atmosphere of secrecy, relating it to notions of shame and protection of the community: 'Informers weren't meant to be spoken of to outsiders [...] local lore has it the names recorded were the wrong names. My father wanted the names to be kept secret to avoid it being cast up to people'.[10] Dooley is also interested in how Carleton's story becomes a cypher, a coded but legitimate way to discuss that which was too terrible to speak about through the medium of fictionalized narrative. This contention is also borne out by fieldwork conducted for the project. As another participant, Pat Drumgoole, put it, 'People were afraid to talk about the story, so Carleton's story became the story'.[11] This atmosphere of secrecy and silence presented a direct problem for the project: how could these accounts be researched in such a silent and close-knit community? In fact, the execution of this project was only possible through my own family links to the area and my access to my family's variant of the story, which afforded me a privileged position as embedded ethnographer in this community of interest.

9 Jim McArt, *Remembering Wildgoose Lodge*, Questionnaire, 2013.

10 Ibid.

11 Pat Drumgoole, *Remembering Wildgoose Lodge* project, Questionnaire, 2013.

## Conducting the Research: Fieldwork as Homework

Using this qualitative approach also necessitated a precise type of fieldwork. This method is best described as 'homework', a concept first introduced by in 1996 by Lavie and Swedenburg as fieldwork in a community which is conducted by a member of that community:

> [Many] essays blur the boundaries that are expected to distinguish authors from their subjects of research, turning authors into subjects or semi-subjects of study. Such insertions of the author as subject are not just pretexts to occupy centre stage in narcissistic self-displays or as omniscient self-controllers [...] In many instances, the interjection of the autobiographical is also concerned with the author's own engagement, and/or belonging or partial belonging to the community under study, which produces a kind of lovingness towards the embattled groups.[12]

Smyth elaborates on this idea of fieldwork that 'refuses the methodological boundaries [...] between "home" and "away", between the place of "legitimate" study – Out There – and the privileged locations – In Here – wherein "legitimate" study may be undertaken'.[13] He examines the classification of fieldwork as 'cognitive fieldwork' (or as he terms it, 'going native'), and asks 'what happens when the fieldworker *is* a native, and when the subject/object dyad that structures disciplinary knowledge is blurred?'.[14] Smyth finds this blurring of boundaries to be vital in creating a successful cross-fertilization of different research methods drawn from autobiography, fiction, historical sources and fieldwork

*Remembering Wildgoose Lodge* is a project intimately connected with the idea of the home as a place where secrets are transmitted. It is significant that the initial meetings of this project, the completion of questionnaires and the recording of stories through interviews all took place in the

12 Smadar Lavie and Ted Swedenburg, *Displacement, Diaspora and Geographies of Identity* (London: Duke University Press, 1996), 22.

13 Gerry Smyth, *Space and the Irish Cultural Imagination* (London: Palgrave, 2001), 128.

14 Ibid., 127.

participants' homes. Using family connections to commence this research was vital. My aunt facilitated a meeting with my initial interviewee, Jim McArt, a journalist, active and influential in his community, who was eager to work with me on this project, and who actively encouraged others to do so. He rapidly became an acknowledged community gatekeeper for this project, and his support was crucial to the recruitment of others. With his help, I ended up working with five final subjects, four retired men and one woman, ranging in age from sixty-eight to ninety-two. With these subjects I completed five sets of structured questionnaires and then recorded interviews with open questions. From these recordings emerged a mixture of fact and fiction as well as odd, diverging details that distinguish the tale's variants. Some of this variance is due to the nature of their oral transmission: 'My generation heard it in homes. Stories were told within people's own families – handed down, not written down, so there are variations from story to story'.[15] The telling and retelling of these stories within family homes all point to Connerton's idea of how social memory is transmitted through rites and repetition.[16]

Building up to recording these variants took time. In 2013, I began with trust-building exercises and mediated meetings with individual one-to-one conversations, and completed the questionnaire with the first participant, McArt. In 2014, I recorded McArt's interview, worked with the second subject, Pat Drumgoole and held a public meeting after which the final three participants were recruited: Ambrose Finn, Gretta Fahey and Enda Matthews. From late 2014 to May 2015, I focused on completing questionnaires with participants, conducting interviews that focused on retelling the story, and recording the divergent tales that ensued. What has emerged from this fieldwork is a series of unique (and previously uncollected) stories that act as cautionary tales. Of these narratives, this chapter explores some specific accounts of curses and ghosts that add to the story of Wildgoose Lodge, by amplifying its ghastly nature.

15 Jim McArt, *Remembering Wildgoose Lodge*, Questionnaire, 2013.

16 Paul Connerton, *How Societies Remember* (Cambridge: Cambridge University Press, 1989).

## Analysis of the Variants: The Curses

There were several allegations in the fieldwork interviews that the story of Wildgoose Lodge itself was cursed. McArt remembers this claim from his childhood: 'There was a curse – if we touched anything to do with it, we would be cursed – cursed if we got involved in a film or any other project. Wildgoose Lodge was a cursed story'.[17] During the course of this fieldwork, by analysing just five different versions of the story, five different curses associated with Wildgoose Lodge were uncovered – the 'Curse of the Four F's', the 'Curse of the 3 R's', the curse of the site, the curse of the gibbets and the curse that McArt cites – a general curse on anything that revived the story of the Lodge. Possibly the most significant curse uncovered through this research is the 'Curse of the 3 R's', described by McArt, Drumgoole and Matthews. This curse first came to light in an initial interview with McArt in 2013. In completing this questionnaire, McArt mentioned, with slight reluctance, the 'Curse of the 3 R's' that his father had told him, which was concerned with the identity and punishment of informers. 'The parish priest (Rev. Marron) wrote a letter to Mrs O'Reilly of Knockabbey [a local Big House] about the incident and he cursed the names of the informers from the altar – the three R's'.[18] His later recorded interview of 2014 reiterates this story:

> The story of the three R's was that they were local informers [names omitted here on request of interviewee] and that they gave information that got people hanged. Their names have never come out when William Pendleton investigated the whole thing, his report, their names have never come out, it's always other informers. But that was said locally.[19]

The three 'R's' are popularly supposed to be the names of the original local informers, names that were not meant to be disseminated to the authorities. McArt was visibly uncomfortable with the story. His father obviously told

17 Ibid.
18 Jim McArt, *Remembering Wildgoose Lodge* project, Questionnaire, 2014.
19 Jim McArt, *Remembering Wildgoose Lodge* project, Interview, 2014.

it to him with strict instructions not to repeat it. In terms of the recording, we agreed on omitting the actual names beginning with 'R' spoken on tape from the transcript. However, McArt was happy to explain that one of these 'R's' was a local miller, who operated Reaghstown Mill. Annie Lynch's oral account of Wildgoose Lodge of 1964–1965 cited in Casey and Dooley[20] also confirms the existence of this character in local lore: 'One man was the name of the Miller Reenan, he lived at the chapel ... he stagged [informed] too.'[21] Lynch's account also provides gory details about another informer (Kavanagh) being shot and left to lie in the road, and another unnamed man who was lashed to a horse's tail 'and galloped round till there wasn't a piece of him left together'.[22]

Moreover, McArt talks of the terrible punishment of informers in the area, and of the fate of this miller-turned-informant: 'But they went home for something and when they came back the miller was dead, someone had put in his head, his skull, with a stone'.[23] Matthews corroborates this story, albeit in a more fragmented manner. He affirms that he heard the story from the same (unnamed) source that McArt heard it from. However, Matthews's account returns to his repeated assertion that Wildgoose Lodge is a dangerous story, a story people tell to implicate others. Like McArt, Matthews will not mention the actual names:

> One of the stories was mentioned was three surnames beginning with 'R' – won't mention them. And the story was that, em, on the scaffold a priest must have been giving last rites or last whatever he gave to the people being hanged and the story was that he took off their shoes – don't know if they took off their shoes before they were hanged? – I don't know – but when he was taking off their shoes he said to the guy 'These three' – he named the three 'R's – 'they will die out. They won't last in this parish.' Now the guy that said that to me would be the guy that Jim McArt would've probably got the story from, but I imagine that guy would have had a slant on his story that he was anxious to ... he would have been anxious certainly that one

20 Daniel Casey, 'Wildgoose Lodge, the Evidence and the Lore'; Terence Dooley, *The Murders at Wildgoose Lodge: Agrarian Crime and Punishment in Pre-Famine Ireland.*

21 Daniel Casey, 'Wildgoose Lodge, the Evidence and the Lore', *Journal of the County Louth Archaeological and Historical Society*, 18(3) (1975), 216.

22 Ibid.

23 Jim McArt, *Remembering Wildgoose Lodge* project, Interview, 2014.

> of the 'R's, maybe two of the 'R's would be implicated in that way. That he wouldn't be a fan of them, d'you know?[24]

Some of these curses were site-specific and acted like cautionary takes, warning listeners to stay away from the places named in the stories. These curses also, of course, include the site of the Lodge itself: 'It was meant to be cursed, of course, the person who owned the land would have bad luck'.[25]

## Analysis of the Variants: The Ghost of Biddy Richards

Of all the accounts recorded, it is Drumgoole's which contains the most stories. Of the five subjects interviewed, only he recalls hearing versions of the story outside the home, from several local elderly people: 'Well the story I was told, a long, long time ago by the old people'.[26] These multiple sources of reception may be the reason why Drumgoole's version of the story is so full of colourful stories. Drumgoole's account tells a story not recorded in any written or oral source, a ghost story featuring the *fetch* or double of Biddy Richards, the servant girl who was burned alive in Wildgoose Lodge.[27] Drumgoole tells this story of Biddy Richards in his recorded interview of 2015:

> Biddy Richard came from Mountrush, Stormanstown. But there's a mass pass from that time … everyone had to walk. There was a mass pass from her place, by her door, it came out at Reaghstown Chapel, still there, the wicker gate's there, still at the back there was a mass pass, eye, people used to go, like when I was serving mass, people'd cross the fields, d'you know, short cut. And an old woman called Mrs Quinn was in hospital, she was ninety years of age if she was an hour [laughs] and me mother could come out to see her, and I'd take out me mother and I'd sit down and I'd ask her. I was interested and she lived beside Richards or her people. And she says 'I'll

24 Enda Matthews, *Remembering Wildgoose Lodge* project, Interview, 2015.
25 Enda Matthews, *Remembering Wildgoose Lodge* project, Questionnaire, 2015.
26 Pat Drumgoole, *Remembering Wildgoose Lodge* project, Interview, 2015.
27 Pat Drumgoole, *Remembering Wildgoose Lodge* project, Questionnaire, 2013.

> tell you a story, the night of the Wildgoose Lodge', that her father handed on down, or that came down through the family. 'He was coming walking down the mass pass and he met his daughter going home, and he said to her 'Biddy, where are you going?' She never spoke. She never spoke. It was her ghost! She was burned that night.[28]

This ghost story of Biddy Richard may be linked to the fact that the house burned down close to the eve of Samhain, a time when ghosts were popularly supposed to walk the land. This was the only ghost story regarding the Lodge that Drumgoole was aware of, although he remarked that '[i]t was a breeding ground for them [ghosts], if ever there was one'.[29]

## Conclusion

These family memories of Wildgoose Lodge, passed on through stories handed down from generation to generation, tell us of the fear felt locally since 1816, a fear preserved through the curses that act as cautionary tales, designed to keep people away from the sites of the story, and through a ghost story that warns of impending death. *Remembering Wildgoose Lodge* plays an important part in documenting the sad history of Wildgoose Lodge – an 'other' version preserved through oral folklore. It also explores the idea of traumatic history as having a profound and lasting effect on rural communities. This project seeks to offer authentic and Othered accounts of the atrocity. These counter-narratives offered by McArt, Drumgoole, Finn, Matthews and Fahey illustrate how gaps in historical narrative are filled in by communities, and how these stories become populated by their own family ancestors, their own familiar places. As Lippard comments:

> When history fails a community, memory takes up the task. If history comes from above and outside, from teachers and governments, and document cultures prove inadequate, grandmothers become the authorities. And the landscape triggers

28 Pat Drumgoole, *Remembering Wildgoose Lodge* project, Interview, 2015.
29 Ibid.

> their memories, becomes symbolic, conveys different messages in different cultural languages.[30]

This project is important because of its societal relevance: it preserves, records, and celebrates memory as a collective and living entity. Social memory or folk memory reflects what is valued by the community, for what survives tells us about the community and its values. It is at this intriguing point that we close off the analysis of the variants. The next project phase involves the creation of a commemoration involving local folk responses to the story of Wildgoose Lodge that will be publicly displayed and performed in the locality. Some of these will be produced by active stakeholders in this project such as McArt and Drumgoole. The final project output in 2016 will be a collaborative album project. This will be released by *Folklore Tapes*, a UK label, which specializes in soundtracks that respond to folkloric stories in Ireland and the UK. The album consists of atmospheric tracks composed by local musician Eamonn Murphy, featuring voice excerpts from the *Remembering Wildgoose Lodge* series interviews. This album will validate and reify these narratives as interesting, complex and creative responses to this story of historical atrocity.

In doing so, the project aims to preserve family memories and to position these oral narratives as a recognized contribution towards the lore of Wildgoose Lodge. Using this mode of research involving interviews, recordings, music and fine art, this project offers an interrogation of this haunting story and attendant collective memories, not to achieve an exorcism of the reality but to create a repository for the local versions of the story. *Remembering Wildgoose Lodge* also attempts to allay the fears of open discussion that surround these stories. In this project, Gothicise hopes to act as a catalyst in facilitating the communication of these stories, the coming together of the community to discuss these and remember them, and ultimately to seek to preserve these unique stories that are bound up in personal, local, regional and national identities.

30 Lucy Lippard, 'Notes from a Recent Arrival', in Clare Doherty, ed., *Situation* (London: Whitechapel Gallery, 2009), 155.

## Bibliography

Anton, James, *Retrospect of a Military Life* (Edinburgh: Lazars, 1846).

Carleton, William, 'Confessions of a Reformed Ribbonman. (An Owre True Tale)', *The Dublin Literary Gazette*, 4 (1830), 49–51.

——, *Traits and Stories of the Irish Peasantry*, Vol. 4 (London: George Routledge and Sons, 1933).

Casey, D. J., 'Wildgoose Lodge, the Evidence and the Lore', *Journal of the County Louth Archaeological and Historical Society*, 18(3) (1975), 211–31.

Cronin, Maura, 'Oral History, Oral Tradition and the Great Famine', in Christian Noack, Lindsay Janssen and Vincent Comerford, eds, *Holodomor and Gorta Mór: Histories, Memories and Representations of Famine in Ukraine and Ireland* (London: Anthem Press, 2012), 231–44.

Dooley, Terence, *The Murders at Wildgoose Lodge: Agrarian Crime and Punishment in Pre Famine Ireland* (London: Four Courts Press, 2007).

Drumgoole, Pat, *Remembering Wildgoose Lodge* Questionnaire 2013, in Tracy Fahey, *Resurrection: The Return of the Gothic Home in Contemporary Irish Art (1990–2015)*, 352–5, unpublished thesis, <https://dspace.mic.ul.ie/handle/10395/2045> accessed 30 March 2016.

——, *Remembering Wildgoose Lodge*, Recorded Interview 2015, in Tracy Fahey, *Resurrection: The Return of the Gothic Home in Contemporary Irish Art (1990–2015)*, 371–6, unpublished thesis, <https://dspace.mic.ul.ie/handle/10395/2045> accessed 30 March 2016.

Fentress, James, and Wickham, Christopher, *Social Memory* from *New Perspectives on the Past* series (London: Wiley Blackwell, 1992).

Gibbons, Luke, *Transformations in Irish Culture* (Cork: Cork University Press, 1996).

Kavanagh, Patrick, 'Introduction', *The Autobiography of William Carleton* (London: MacGibbon & Kee, 1968).

Kiely, Benedict, 'Return to Wildgoose Lodge –: Leapfrog and a Magic Ash Tree', *The Irish Times* (28 July 1972).

Lavie, Smadar, and Swedenburg, Ted, *Displacement, Diaspora and Geographies of Identity* (Durham, NC, and London: Duke University Press, 1996).

McArt, Jim, *Remembering Wildgoose Lodge* Questionnaire 2013, in Tracy Fahey, *Resurrection: The Return of the Gothic Home in Contemporary Irish Art (1990–2015)*, 350–1, unpublished thesis, <https://dspace.mic.ul.ie/handle/10395/2045> accessed 30 March 2016.

——, *Remembering Wildgoose Lodge* Recorded Interview 2014, in Tracy Fahey, *Resurrection: The Return of the Gothic Home in Contemporary Irish Art (1990–2015)*,

363–70, unpublished thesis, <https://dspace.mic.ul.ie/handle/10395/2045> accessed 30 March 2016.

Matthews, Enda, *Remembering Wildgoose Lodge* Questionnaire 2015, in Tracy Fahey, *Resurrection: The Return of the Gothic Home in Contemporary Irish Art (1990–2015)*, 356–7, unpublished thesis, <https://dspace.mic.ul.ie/handle/10395/2045> accessed 30 March 2016.

——, *Remembering Wildgoose Lodge* Recorded Interview 2015, in Tracy Fahey, *Resurrection: The Return of the Gothic Home in Contemporary Irish Art (1990–2015)*, 380–2, unpublished thesis, <https://dspace.mic.ul.ie/handle/10395/2045> accessed 30 March 2016.

Murray, Raymond, *The Burning of Wildgoose Lodge: Ribbonism in Louth, Murder and the Gallows* (Armagh: Armagh Diocesan Historical Society, 2005).

Patterson, T. G. F., 'The Burning of Wildgoose Lodge', *County Louth Archaeological Journal*, 12(2) (1950), 159–80.

Smyth, Gerry, *Space and the Irish Cultural Imagination* (London: Palgrave, 2001).

*Digital sources cited*

Gothicise website <http://www.gothicise.weebly.com> accessed 31 March 2016.

GAIL BAYLIS

# 3 The Easter Rising 1916: Photography and Remembrance

The Rising was conceived and implemented by a minority within a minority of Irish radical politics who did not hold a popular mandate. Originally planned as an all-Ireland affair it quickly dwindled into a Dublin-based insurrection that was quashed by British military forces within a week. However, in terms of collective memory it became a mythic event and one positioned as foretelling the creation of the Irish State. In order to understand this transformation, it is necessary to recognize that the Rising was a highly mediated affair: indeed, it has been claimed that it 'was a media event as much as a military operation'.[1] This chapter aims to elucidate the role of photography in the immediate aftermath of the insurrection in order to point to the emergence of modes of remembrance that were only made possible through the repositioning of photographs. It focuses on portrait photographs of the executed leaders that appeared within weeks of the events of Easter Week (1916) and considers their significance in creating a mediated context of remembrance. The coverage of the insurrection was modern in that it employed modern technologies of media and also in that these technologies created externalized remembrance, a key feature of modern memory.[2] It was 'the convergence of photography within

1 Christopher Morash, *A History of the Media in Ireland* (Cambridge: Cambridge University Press, 2010: 2012), 127.

2 Modernist theories of memory prioritize technologies of visualization, particularly photography and cinema. See Esther Leslie, 'Absent-Minded Professors: Etch-A-Sketching Academic Forgetting', in Susannah Radstone and Katharine Hodgkin, eds, *Regimes of Memory* (London and New York: Routledge, 2003), 172–85; and Morash, *A History of Media*, 103–15 for technologies of visualization in Ireland in the run-up to the Rising.

new communication network in the late nineteenth century that enabled its transformation into the dominant visual technology of modernity'[3] and, additionally, the preeminent mode of memorializing. A photograph, however, is not a memory in itself but it can become a means to that end through how it is used: as has been noted, 'the language of memory does seem to be above all a language of images'.[4] However, it needs to be stressed that media do not control remembrance, as the '"mediation of memory" equally refers to the perception of media in terms of memory as well as to the perception of memory in terms of media'.[5]

## Production

The overtly political postcard begins with the second Boer War (1899–1902), and its utility value as a commercially viable means to cater to markets eager for accessible news was not missed in the wake of the Rising. The Powell Press of 22 Parliament Street, Dublin, produced a series of postcards based on photographs of the Easter Rising leaders, those who were condemned to penal servitude and figures associated with the event. These postcards were printed using the halftone process, a mechanical reproduction technology attributed with encapsulating the 'rapid and ephemeral nature of modernity'.[6] The series was titled 'Irish Rebellion, May 1916' and in each postcard the sitter's name is given, a summary of his role and sentence. The dating of the series to May rather than April (when the insurrection took place) already acts as a mnemonic signifier for both the event and its consequences, the executions of the leaders which occurred in Dublin

3 Martin Hand, *Ubiquitous Photography* (Cambridge: Polity Press, 2012), 132.

4 Annette Kuhn, 'A Journey through Memory', in Susannah Radstone, ed., *Memory and Methodology* (Oxford and New York: Berg, 2000), 188.

5 José van Dijck, 'Mediated Memories: Personal Cultural Memory as Object of Cultural Analysis', *Continuum: Journal of Media and Cultural Studies*, 18(2) June (2004), 272.

6 Gerry Beegan, *The Mass Image: A Social History of Photomechanical Reproduction in Victorian London* (Basingstoke: Palgrave MacMillan, 2008), 2.

between 3 and 12 May. At this point, it is worth noting that the leaders were unknown as such to most in Dublin prior to and during the Rising and that the insurrection was not popularly received. The secrecy with which it was planned added to an air of bafflement and confusion as to the cause of the fighting. Few people knew which group was behind the insurrection and the attribution of it to Sinn Féin, while incorrect, indicates the lack of credible news available at the time.[7] My suggestion is that this left a memory void that required filling. In this respect it was the executions and not the actions of the insurgents that brought about a sea-change in public opinion. In large measure the circulation of portrait postcards afforded the means to give a face to a name and thereby a point of identification. The portrait is a 'sign whose purpose is both the description of an individual and the inscription of social identity'.[8] The choice by the Powell Press to employ the photographic portrait format was therefore affective: it invited a personalized relationship between viewer and subject represented and thereby a means for the portrait to enter a culture of popular remembering.

Information on the Powell Press is notable by its absence; however, by tracing the postcard image to its source it is possible to reveal the process by which it came about. The majority of the Powell Press series are based on photographs produced by the Keogh Brothers, a commercial photographic firm with, in 1916, premises at 75 Lower Dorset Street, Dublin. The firm had photographed all of those who later became recognized as the leaders of the Rising: the seven signatories to the Proclamation (Patrick Pearse, James Connolly, Tom Clarke, Seán MacDermott, Thomas MacDonagh, Joseph Plunkett and Éamonn Ceannt) and the other combatants who were executed in Dublin (Michael Mallin, Michael O'Hanrahan, Sean Heuston, Cornelius Colbert, Edward Daly, John McBride and Willie Pearse).[9] For some of the Powell postcards the Keogh source is a direct portrait, as in the case of MacDermott; for the majority, the individual head and shoulder image has been produced by 'picking out' from a larger group portrait.

7 Brian Feeny, *16 Lives: Sean MacDiarmada* (Dublin: O'Brien Press, 2014), 287.

8 John Tagg, *The Burden of Representation: Essays on Photographies and Histories* (Amherst: University of Massachusetts Press, 1988), 37.

9 While the majority of the postcards are based on Keogh photographs for Patrick Pearse, Willie Pearse and Connolly the source is Lafayette photographs.

This working method is visually recorded in the Keogh archive, an example of which is the process behind the single portrait postcard of Clarke. The source image is a large group portrait in which men appear seated or standing in rows. In the glass plate prototype for the postcard, tight cropping and enlargement have been employed and the effect, visually, is that the heads of the men in the row behind Clarke have been removed and the bodies of the men who appeared on either side of him are now only partially visible. The plate has also been worked on through over-drawing, shading and cross-hatching. Clarke's shoulders have been outlined in pencil; part of his arm, which was not visible in the original photograph, has been drawn in, and the presence of another man's hand has been eradicated by shading. Most noticeable is the over-drawing on a man's head [Figure 1]. Interventions such as these reveal the degree of manual manipulation involved in the reproduction of photographs in the early twentieth century.[10] In postcard form a further darkening of the backdrop, an oval mount, header title and caption direct the viewer's interpretation and render the single portrait as the apparently unmediated presence of Clarke [Figure 2].

A postcard of Heuston, another in the 'Irish Rebellion, May 1916' series, also derives from a large group portrait, in this instance of the Fianna Éireann (the Irish nationalist boy scouts) taken by the Keogh Brothers in 1913. In the original, Heuston appears in the second row, forth from the left wearing a Fianna uniform; his head is tilted slightly sideways. As with Clarke, the glass plate helps explain the process of transference to single portrait. In this instance the outline of burning in (darkening) reveals what will become the oval single portrait. Heuston's execution, given his relative obscurity prior to the Rising and the fact that he was the youngest of the condemned men to be shot, caused public disquiet. In these circumstances the postcard portrait served to individualize and personalize his fate as the tragedy of lost youth. Likewise, the postcard portrait of Colbert derives from another large group photograph of the Fianna taken by the Keogh Brothers and the effect is similar.[11] Here masking (bleaching out) and over-drawing have been employed to eradicate unwanted traces of the original

10 Beegan, *The Mass Image*, 177.

11 Gail Baylis, 'Boy Culture and Ireland 1916', *Early Popular Visual Culture*, 13(3) (2015), 202.

[Figure 3]. As with the other examples discussed, further cropping and darkening of the backdrop serves to render the image as a single portrait. Glass plate prototypes for the postcards reveal other instances of these procedures. In the instance of MacDonagh and Éamon de Valera the source photograph is a large group image of the 1915 O'Donovan Rossa Funeral Committee, again a Keogh photograph.

What these interventions with the photograph reveal is that photography in print form is far from the transparent medium of communication that it has been credited with; however, it becomes effective as such because by the beginning of the twentieth century it becomes easier to hide the signs of mediation. The process of producing the Powell postcards shared commonalities with the process of printing photographs in magazines: from the 1890s the cropping of photographic images began to 'exclude content and alter their meanings'; in other words, it realigned meaning.[12] The repositioning of photographs to individualize them occurred during a period when subjectivity enters the reading of the image through the ability of technologies of reproduction 'to rapidly circulate the appearance, personal details and opinions of personalities on a huge scale'.[13] In this context, the production of the Powell Press postcards as individual portraits acts to personalize an event and in the process afford the means for the making of heroes of those who partook in it.

## Display

One of the notable features of the Rising was the speed with which it was commodified: within weeks, souvenir brochures, postcards and a raft of ephemera appeared.[14] Postcards were displayed for sale in shop windows. As one observer noted in May 1916:

12 Beegan, *The Mass Image*, 15.

13 Ibid., 11.

14 Stephanie Rains, *Commodity Culture and Social Change in Dublin 1850–1916* (Dublin: Irish Academic Press, 2010), 206–7.

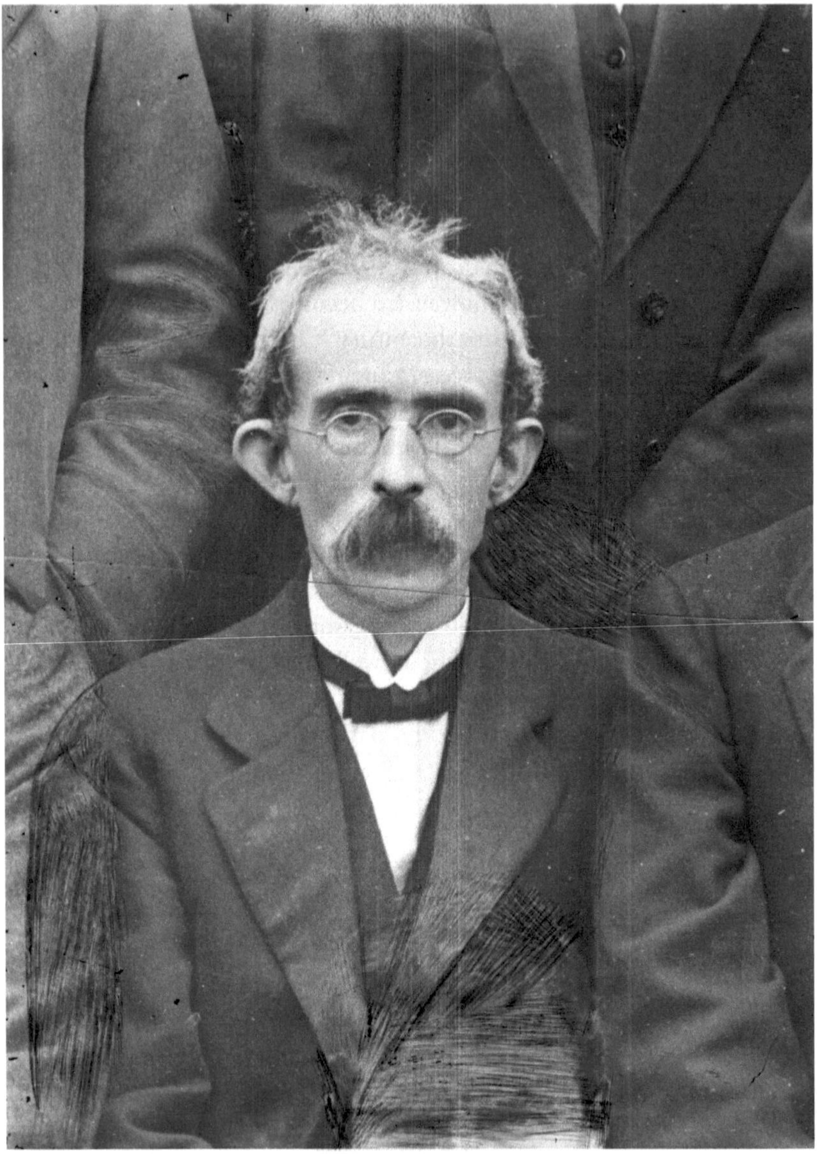

Figure 1: 'Working' glass plate for the repositioning of Tom Clarke's portrait. Keogh Brothers. Courtesy of the National Library of Ireland.

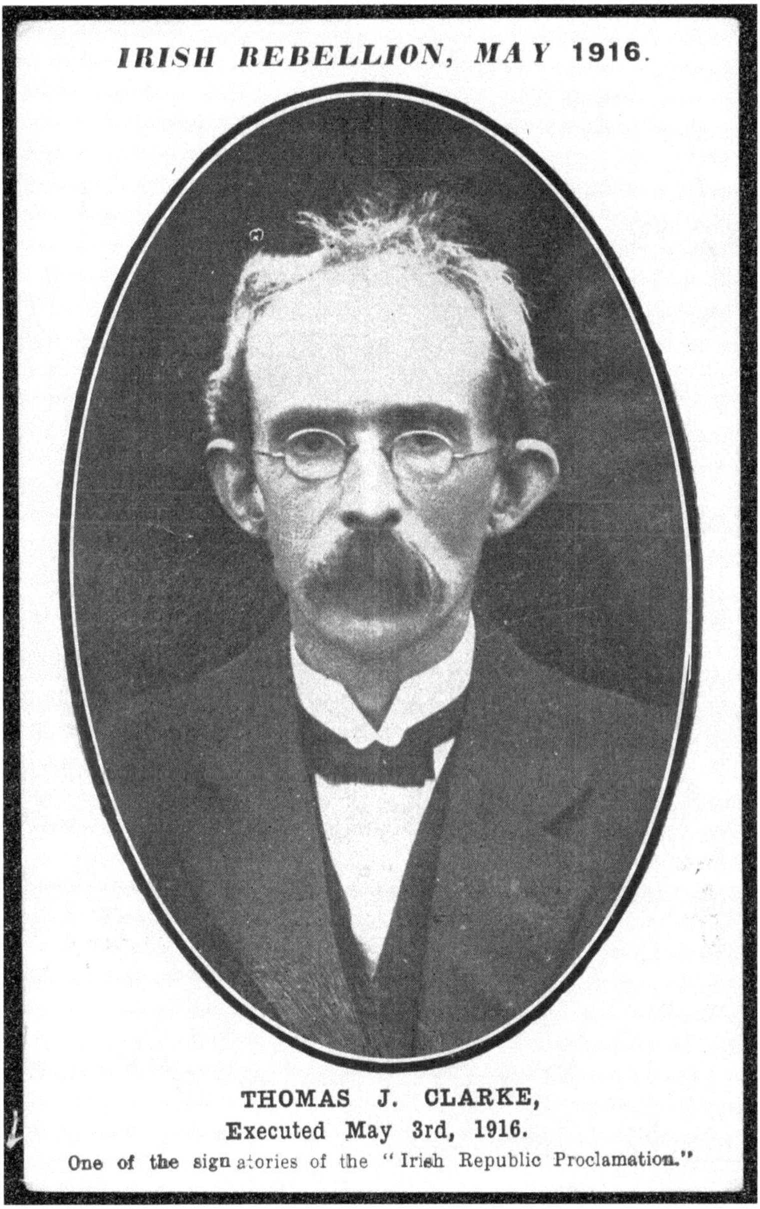


Figure 2: 'Irish Rebellion, May 1916' postcard of Tom Clarke, 1916. Powell Press. Courtesy of the National Library of Ireland.

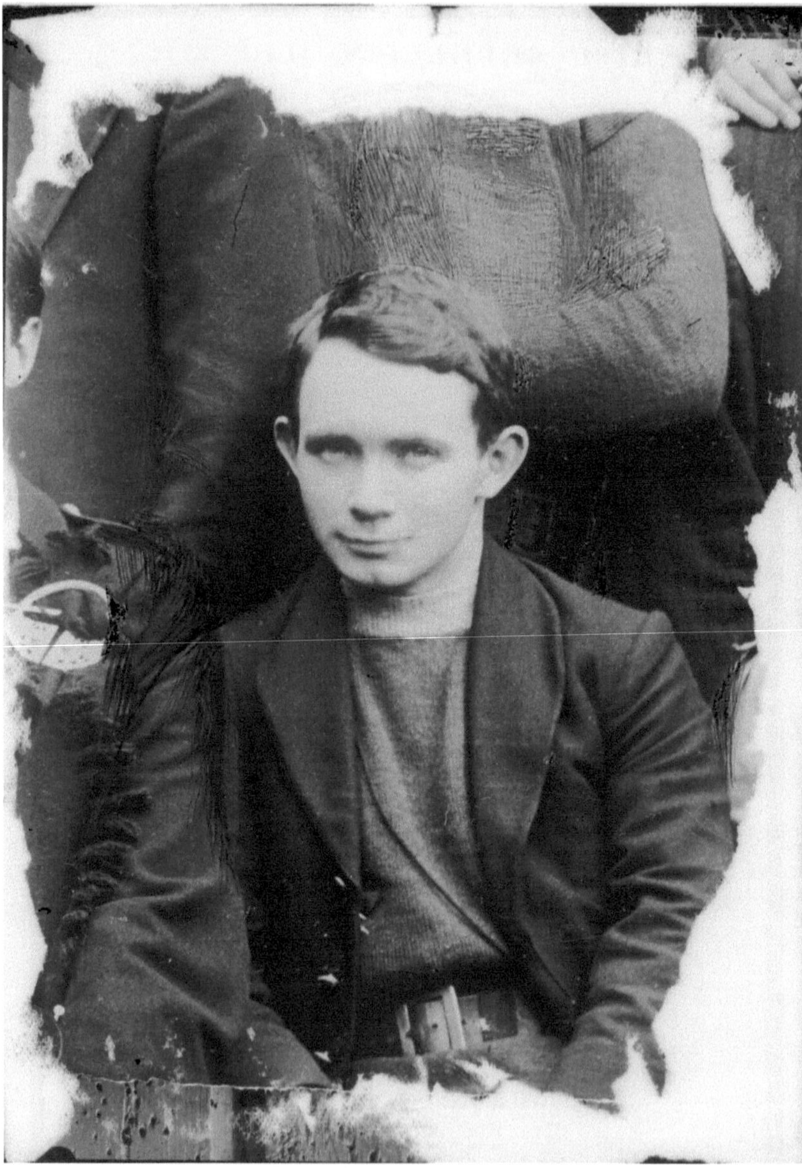

Figure 3: 'Working' glass plate for Cornelius Colbert. Isolating the figure interpolates the viewer to identify with Colbert's boyish looks and direct gaze. Keogh Brothers. Courtesy of the National Library of Ireland.

> Picture postcards of the executed rebels were displayed in almost every shop window, and their faces were gazed upon with silent veneration by the passer-by. A large photogravure of P. H. Pearse, produced by some enterprising firm, attracted crowds whenever it was displayed. Up and down Sackville Street [now O'Connell Street] urchins ran selling broad sheets purporting to contain 'The last and inspiring speech of Thomas MacDonagh'... So far as one could tell, except among the shopkeepers who had not received compensation for their losses and among the upper classes, all resentment against the Sinn Féiners had died away.[15]

Such advertising is evidence of the commodification of the Rising, how the postcard portrait came to be interpreted in a politicized context, how it served to give meaning to an event and how interpretation of the postcards occurred within a verbal and visual repertoire. And in this context, as Mieke Bal reminds us, 'we need to consider verbal imagery as part of visual culture'.[16] The significance of extra-textual referents is particularly apposite when considering how the Powell postcards initially communicated. The literary and theatrical dimensions of the Rising are now well recognized and indeed were at the time; the letters and poems penned by the condemned men were written with a view to posthumous effect. The photographic portrait postcard showed the man literally as in life but viewing relations were modulated by knowledge both of his execution and also by the context of the circulation of his last written testimony, which exalted willing sacrifice in the cause of nation. A photograph emanates presence, even when it is repositioned in a different content. What the Powell Press produced, albeit under commercial dictates, was a visual reference *in-anticipation* of memory.

It has been advanced that the 'patriotic cult' engendered after the executions depended in no small measure on the 'flood of rebel memorabilia' produced in its immediate wake, which was 'probably more influential than revolutionary ideas or texts'.[17] The Powell Press were not the only publishers to produce commemorative postcards that represented the executed

15 Douglas Goldring, *Dublin Explorations and Reflections by an Englishman* (Dublin: Maunsel Co. Ltd, 1917), 33.

16 Mieke Bal, 'Visual Essentialism and the Object of Visual Culture', *Journal of Visual Culture*, 2(5) (2003), 21.

17 Peter Hart, *The IRA and its Enemies: Violence and Community in Cork, 1916–1923* (Oxford: Oxford University Press, 1998), 207.

leaders. Sinn Féin was particularly active in producing commemorative ephemera: '[m]any of these items were mass-produced [...] making the most of the opportunities for propaganda'.[18] However, their postcards differed in being primarily autographic and clearly political. In contrast, the Powell Press postcards become political due to interpretative repertoires and context. There is nothing inherently political in these portraits, for all were initially produced from photographs taken some time before the events; titling and caption indicate sparse factual information and, possibly, shrewd marketing to a number of potential audiences. The postcards become political because of cultural needs. And here the medium becomes salient: photography's visual credentials as an aid for remembrance stems from its indexical nature. A portrait photograph offers a trace of the living person caught in time and space through a chemical process of fixing an image. A halftone, while offering a different material relationship, still carries a ghosting of that presence. A notable feature of these postcards was the cropped head and shoulder view which puts the 'viewer into an intimate position with the person seen ... and offers a point of psychological identification'.[19] The modulation between what the viewer sees and knows proved particularly effective for how the Powell postcards were understood in a culture of popular remembering.

The authorities were not blind to these effects: by June 1916 concerns were being voiced about how 'public sympathy' was being 'stimulated by the sale of photographs of the rebel leaders, and letters written by some of them on the eve of execution, together with mourning badges of green and black ribbon' which were readily available for purchase.[20] Colonel William Johnstone, Head of the Dublin Metropolitan Police, elaborated the nature of these concerns: 'there can be no doubt that the publication and sale of the leaflets, photographs etc. [...] has a tendency to keep alive the feeling of disloyalty and unrest'. He explained to the Under Secretary that the sale of pamphlets and leaflets 'can easily be dealt with', whereas the prohibition of the sale and exhibition of photographs proves more problematic, because

18 Clair Wills, *Dublin 1916: The Siege of the GPO* (London: Profile Books, 2009), 105.

19 David Bate, *Photography: The Key Concepts* (Oxford and New York: Berg, 2009), 75.

20 Royal Irish Constabulary Reports, June 1916 quoted in Charles Townshead, *Easter 1916: The Irish Rebellion* (London: Allen Lane, 2005), 308.

the 'photographs have already been published in newspapers whose loyalty cannot be questioned and are being sold by traders of the utmost respectability [...] The general public are buying them'.[21] Johnstone's observations indicate not only the difficulties of policing photographs under the Defence of the Realm Act (DORA), but also how the portrait photographs of the rebel leaders became incorporated into the popular culture of the Rising. His observations also make evident the speed with which the 'Irish Rebellion, May 1916' postcards were brought into circulation.[22]

Johnstone was not the only one to perceive that there was nothing political *per se* in the portrait photograph but that it could function in particular ways in politicized contexts. Publishers did not miss this opportunity and a plethora of posters commemorating the leaders were produced, most of which reproduced Powell postcards. The Powell Press was itself in a favourable position to take advantage of the mood for commemoration. It produced both broadside posters of the leaders and of the lesser figures. Both employed a grid arrangement of sixteen image layout that directly reproduced the portraits as they appeared in postcard format. Another poster was produced by Francis Rigney. In it a direct analogy is made between a popular republican song that commemorated the abortive 1798 rebellion by the United Irishmen and 1916. This poster includes a copy of the Proclamation and the music and words of the 1798 rebel song, rewritten for the Rising. Surmounting the song lyrics is a tri-part arrangement of Pearse, Connolly and Roger Casement's portraits that form a shamrock shape. Major locations of the rebellion are presented along with autographic figures of a 1798 rebel and 1916 Volunteer. Portraits of the other executed leaders flank the sides of the layout. Countess Markievicz appears presiding in the lower central section surrounded by flags, rifles and

21 Letter from Colonel William Johnstone to the Under Secretary (2 June 1916). In Joseph Brennan Papers, 'Miscellaneous Materials ... relating to the suppression of seditious literature', MS 26, 154 (National Library of Ireland).

22 Johnstone's evaluation was made at the beginning of June, which indicates that the postcards must have been produced some time after 12 May (the last executions in Dublin) and well before the end of that month to gain the sort of circulation that warranted police comment.

pikes; the symbols of both insurrections. The visual and literary arrangement serves to emphasize commonalities of pasts, not failure or death: the effect is to immortalize the rebels. All the portrait images are replications of Powell postcards.

Numerous copies of the Proclamation were produced with decorative Celtic borders. Powell postcard images were inserted within these borders, either by direct photographic reproduction or as woodblock print. A hierarchy begins to be confirmed with Pearse's portrait featuring at the top from which all the other portraits descend. The combination of reproduced photographs inserted within decorative patterning set around written text imbues to the modern print typeface of the Proclamation the quality of a precious illuminated manuscript. The popularity of last letters and speeches did not escape this type of embellishment. One of MacDermott's last letters is a case in point. In contrast to the other leaders, MacDermott was economical in the numbers of letters he wrote while awaiting execution, writing only two: one to his family and another to John Daly, a close associate. A facsimile of the letter to Daly was produced, in which stylistic flourishes in phraseology suggest a conscious awareness of posthumous effect: 'We die that the Irish nation may live. Our blood will – baptize and reinvigorate the old land. Knowing this it is superfluous to say how happy I am'. It has been claimed that the symbolic language MacDermott employs rather than being evidence of his religious fervour (he was highly critical of the Church) was a rhetorical style learnt from Pearse.[23] In facsimile form his words are encased in a loose flowing Arts and Crafts' style Celtic border with portraits of MacDermott (top left), John Daly (top right), Edward Daly (bottom left) and Clarke (bottom right) forming the four corner points of the design.

The border is clearly an additional adornment to the original letter; with the exception of the image of John Daly, all are reproductions of the 'Irish Rebellion, May 1916' postcards. John Daly and Clarke were veteran Irish Republican Brotherhood (IRB) men with whom MacDermott had close political and personal relationships. All three maintained close links with the Clan na Gael in North America. Clarke was instrumental

23 Feeney, *16 Lives*, 313.

in nurturing youthful republicans, with MacDermott being his closest and most single-minded protégé; between them they were the masterminds behind the Rising. MacDermott would have known Edward Daly through the IRB, from his numerous holidays at the Daly home in Limerick and his visits to the Clarkes.[24] The insertion of photographic portraits in the design signifies the production of a type of 'family tree' of advanced nationalism. An earlier group photograph of John Daly, Clarke and MacDermott had been sent to an associate in North America; postcard copies of it were in circulation. This suggests a possible intended market for the facsimile. The ability to mass reproduce photographs, together with the speed with which they could be disseminated to a number of audiences and in a range of formats, as in this example, makes evident the mediated context of the Rising's cultural positioning.

The Catholic Church in Ireland was not slow to co-opt the Rising to its own ends. Testimony from the Capuchin Friars who ministered to the condemned men in their final days was reported in the press. These accounts stressed the piety and suffering of the men and 'helped to shape the nationalist perception of the executions as a latter-day Passion'.[25] Such accounts whitewashed the republicanism and anti-clericism of the likes of Clarke and MacDermott, as well as the socialism of Connolly and Ceannt's refutation of the surrender to the end. In other word, the political and military underpinning of the Rising came to be replaced by a hagiographic account of saintliness. Masses were held for the executed leaders and these became a forum for the display of patriotic sentiments. Mass cards 'were printed with halos round the heads of the leaders'.[26] Requiem Masses continued well into 1917 and were timed to accord with the return of the detainees.

Memorial cards were produced carrying the photographic portrait of the executed leader, often a cropped image of the Powell postcard or one that shared commonalities with it, being a close-up head and shoulders

24 Edward Daly was a nephew of John Daly, brother of Kathleen Clarke who was married to Tom Clarke.

25 Fearghal McGarry, *The Rising Ireland: Easter 1916* (Oxford: Oxford University Press, 2010), 272.

26 Wills, *Dublin 1916*, 112.

view. These cards, combining image and text, exhorted the owner to pray for the soul of the departed who, it is stated, had given his life for Ireland. The conflation of image, text and religious incantation imbued to the reproduced portrait the status of the religious icon, which in turn came to be attributed to the postcards. One adolescent, radicalized by the executions, makes evident how this connection was formed: 'In the early mornings Mother and I went into town to the Franciscan or Augustinian church where Mass was said for the dead rebels, and on the way back we bought picture postcards of them.'[27] The mere sight of the photographic postcards induced powerful religious connotations as one observer in Dublin recorded.

> I remember once when I was standing outside a shop in which photographs of all the Sinn Féin leaders were on view, a very old man came and stood beside me and looked at the faces of the dead, with tears streaming down his cheeks. Suddenly he turned to me and caught my arm. 'And they were all Sodality boys …' he said, in a voice broken with emotion, 'all Sodality boys! All near to the Sacred Heart of Jesus!'[28]

The repositioning of photographs as current visual perception, made possible by technologies of reproduction, induce the capacity for highly charged responses. Memory 'essentially emerges from the present ascriptions with their respective conditions of knowledge-production and usage of the past'.[29] How the past, even as a relatively recent event, is remembered does not necessarily correspond with the historical facts. Window-display, in this instance, prompts visual metaphrase: in the interpretation of the old man mass-produced imagery translates as evidence of the pious brotherhood ('sodality boys') and the iconic. As this encounter indicates the portrait photographs of the dead leaders could act as powerful signifiers: in photography 'presence [at a certain past moment] is never metaphoric'.[30]

27 Frank O'Connor, *An Only Child* (London: Readers Union/Macmillan, 1962), 110.

28 Goldring, *Dublin Explorations*, 33–4.

29 Gregor Feindt, Félix Krawatzek, Daniela Mehler, Friedemann Pestel and Rieke Trimçev, 'Entangled Memory: Toward a Third Wave in Memory Studies', *History and Theory*, 53 (2014), 28.

30 Roland Barthes, *Camera Lucida: Reflections on Photography*, trans. Richard Howard (London: Vintage, 1993), 78.

## Uses

Photographs are not only two dimensional visual images; they are also artefacts that carry physicality in how they communicate, can be experienced and used. Photographs 'exist in time and space and thus in social and cultural experience [...] they occupy spaces, move into different spaces'.[31] The Powell postcards were handled and collected but rarely sent. The postcards' popularity was indicated by their display in houses all over Ireland[32] and by their incorporation into other display formats, such as the poster and facsimile. The transference of photographs into physical objects brought a haptic dimension to remembrance: 'when fitted into a frame, photographs become three-dimensional; and that framing makes us aware that the photograph is a thing as well as an image'.[33] It elicits touch as well as sight as the mode of remembering.

The Powell postcards were transferred onto a plethora of ephemera ranging from badges and lockets to decorative household objects. Such artefacts held physical presence and demanded an embodied relationship. These shifts in material form acted as both 'signals' and determinants for 'different expectations and use patterns';[34] they also called into play different patterns of remembering. Referring to 'the modification of pieces of mass-produced material', Jack Elliot discusses how cheap tin lockets with flimsy cardboard backings into which newspaper reproductions of the 1916 leaders' portraits had been inserted were 'integral to the ways in which individual responses to the Rising were modified'.[35] The newspaper clippings were of Powell postcards, which were inexpertly in-laid,

31 Elizabeth Edwards and Janice Hart, 'Introduction: Photographs as Objects', in Elizabeth Edwards and Janice Hart, eds, *Photographs, Objects, Histories: On the materiality of images* (London: Routledge, 2004), 1.

32 Maurice Walsh, *The News from Ireland: Foreign Correspondents and the Irish Revolution* (London: I. B. Taurus, 2008), 54.

33 Geoffrey Batchen, *Forget Me Not: Photography & Remembrance* (New York: Princeton Architectural Press, 2004), 40.

34 Edwards and Hart, 'Introduction', 3.

35 Jack Elliott, '"After I hang my portrait will be interesting but not before": Ephemera and the Construction of Personal Responses to the Easter Rising', in Lisa Godson

leaving some of the frame backing exposed. Such items reveal a level whereby hand-craft and mass-produced image integrate in the making of memory. The ring on the locket highlighted that it was meant to be worn, either around the neck by a chain or on clothing by the insertion of a pin. Such hand-crafted objects brought the photograph literally to the body and made the experience of the photograph physical. The wearing of lockets, brooches and badges with the transferred photographic likeness of one of the leaders on it acted both as a statement of patriotic fervour and also as a more personalized, physical experiencing of remembering.

The postcards were also embellished in other ways that brought with them their own memorializing contexts. These involved hand-crafted additions to the display of the photograph in a frame and the decorating of the postcard. Such additions inserted a personalized relation to the image of the dead. A photograph of Plunkett, familiarized through the circulation of the Powell postcards, appears encased in a crocheted tracery. The decorative addition indicates a memorial context, one personalized by the craft work. The crochet design uses three colours, gold, white and green, and it fans out to form a casing to the photograph that interjects a three-dimensional aspect to the image. The colours chosen were ones that became emblematic of popular nationalism after 1916. The inclusion of a bow positioned above Plunkett's head references a tradition of the wreath knot [Figure 4]. This type of craft adornment exemplifies the hybrid nature of photography and how it could be translated from mass produced imagery into a personalized familial memory talisman. The labour involved in this type of handcraft, as with embroidery, as Geoffrey Batchen notes in a wider context, 'ensured that the act of remembrance would be painstaking, extended through time, deliberated. The object also thereby manages to simultaneously refer inside and outside itself, to the time when the photograph was taken and to the later act of encircling it'.[36]

Visible on the verso is a diamond shape produced from the crochet skeins that hold the hand-crafted addition to the photograph in place. It

---

and Joanna Bruck, eds, *Making 1916: Material and Visual Culture of the Easter Rising* (Liverpool: Liverpool University Press, 2015), 92.

36 Batchen, *Forget Me Not*, 87.

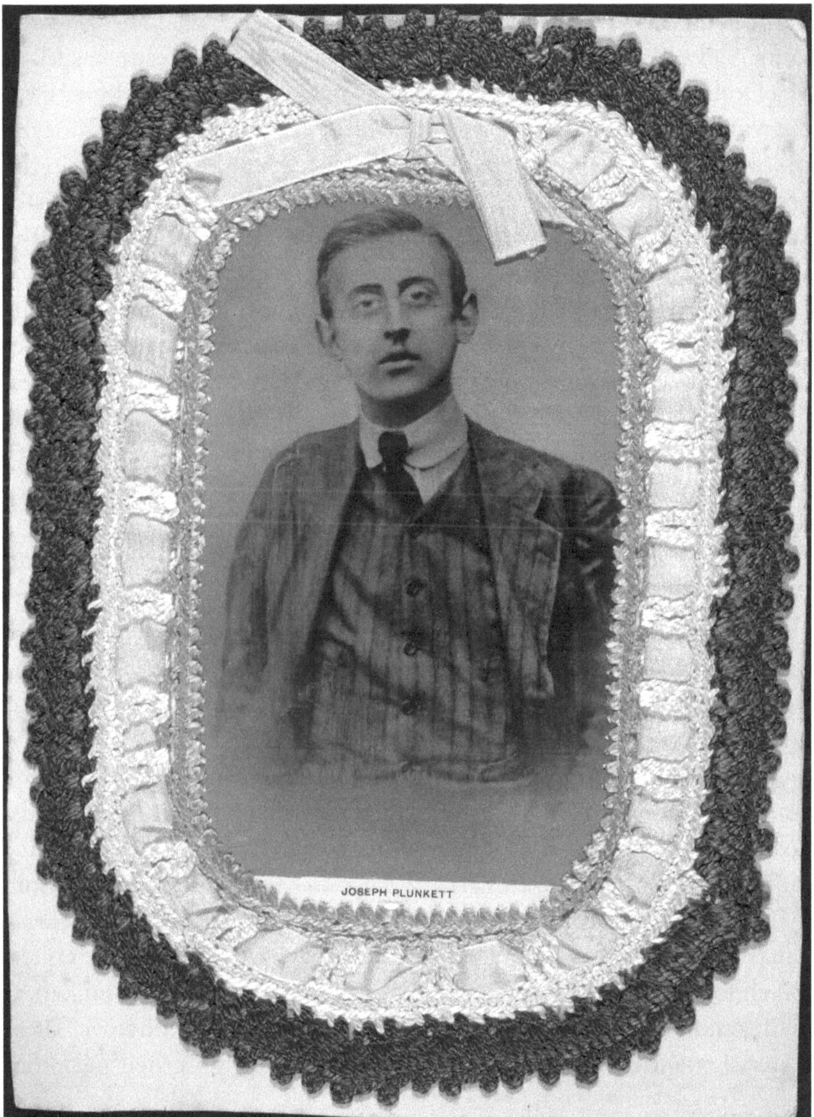


Figure 4: Joseph Mary Plunkett. The hybrid image shifts relations of mediation. Courtesy of the National Library of Ireland.

has been used as a frame in which text has been placed, which reads: 'given to me S[ister] Francesca by Grace Plunkett.' Sister Francesca was Mary MacDonagh, the religious sister of MacDonagh; Grace (nee Gifford) was the widow of Plunkett[37] and sister of Muriel MacDonagh, MacDonagh's widow.[38] The written inscription inserts the death of Plunkett within the orbit of a family tragedy. At the lower frame of the mount verso Sister Francesca has added: 'Shot at Kilmainham May $4^{\text{th}}$ 1916/ He died for the Land he loved'. This written commentary reiterates the standard phraseology found on memorial cards. The use here of both personal and public address highlights how a photograph can inhabit two spaces simultaneously. The diamond outline, a product of hand labour, encases an expression of personal loss while the mass produced photo mount that of public memorializing. The faded handwriting adds another level of materiality to the image, that of age and it speaks of what was then and how the photograph communicates now.

Another mounted photograph of Plunkett has been inscribed by Sister Francesca; this time she is the sender not the receiver of the image. This photograph is another hybrid, an enlarged reproduction of the image that appears on a memorial card for Plunkett, which is itself taken from a portrait photograph of him. The mount to the photograph hides most but not all of the imperative address found on the memorial card to 'Pray for the Repose of the Soul of Joseph Mary Plunkett'. Inscribed on the verso is a similar public address as on the crochet-adorned photograph. It reads: 'Joseph Mary Plunkett shot at Kilmainham Jail May $4^{\text{th}}$ 1916/ He died for the Land he loved'. Below this Sister Francesca has added, 'To his Godson. Donagh MacDonagh/ from his Godmother and aunt, Sister Francesca/ Xmas 1916'. Plunkett was the godfather of Donagh MacDonagh, the fatherless child of Thomas MacDonagh. The portrait photograph (Plunkett as in life) and text imbue in this one artefact a message from the grave. These material qualities evidence how photographs can shift in their meanings

37 Her marriage to Plunkett in Kilmainham gaol hours before his execution became one of the romantic tales of the Rising.

38 Thomas MacDonagh Family Papers, <http://www.nli.ie/pdfs/mss%20lists?131_ThomasMacDonagh.pdf>.

and occupy more than one space. Here there is both public articulation and also a private register of memory meaning.

The Rising occurred at the height of the postcard collecting craze and the vogue for displaying postcards in albums. Albums afford their compilers the means to store, organize and determine the meaning of photographs. They also add another layer of materiality to the photograph: albums are 'tactile objects with moveable parts, and to be experienced fully [...] demand that we add the physical intimacy of touch to the more distanced apprehension of looking'.[39] The sequencing of photographs, hand-written commentary and the addition of other materials all further affect how photographs will communicate in an album. One Irish postcard album that documents the Rising was produced by Art O'Murnaghan (1872–1954). He was an artist who was active in the Gaelic League, friend of Arthur Griffith, leader of Sinn Féin, and producer of *Leabhar na hAiseirghe* (*Book of Resurrection*), a folio decorative manuscript commemorating those who had died in the struggle for Irish independence.

The hand written inscription on the fly-leaf reads, 'Postcards illustrating Irish Rising, 1916. The Gift of Art O'Murnaghan (1927)'.[40] The dating of the album is significant as it corresponds with the completion of *Leabhar na hAiseirghe* and it indicates that O'Murnaghan was framing his dedication to 1916 within the context of later revolutionary events. Measuring twenty-two by twenty-nine centimetres this sparsely embossed dark blue covered album holds all the Powell postcards, views of Dublin in ruins, scenes of the roundup of prisoners, scenes at the prison gates, as well as postcards of the return of detainees at Westland Row, Dublin in 1917. The sequencing of the postcards creates a narrative of meaning. The first numbered page displays 'Irish Rebellion May 1916' postcards of Pearse and Clarke, followed by a sequencing of the executed, those killed in action, those condemned to penal servitude, as well as images of those who were released from prisons in 1917.[41] On the opening pages O'Murnaghan has added his own

39 Batchen, *Forget Me Not*, 49.

40 Album 113, National Library of Ireland.

41 On the opening cover leaf appear portraits of Connolly and MacDermott. It is unusual to have materials pasted onto the cover leaf of an album, which suggests that these images may well be an addition to the initial compilation of the album.

commentary on some of the postcards: he has scored through the word 'Rebellion' in the Powell title and written above it in capitals 'RISING'. This iteration makes clear how the owner of the portrait postcard could modify its meaning and interject a personal reading of significance. Loose items placed at the back of the album are accounts of memorials for Monsignor Hickey, Dean of University College Dublin who died in 1925. All these items, through their material display format and the personalized traces evident on them indicate that media may be a source for remembrance but within this, modes of remembering create object-ness from such sources. This makes apparent the physicality of photographs; they are material presences that are used and made relevant by personal ordering and additions and, additionally, are continually modified by shifts in remembering, as revealed in this album where we witness memory in process.

## Conclusion

It is the case that media texts of memorializing affect memory as a modern phenomenon but equally memory inhabits those texts and in the process changes their meaning. The 'Irish Rebellion, May 1916' postcards were produced to cater to a market that needed to position a recent event that was little understood at the time. In this sense, the postcards served as a remembrance in place of the event, and one that drew on earlier photographs, which were modified to suit a specific cultural space. Standardized titles and captioning led viewers in how to read the postcards but it did not fully constrain their meaning. This is because the meaning of a photograph is dialogic: 'it comes about through rather than existing prior to interpretation'.[42] Charting the use of the portraits reveals that they moved into different spaces. Sometimes, as outlined, they were adapted to occupy two spaces at once: that of public declaration and that of personal recall.

42 Bal, 'Visual Essentialism', 24.

This study has focused on some of the now lesser remembered executed leaders in order to reveal how in remembering 'people, media, mnemonic forms, contents, and practices are in constant, unceasing motion'.[43] It has indicated some of the ways in which the Easter Rising became invested with personal needs through how photographs were displayed, added to and decorated. The outcome was that the photograph becomes a hybrid object that was experienced as an embodied memory form, and one that involved touch as well as sight. Usage indicates how photographs in remembrance extend their two-dimensionality to gain material and object presence. The significance of the repositioning of photographs to meet a specific cultural need has also been foregrounded. The isolating of the figure in postcard format served to afford a means whereby people could both individualize and collectivize the meaning of the Rising and invest it with heroic status. In this, the portraits of the leaders can be seen as serving in the production of what would come to be collective memory while at the same time revealing submerged micro-histories of personal relevance that have been overwritten by the production of a hegemonic nation narrative.

## Bibliography

Bal, Mieke, 'Visual Essentialism and the Object of Visual Culture', *Journal of Visual Culture*, 2(5) (2005), 5–32.

Barthes, Roland, *Camera Lucida: Reflections on Photography*, trans. Richard Howard (London: Vintage, 1993).

Batchen, Geoffrey, *Forget Me Not: Photography & Remembrance* (New York: Princeton Architectural Press, 2004).

Bate, David, *Photography: The Key Concepts* (Oxford and New York: Berg, 2009).

Baylis, Gail, 'Boy Culture and Ireland 1916', *Early Popular Visual Culture*, 13(3) (2015), 192–208.

43 Astrid Erll, 'Travelling Memory', *Parallax*, 17(4) (2011), 12.

Beegan, Gerry, *The Mass Image: A Social History of Photomechanical Reproduction in Victorian London* (Basingstoke: Palgrave MacMillan, 2008).

Dijck, José van, 'Mediated Memories: Personal Cultural Memory as Objects of Cultural Analysis', *Continuum: Journal of Media and Cultural Studies*, 18(2) (June 2004), 261–77.

Edward, Elizabeth, and Hart, Janice, 'Introduction: Photographs as Objects', in Elizabeth Edwards and Janice Hart, eds, *Photographs. Objects, Histories: On the Materiality of Images* (London: Routledge, 2004), 1–15.

Elliott, Jack, '"After I am hanged my portrait will be interesting but not before": Ephemera and the Construction of Personal Responses to the Easter Rising', in Lisa Godson and Joanne Bruck, eds, *Making 1916: Material and Visual Culture of the Easter Rising* (Liverpool: Liverpool University Press, 2015), 91–8.

Erll, Astrid, 'Travelling Memory', *Parallax*, 17(4) (2011), 4–18.

Feeney, Brian, *16 Lives: Sean MacDiarmada* [MacDermott] (Dublin: O'Brien Press, 2014).

Feindt, Gregor, Félix Krawatzek, Daniela Mehler, Friedemann Pestel and Rieke Trimçev, 'Entangled Memory: Toward a Third Wave in Memory Studies', *History and Theory*, 53 (2014), 24–44.

Golding, Douglas, *Dublin Exploration and Reflections by an Englishman* (Dublin: Maunsel and Co. Ltd, 1917).

Hand, Martin, *Ubiquitous Photography* (Cambridge: Polity Press, 2012).

Hart, Peter, *The IRA and its Enemies: Violence and Community in Cork, 1916–1923* (Oxford: Oxford University Press, 1998).

'Joseph Brennan Papers', 'Miscellaneous Materials … Relating to the Suppression of Seditious Literature', MS26, 154, National Library of Ireland.

Kuhn, Annette, 'A Journey Through Memory', in Susannah Radstone, ed., *Memory and Methodology* (Oxford: Berg, 2000), 179–96.

Leslie, Esther, 'Absent-Minded Professors: Etch-a-Sketching Academic Forgetting', in Susannah Radstone and Katharine Hodgkin, eds, *Regimes of Memory* (London and New York: Routledge, 2003), 172–85.

MacDonagh, Thomas, Family Papers, <http://www.nli.ie/pdfs/mss%20 lists?131_ThomasMacDonaghpdf>.

McGarry, Fearghal, *The Rising Ireland: Easter 1916* (Oxford: Oxford University Press, 2010).

Morash, Christopher, *A History of the Media in Ireland* (Cambridge: Cambridge University Press, 2010: 2012).

O'Connor, Frank, *An Only Child* (London: Readers Union/MacMillan, 1962).

Rains, Stephanie, *Commodity Culture and Social Change in Dublin 1850–1916* (Dublin: Irish Academic Press, 2010).

Tagg, John, *The Burden of Representation: Essays on Photographies and Histories* (Amherst, MA: University of Massachusetts Press, 1988).
Townshead, Chris, *Easter 1916: The Irish Rebellion* (London: Allen Lane, 2005).
Walsh, Maurice, *The News from Ireland: Foreign Correspondents and the Irish Revolution* (London: I. B. Taurus, 2008).
Wills, Clair, *Dublin 1916: The Siege of the GPO* (London: Profile Books, 2009).

# PART II

# Contested Politics

EVE MORRISON

# 4 Hauntings of the Irish Revolution: Veterans and Memory of the Independence Struggle and Civil War

Separatist nationalist veterans of the Irish independence struggle and civil war (1913–1923) were a generation of transition, their memories stretching from the old Ireland in which they were reared – the Ireland they fought against – through the independence struggle itself and continuing into the new, post-revolutionary, partitioned jurisdictions created by the conflict. A host of official and independent projects were initiated in the decades from 1922 up to the 1980s to gather testimony and documents from them. The vast majority of contributors were former members of the Irish Volunteers or Irish Republican Army (IRA).[1] Much of the material has only recently come to light or become easily accessible, and there is almost certainly more awaiting discovery. Personal testimony from separatist veterans is a rich source for exploring both the autobiographical and social memory of the Irish Revolution. Veterans giving or conducting interviews, writing statements or articles, memoirs and letters were active agents of memory in a public sphere of well-known published accounts and private disputes. Heroic public or published narratives and 'fighting stories' had the highest profile, but they were rarely representative of its totality.[2] Veterans also influenced, and were influenced by, local memory.

1 Eve Morrison, 'Bureau of Military History Witness Statements as Sources for the Irish Revolution', (Military Archives of Ireland [hereafter MAI] website, 2010) <http://www.bureauofmilitaryhistory.ie/files/Bureau_of_Military_witness_statements%20as_sources%20for_the_Irish%20Revolution.pdf>.

2 Eve Morrison, 'Kilmichael Revisited: Tom Barry and the "False Surrender"', in David Fitzpatrick, ed., *Terror in Ireland: 1916–1923* (Dublin: Lilliput Press/Trinity History Workshop, 2012), 158–80.

Oral history, personal testimony and received tradition (folklore) are not interchangeable categories, but there is no neat dividing line between them either. This chapter explores the architecture of Irish revolutionary memory. It examines the interfaces, cross-overs and interactions between veterans' personal accounts, written sources and orally transmitted knowledge systems, and the interrelationship between different sources of information: local, autobiographical, oral, written, contemporary and retrospective. Also considered is the context and circumstances in which testimony was given and the role, interests and expectations of interviewers.

## Survival of the Supernatural

One of the stand-out characteristics of the Irish Folklore Commission's (IFC, in operation from 1935 to 1971) archive is the sheer volume of supernatural legend it contains.[3] In veterans' accounts this aspect of Irish popular culture is much less evident, although there are indications that it informed their life experience. In February 1968, Gerry Mulhern, a former member of Dundalk Fianna Eireann, wrote a letter to Fr. Louis O'Kane describing his first trip back in fifty years to County Tyrone:

> I was afraid to go because I thought the images I had of it would be shattered ... I was pleasantly surprised ... the atmosphere and the structure of impressions remained unchanged. I remember Fivemiletown, the bigotry aside, as one of the most typically Irish towns in the traditional sense. It had all the old-time pishogues, superstitions and beliefs in the occult, mumbo-jumbo cures and ghosts. Every field had a history, and in the margin of every country road was the evidence of long ago – the ruins and the marks of the people plain and strange who dwelled there. The route of the Black Pig was there too. The sites of the *Feir Gorta* [Hungry Grass] were well known, & my Uncle never went out where they were without oaken bannock in his pocket. ...

3 Niall Ó Ciosáin, 'Approaching a Folklore Archive: The Irish Folklore Commission and the Memory of the Great Famine', in *Folklore*, 115 (2004), 223; Angela Bourke, *The Burning of Bridget Cleary: A True Story* (London: Pimlico, 1999), 27.

> The Cunnion ['Cooneen' or 'Coonian'] Ghost occurred when I was there in 1918 & I've heard the BBC had it on TV recently. My Grandmother was the only survivor of her family during the Famine.[4]

Folklore is a term coined in the mid-nineteenth century to describe the 'non-written cultural inheritance and exchanges of a group of people'.[5] One of its defining characteristics is fluidity. It reflects how communities remembered and understood their past and is at the same time 'alive and real in the present, contemporary world'.[6] Studies of popular culture as well as many memoirs reflect the persistence of a mix of pre-Christian and Christian supernatural beliefs in Ireland, confirming that Mulhern's description of a 'typical' Irish town was just that.[7] The Cooneen poltergeist case occurred in 1913 (not 1918), but this famous Fermanagh ghost story is told to this day.[8] Whether or not people actually believe it to be literally true is a moot point.

Ireland was, in this respect, far from unique. Studies of British and European popular culture suggest that the extent to which 'dechristianization',

4 Gerry Mulhern to Louis O'Kane, 29 February 1969 (Cardinal Tomás Ó Fiaich Library and Archive [hereafter CÓFLA], L. O'Kane interviews [hereafter LOK] IV.C.48, Box 0003, Folder 04); Fianna Eireann, Dundalk, 1st Critical Date (MAI/Military Service Pension Project [hereafter MSPC]/FE/23/1).

5 Anne Markey and Anne O'Connor, *Folklore and Modern Irish Writing* (Sallins: Irish Academic Press, 2014), 6.

6 Ibid.; Clodagh Tait, 'Wandering Graveyards, Jumping Churches and Rogue Corpses: Tolerance and Intolerance in Irish Folklore', in James Kelly and Mary Ann Lyons, eds, *Death and Dying in Ireland, Britain and Europe: Historical Perspectives* (Sallins: Irish Academic Press, 2013), 277–96, 278.

7 Anne O'Connor, *The Blessed and the Damned: Sinful Women and Unbaptised Children in Irish Folklore* (Oxford; New York: Peter Lang, 2005), 55; Rìonach uí Ógáin and Tom Sherlock, *The Otherworld: Music & Song From Irish Tradition* (Dublin: Comhairle Bhealoideas Eireann, 2012), 7, 139; 'Extracts from the memoirs of Florence John Crowley', in *Dunmanway Doings*, Journal IV, Dunmanway Historical Society (2010), 16–29, 18–20; Cathal Ó Searcaigh, *Light on Distant Hills* (London: Simon & Schuster, 2009), 87–99; Benedict Kiely, *Drink to the Bird: a Memoir* (London: Methuen, 1991), 26–9; David Thompson, *Woodbrook* (London: Vintage, 1994; 1st edn 1974), 267.

8 Frank Galligan, 'Spooked by the Cooneen Ghost', in *Tyrone Herald*, 24 September 2007; John J. Dunne, *Irish Ghosts* (Appletree: Belfast, 1999), 54.

'secularism', 'modernism' and the decline of religious institutions had eroded belief in the supernatural in Western societies was greatly exaggerated at the time. Evangelical revivals in Scotland and Ireland, Great War trenches, rural France and interwar Britain were awash with visions, miracles, faith healers, spiritualists, fairies, animate statues of saints and apparitions.[9] According to one local British folklorist's calculation in the 1940s, every mile of the Warwickshire countryside had its own ghost.[10]

Ireland is similarly rich in 'remembered' history, but it remains a loaded concept, weighted down by polemic. Popular memory of colonialism and Irish attempts to challenge British rule is either defended as the authentic nationalist tradition of the risen people or derided as a dangerous mythology promoting (or eliding the bloody reality of) political violence.[11] The narrowing confines of the 'revisionist' debate tend to lock historical analysis of memory and oral testimony into the most limiting kind of positivism. It has been over twenty years since Gearóid Ó Tuathaigh pointed out that 'revisionists' and 'anti-revisionists' alike were making too many assumptions about popular historical consciousness and *mentalité*, and that Irish historians needed to pay 'careful attention to a wide range of sources and texts

9 Judith Devlin, *The Superstitious Mind: French Peasants and the Supernatural in the Nineteenth Century* (New Haven, CT; London: Yale University Press, 1987), 5–6, 15–18, 43; Joseph Clarke, 'Rethinking Death in the Year II: The Dechristianization of Death in Revolutionary France', in James Kelly and Mary Ann Lyons, eds, *Death and Dying in Ireland*, 143; Owen Davies, *The Haunted: a Social History of Ghosts* (Basingstoke: Palgrave Macmillan, 2009), 249; Daniel Richie, 'The 1859 Revival and its Enemies: Opposition to Religious Revivalism within Ulster Presbyterianism', *Irish Historical Studies*, 40 (2016), 69; D. W. Bebbington, *Victorian Religious Revivals: Culture and Piety in Local and Global Contexts* (Oxford: Oxford University Press, 2012), 47, 174–5; J. M. Winter, *Sites of Memory, Sites of Mourning: The Great War in European Cultural History* (Cambridge: Cambridge University Press, 1996), 64–77; Jenny Hazelgrove, *Spiritualism and British Society Between the Wars* (Manchester: Manchester University Press, 2000), 13–14.

10 Davies, *The Haunted*, 1.

11 Desmond Fennell, 'Against Revisionism', in Ciaran Brady, ed., *Interpreting Irish History: the Debate on Historical Revisionism 1938–1994* (Blackrock, Co. Dublin: Irish Academic Press, 1994), 183–90; R. F. Foster, 'History and the Irish Question', *Transactions of the Royal Historical Society*, 33 (1983), 169–92, 191.

of a kind that many [...] are not accustomed to handling', and to make a 'more systematic, and a more subtle, examination of various historiographical currents'.[12] Almost a decade has passed since Guy Beiner's *Remembering the Year of the French* urged historians to accept the 'radical challenge' presented by folk memory.[13] Yet mainstream historians of modern Ireland still generally ignore broader scholarship relating to oral history, memory studies, historical geography and anthropology. 'Folklore' remains largely a term of dismissal and abuse.

In reality, as we shall see, the surviving oral and/or written evidential fragments of personal and social memory relating to the revolutionary period are much too diverse either to conform or to condone any one set of beliefs. The inherent subjectivity and dynamism of memory demands a more subtle approach. Folk narratives of historical events and oral testimony retaining folkloric characteristics are ill-served by conventional methodologies. In an essay on the memory of 1798, for instance, Roy Foster used an anecdote about superstitious locals refusing to cut down a certain tree associated with 1798 executions to illustrate how historical memory and commemoration represses and sanitizes violence. However, the story, even as Foster tells it, can be read as an example of how local tradition actually preserved and protected a *lieu de mémoire*. The tree proved to be so full of lead from bullets fired into rebels tied to it in 1798 that the saw blades ground to a halt when men (from outside the area) were brought in to cut it down a century later.[14] Popular historical memory often collected around particular sites and areas, in what Guy Beiner refers to as the 'vernacular

12 M. A. G. Ó Tuathaigh, 'Irish Historical Revisionism', in Brady, ed., *Interpreting Irish History*, 324.

13 Guy Beiner, *Remembering the Year of the French: Irish Folk History and Social Memory* (Madison, WI; London: University of Wisconsin Press, 2007); Guy Beiner, 'Recycling the Dustbin of Irish History: The Radical Challenge of "Folk Memory"', *History Ireland*, 14 (2006), 42–7.

14 R. F. Foster, 'Remembering 1798', in McBride, ed., *History and Memory in Modern Ireland* (Cambridge: Cambridge University Press, 2001), 67, 93; For a more nuanced 'revisionist' examination of 1798 see Tom Dunne, *Rebellions: Memoir, Memory and 1798* (Dublin: Lilliput Press, 2004).

landscape'.[15] The tree was a tangible marker of past violence. Removing it was the equivalent of book burning.

'Reading' personal testimony and social memory effectively needs more than a sound knowledge of contemporary printed and archival primary sources (essential though this is). There are numerous historical, historiographical and analytical contexts to be negotiated. Being overly critical of oral sources is just as unhelpful as is using them too naively. Veterans' recollections are not always factually accurate, but to *assume* they are not without proof is not evidence-based reasoning (the baseline of all historical enquiry).[16] Yet all too frequently, the rhetoric of 'rational' discourse is employed to discredit oral testimony without considering the extent to which that testimony conforms or deviates from the cultural norms and *mentalities* of the society in which the subject lived, or the circumstances in which the information was imparted. Should Cork IRA Captain Martin Corry's account of 'Sing Sing' (the infamous underground vault in a Knockraha graveyard used by the IRA as prison for captured prisoners) really be ignored or discounted because at one point he described an encounter with a phantom black dog while on the run?[17] Black dogs are a well-documented motif in Irish and British folklore.[18] Jim Fitzgerald,

15 Guy Beiner, 'Mapping the "Year of the French": the Vernacular Landscape of Folk Memory', in C. E. J. Caldicott and Anne Fuchs, *Cultural Memory : Essays on European Literature and History* (Oxford: P. Lang, 2003), 191–208, 198; Erika Kuijpers, Judith Pollmann, Johannes Müller, Johannes and Jasper van der Steen, *Memory Before Modernity : Practices of Memory in Early Modern Europe* (Leiden: Brill, 2013), 153; Patrick J. Duffy, *Exploring the History and Heritage of Irish Landscapes* (Dublin: Four Courts, 2007), 209–18.

16 For an example of an unreliable personal testimony, see Frank McNally, 'Changed Utterly – An Irishman's Diary about Revisionism and 1916, Edward Kelly's Change of Mind', *Irish Times* (19 March 2015).

17 See John Borgonovo's comments in relation to Martin Corry in 'Separating Fact from Folklore: Review of "In the Name of the Republic"', *Irish Examiner* (27 March 2013) <http://www.irishexaminer.com/viewpoints/analysis/separating-fact-from-folklore-226559.html> accessed 24 April 2016.

18 St John D. Seymour and Harry Neligan, *True Irish Ghost Stories: Haunted Houses, Banshees, Poltergeists, and Other Supernatural Phenomena* (Mineola, NY; Dover; Newton Abbot: David & Charles, 2005; 1914).

the local historian who interviewed Corry in the 1970s, was interested in supernatural lore, and there was a lot of it about Knockraha.[19]

As we shall see, when interviewers sought or expressed an interest it was not so uncommon for interviewees to relate this kind of information. Corry's narrative may or may not be credible, but the black dog seems less an indication of his unreliability as a narrator than an example of a local story told in a familiar way. Corry did not mention the dog to Ernie O'Malley (the writer and anti-treaty republican who interviewed Corry in the 1950s) or if he did, O'Malley chose not to write it down.[20]

O'Malley interviewed over 400 (mostly anti-treaty) veterans in the 1940s and 1950s for a planned history of the civil war. The wider context in which testimony was collected and the motivation and interests of collectors had a significant impact on how and what was said. In the decades after the Second World War, both official projects and independent collectors like O'Malley tended to concentrate primarily on sourcing operational accounts of military actions and military organization with a view to legitimizing historic claims to independence and the war waged to achieve it, or sides taken in the civil war.[21]

The Irish government established institutions to gather both written and oral heritage, as did other newly established states in Central and Eastern Europe where post-Great War independence struggles took place.[22]

19 James Fitzgerald, *Knockraha 1977: Macra na Feirme Foras Feasa na Paroiste* (Knockraha, Cork, 1977), 88–90, 91–2,137, 144; James Fitzgerald interviews [hereafter Fitzgerald interviews] (Collins Barracks [hereafter CB], Cork).

20 Interview notes [hereafter 'Interview'], Martin Corry (Ernie O'Malley notebooks [hereafter EOMNbks] (University College Dublin Archives [hereafter UCDA], Ernie O'Malley Papers [hereafter EOMP], P17b/112), 142–5.

21 Eve Morrison, 'Witnessing the Republic: The Ernie O'Malley Notebook Interviews and the Bureau of Military History Compared', in Cormac K. H. O'Malley and John P. Waters, eds, *Modern Ireland and Revolution: Ernie O'Malley in Context* (Irish Academic Press, Dublin, 2016), 178–203, 188.

22 Michael Kennedy and Deirdre McMahon, *Reconstructing Ireland's Past: A History of the Irish Manuscripts Commission* (Dublin: Irish Manuscripts Commission, 2009); Micheál Briody, *The Irish Folklore Commission: 1935–1970, History, Ideology, Methodology* (Helsinki: Finnish Literature Society, 2008); Diarmuid Ó Giolláin, *Locating Irish Folklore: Tradition, Modernity, Identity* (Cork: Cork University Press, 2000) 63, 76–7, 146.

Eamon de Valera played a pivotal role in inaugurating two major official projects documenting oral culture, the Bureau of Military History (the Bureau, from 1947 to 1957) and the IFC.[23] The Bureau was the largest and most significant government-sponsored initiative to collect personal testimony from veterans in Ireland. Run by the Department of Defence, the Bureau's main control and source for locating potential interviewees was the department's records relating to its administration of various disability and military service pensions for separatist veterans, a process which had generated a massive reinvestigation into the 1916–1923 period.[24] This archive is another important source for veterans' testimony.

While earlier efforts were generally initiated and conducted by the veterans themselves, from the 1960s onwards, projects were often compiled as an adjunct to wider commemorative practices organized by local history societies and libraries, or undertaken as research by local or non-academic historians, radio and television documentary makers, and sometimes by veterans' relatives. Priests played a prolific and conspicuous and role from early on as establishers of local history groups, writers, ghost-writers, collectors and interviewers.[25] In the early 1960s, three priests and a local school teacher involved in the Clogher Historical Society (CHS) devised a scheme to interview Monaghan veterans.[26] From 1963 to his death in September 1973, Fr. Louis O'Kane conducted over four hundred hours of interviews and corresponded with and collected statements from Northern veterans, mostly from Armagh, Derry and Tyrone.[27] In 1981, Seán Ó Súilleabháin,

23 Ibid.; Eve Morrison, 'Witnessing the Republic', 185; Briody, *Irish Folklore Commission*, 107–16.

24 For an overview of this collection see Catriona Crowe, ed., *Guide to the Military Service (1916–1923) Pensions collection* (Department of Defence: Dublin, 2012).

25 Among other clergy involved in history-writing and collecting (in addition to those discussed in this chapter) were Colm Ó Labhra, Patrick J. Twohig, J. Anthony Gaughan and F. X. Martin. For details see bibliography.

26 Father Laurence Marron Collection (Monaghan County Museum). The clergy involved were Peadar Livingston, Laurence Marron and Joseph Duffy (later to be bishop of Clogher).

27 Proof copy of Dónal McAnallen, 'Rev. Louis O'Kane (1905–1973) and his revolutionary recording of Irish Volunteers: Aspects of South and East Tyrone

then Leitrim's County Librarian, established the (still ongoing) Leitrim Library Oral History Project which interviewed a number of veterans.[28]

Very little of the early material conforms to modern oral history practice although it was generally derived from some kind of interview process. Bureau witness statements, O'Malley interviews and MSP claim files dating from the 1930s to the 1950s, as well as the accounts collected by the CHS, were assembled from interviewer's or stenographer's notes, or personal correspondence. Of these, only MSP transcripts include the questions asked by interviewers, and O'Malley's handwritten notes are notoriously illegible. Most of the later interviews were recorded, and interviewers often quizzed interviewees about a wider range of subjects. The most extensive collection of recorded interviews was O'Kane's. Although concentrating primarily on the independence struggle, Fr. O'Kane was also keen to document stories, beliefs and religious practices. He quizzed Protestant IRA man Alf Cotton about local families and folklorists, customs like 'Blackberry Sunday', bonfires on St. John's Eve, divining fortunes on Halloween, and stories of Protestant and Catholic unity in 1798 and 1860 'before the Presbyterians became Orangemen'. O'Kane remarked at one point in an interview: 'I am fey enough still to believe in fairies and to consider that whatever their nature or origins they were usually kind'. Born into the Plymouth Brethren, Cotton's interview suggests that the 'Otherworld' could be as much a part of a rural South Derry Protestant upbringing as a Catholic one:

on Tape', *Dúiche Néill: Journal of the O'Neill Country Historical Society*, 23 (2016).

28 Leitrim Library Oral History Project (Ballinamore County Library). Special thanks to Mary Conefrey, Assistant Librarian, for information and help in locating relevant material. Also of interest are: Oliver Snoddy Survey (National Museum of Ireland, AI/080); Kenneth Griffith and Timothy E. O'Grady, *Curious Journey: An Oral History of Ireland's Unfinished Revolution* (London: Hutchinson, 1982); Kenneth Griffith, 'A Curious Journey', Tempest Films, release date: 1979; Raidió Teilifís Éireann's [hereafter RTE] Archives also hold extensive interviews.

| | |
|---|---|
| Cotton: | I remember my own mother telling me of the existence of a fairy-forth [sic] which was around her native place [...] and that she and the other children use to hear fairy music. |
| O'Kane: | Well, did you believe in fairies? |
| Cotton: | I don't know whether I did or not.[29] |

Cotton was not the only O'Kane interviewee to mention the supernatural. As a child in Dublin, James T. Donnelly encountered not one, but two poltergeists and was also exorcized as a baby.[30] When Paddy Diamond and his fellow IRA men raided Mayola Castle in Castledawson for arms, they sent their 'friendly Protestant' guide (who worked on the estate) into its famous 'ghost room': 'You see, as Catholics we were more superstitious than he'.[31] Likewise, when the Leitrim library interviewed Francis Joseph Bohan, a member of 'B' (Aughavasa) Company, 1st Battalion, South Leitrim Brigade, he described cures by local women which extracted thorns that x-rays could not find. The ghosts of drowned children, strange women and a man with a cloven hoof wandered Leitrim roads and crossroads. Some of these sighting Bohan heard about, others he saw himself, although by 1982 the hauntings were in decline:

| | |
|---|---|
| Interviewer: | ... It's funny alright, you know, what people experienced that time. |
| Bohan: | Oh, there was things that time surely ... then it all died away since the motors came.[32] |

This kind of material is not entirely absent from the early collections. A few Bureau witnesses described encounters with demonic entities, willow-the-wisps and fairies.[33] One man recalled the 'eerie feeling' that came over

29 Interview, Alf Cotton (transcript), (20 June 1966) (CÓFLA, LOK II AO7, box 0001, folder 12).

30 Interview, Dr James 'Laddy' Donnelly, (2 September 1965) (CÓFLA, LOK IV.A 01A, box 0009, folder n.a.).

31 Interview, Patrick Diamond (transcript), n. d. (CÓFLA, LOK IV.B.01, box 0002, folder 08).

32 Interview with Francis Joseph Bohan by Seán Ó Súilleabháin and Michael Whelan, (20 June 1982); <http://www.census.nationalarchives.ie/pages/1911/Leitrim/Aghavas/Killyfea/658141/>; 1st Battalion, South Leitrim Brigade (MAI, MSPC/RO 582).

33 Michael V. O'Donoghue (MIA, Bureau of Military History [hereafter BMH], Witness Statement [hereafter WS] 1741, part 2 and investigator's notes), 215–17.

him after a mysterious lady on the road warned him about an approaching RIC patrol, and vanished. He eventually concluded she had been an ordinary woman secretly waiting for her lover: 'To explain the feeling I got, I thought I had seen a ghost but to-day I have a more mundane explanation'.[34] Generally though, for Bureau witnesses, ghost-stories were for gullible folklorists. Superstition was a cloak to hide their activities. Haunted houses were useful for storing arms. Only 'West of Ireland' Volunteers believed in ghosts.[35]

These efforts by witnesses to distance themselves from traditional belief systems are to be expected. The preferred focus and cultural reference points of Irish modernizers were historical and related to the old Gaelic world. The '"fairification" of popular culture' in the Victorian period (and earlier) had been largely a conservative, unionist undertaking.[36] Many witnesses were genuinely sceptical and in the nineteenth and early twentieth centuries, establishment figures opposed to self-government for Ireland maintained that the prevalence of Irish 'superstition' made the country unfit for home rule.[37] Supernatural experiences were also increasingly characterized as hallucinations or indicators of 'pathological imbalance'.[38] It is also important to bear in mind, however, that unlike Father O'Kane and the Leitrim librarians, neither Ernie O'Malley nor the Bureau sought to document popular culture.

That said, both the O'Malley and Bureau collections contain interesting second-hand descriptions of the supernatural events and miraculous

---

O'Donoghue's Bureau investigator described him as 'rather flamboyant' and 'with a marked tendency to over-elaboration'; Sean MacEoin (MAI, BMH, WS 1716), 124, 126; Andrew McDonnell (MAI, BMH, WS 1768) and (UCDA, EOMNbks, P17b/100), 58–9.

34 Thomas Hevey (MAI, BMH, WS 1668), 33.

35 Alfred White (MAI, BMH, WS 1207), 23; Patrick O'Sullivan (MAI, BMH, WS 878), 6; James Cullen (MAI, BMH, WS 1343), 3; Moira Kennedy O'Byrne (MAI, BMH, WS 1029), 5; Thomas Pugh (MAI, BMH, WS 397), 17.

36 Joseph Leerssen, *Remembrance and Imagination: Patterns in the Historical and Literary Representation of Ireland in the Nineteenth Century* (Cork: Cork University Press in association with Field Day, 1996), 164, 170.

37 Ó Giolláin, *Locating Irish Folklore*, 112.

38 Shane McCorristine, *Spectres of the Self: Thinking about Ghosts and Ghost-Seeing in England, 1750–1920* (Cambridge: Cambridge University Press, 2010), 31.

occurrences that ensued as the pace of the IRA's campaign increased over the summer and autumn of 1920 and Crown Forces responded by increasingly targeting the civilian population. In an era before analytical vocabularies for describing emotions were common, and during events so traumatic that finding words might be difficult anyway, supernatural explanations were a means of expressing anxiety, fear and trauma for both perpetrators and victims. Several Bureau witnesses and one O'Malley interviewee mention the bleeding statues and other miraculous events in Templemore, County Tipperary in 1920 which commenced shortly after Crown Forces engaged in widespread 'official' and 'unofficial' reprisals against the local population.[39] 'I saw the blood on them', Paddy Kinnane told O'Malley.[40] Thousands of pilgrims descended on the town, and according to one Bureau account, the previously aggressive Crown Forces suddenly became 'most inoffensive'. Some even took part in pilgrimages.[41] Local Volunteers took over the policing of the town, carried out collections and imposed a levy on jarveys and side cars transporting visitors to and from the site.[42] Despite disciplinary problems, the episode provided a 'windfall' for the brigade.[43]

Similarly, two incidents in Galway in November 1920 suggest that supernatural explanations also provided a way of conveying information about extrajudicial killings by Crown Forces that might otherwise attract retribution from the perpetrators. Shortly after parishioners told a local curate that their cattle behaved strangely when passing a certain spot on a bye-road outside Barna the body of Father Michael J. Griffin (almost

39 David Leeson, *The Black and Tans: British Police and Auxiliaries in the Irish War of Independence, 1920–1921* (Oxford: Oxford University Press, 2012), 215; Reynolds, John, 'The Templemore Miracles', *History Ireland*, 1 (January/February 2009), 41–4.

40 Paddy Kinnane interview (UCDA, EOMNbks, P17b/125), 66. He doesn't mention the statues in his witness statement, Patrick Kinnane (MAI, BMH, WS 1475).

41 James Leahy (MAI, BMH, WS 1454), 42.

42 Paddy Kinnane (UCDA, EOMNbks, P17b/125), 66; James Leahy (MAI, BMH, WS 1454), 41–3.

43 Ibid., 44; Sean Scott (MAI, BMH, WS 1486), 7–8.

certainly killed by Auxiliary police) was found nearby, buried in a bog.[44] Likewise, the blackened, mutilated bodies of Patrick and Harry Loughnane (also tortured and killed by Crown Forces) remained undiscovered until the Loughnanes' cousin, Michael Loughnane (who had been reluctant to cooperate) testified at the Court of Inquiry on 8 December 1920 as to their whereabouts.[45] Both contemporary newspaper reports and subsequent accounts say the location came to Michael Loughnane in a dream.[46] There is also a heavy dose of Catholic mysticism in the Pádraig Ó Fathaigh (an IRA veteran, teacher and *Gaeilgeoir)* account of the Loughnane brothers' deaths, but this may be retrospective embellishment. The statement was written for the Bureau at the behest of the Loughnanes' sister Nora, who was by then a nun and prone to visionary dreams herself.[47]

> The brothers, wrapped in linen cloths, were laid side by side on the floor. The Rosary was recited in Irish. The blood began to trickle from Harry's wounds, and the responses were interspersed with the sobs and wails of a grief-stricken people, whose handkerchiefs were dipped in the martyrs' blood.[48]

O'Malley transcribed a version of the Ó Fathaigh account, but his interviews generally are more prone to irreverence.[49] Michael Fitzpatrick,

44 Mary Leech (nee O'Meehan) (MAI, BMH, WS 1034) 3; Summary of police reports November 1920 ((The National Archives of the United Kingdom [hereafter NAUK], Colonial Office [hereafter CO] 904/143); Rev. Michael Griffin, CC, Court of Inquiry (NAUK, War Office [hereafter WO] 35/151A/9).

45 Patrick and Harry Loughnane, Court of Inquiry (NAUK, WO 35/153a).

46 *Irish Independent*, 9 December 1920; *Connacht Tribune*, 11 December 1920; Henry O'Mara (MAI, BMH, WS 1652), 8, 9; Pádraig Ó Fathaigh (MAI, BMH, WS 1517), 3; Pádraig Ó Fathaigh and Timothy G. McMahon, *Pádraig Ó Fathaigh's War of Independence: Recollections of a Galway Gaelic Leaguer* (Cork: Cork University Press, 2000), 64.

47 *Connacht Tribune* (3 July 1981); Ó Fathaigh's account was also published in the *Connacht Tribune*, 15 December 1978; Pádraig Ó Fathaigh (MAI, BMH, WS 1517, investigator's notes). A former *Irish Bulletin* staff member gave the Bureau an earlier, plainly written deposition by Nora Loughnane and her mother. (Liam McKenna Collection (BMH CD 230/3).

48 Pádraig Ó Fathaigh (MAI, BMH, WS 1517), 5.

49 Nora Loughnane (UCDA, EOMNbks, P17b/68 and 136), 32–4.

for instance, whose Bureau account contains a lengthy description of how his wounds were miraculously cured by a priest after an ambush, took a particularly earthy tone with O'Malley: 'There was a clerk of the chapel then who used to look after stuff for us. Barlow's in Shragh used to give it to him, and the clerk would "go over to St Joseph's statue": and the directions for finding it were: "put your hand up his hole and you'll find a revolver. Then go over to the Blessed Virgin's hole and you'll find 400 rounds of 45 ammunition"'.[50]

## Veterans' Accounts and Popular Memory

Acts of personal and communal remembering are invariably repositories of information, documenting occurrences, attitudes, modes of expression and narrative, mentalities, truths, lies and feelings. 'Rememberers' invest their accounts with meaning by referencing events and cultural norms from every stage of their lives. Stories can change with the passing of time or to suit a particular audience. Autobiographical memories almost always incorporate second-hand information from stories and conversations, contemporary and subsequent public or published sources. Like all personal memories – even very mundane ones – veterans' accounts are complex, inherently dynamic, unstable and chronologically fluid fusions of facts and representation. The more aware a reader is of these various contexts, the more what at first seemed like simple, operational accounts of ambushes and attacks reveal themselves as densely layered masses of symbolism and story in need of careful negotiation. These qualities do not make personal testimony particularly easy for historians to employ, but they are part of what makes it fascinating and worth using.

50 Michael Fitzpatrick (MAI, BMH, WS 1443), 14–15; Interview, Micky Fitzpatrick (UCDA, EOMNbks, P17b/114), 11.

## Patrick Boland and Michael Coen

By the 1950s, when Ernie O'Malley interviewed John Greally from Mayo, Greally's narrative could draw on a range of private and public information. At one point, he described the deaths of Patrick Boland and Michael Coen, both local IRA men notoriously murdered by Crown Forces in 1921 (27 May and 2 April, respectively):

> [...] they bayoneted poor Paddy Boland who was an old [only] son. They bursted the bayonet in him and they almost cut his nose off with a bayonet also. Mick Coen, they found in his own house, and they cut off his private parts and pinned them on his breast, and they cut the flesh of his legs and arms. They cut off his ears and they left him out on the road [...] They captured him in his own home [...] That had a terrible effect on the people [...].[51]

The gruesome (if slightly exaggerated) detail in Greally's account reflects the efforts of O'Malley to capture the original idioms and speech patterns of his subjects. Greally was not an eyewitness to either death, though he, Boland and Coen lived within a few miles of each other in and around Ballyhaunis, and he had been a member of Boland's company. Contemporary parliamentary debates, compensation claims and courts of enquiry were public events and often reported in the press.[52] Sir Hamar Greenwood, the notorious Chief Secretary for Ireland, insisted to the House of Commons that Boland had been shot after making a 'determined attempt to escape.'[53] Coen's father applied for compensation, and details of his case were published in local newspapers. Both he and a local RIC sergeant who testified in support confirmed that his body, found near the house, was in a 'terrible condition'.[54] A couple of weeks afterwards, when

51 Interview, John Greally (UCDA, EOMNbks, P17b/113), 3; Coen was a member of Holywell Company and Boland had been Captain of Crossard Company, both in 4th Battalion, East Mayo Brigade. (MAI, MSPC, RO/296).

52 Michael Cohen, Court of Inquiry (NAUK, WO 35/147A).

53 House of Commons Debate, 16 June 1921, vol. 143 cc576–8.

54 *Mayo News* (5 November 1921).

addressing a crowd in Ballyhaunis, William Sears, a Sinn Féin TD, said that local people 'knew what happened [to] poor Boland, whose body had been mangled after he had been killed, and Coen and the others.'[55]

Some events of the Irish revolutionary period remain imbedded in social memory. Colum Cronin, West Cork folklorist and local historian, for instance, does not conduct interviews specifically focussed on the period, but says references to it 'seep' into the interviews all the time.[56] The deaths of Boland and Coen fit easily into the canon of republican martyrs, and their deaths were regularly commemorated over the years.[57] However, the relationship between veterans' accounts and the stories surviving in local areas was more nuanced than is often acknowledged. Sometimes their accounts display an emotional range, complex morality and empathy rarely evident in official commemorations or fighting stories. Arguably, these were as important a locus for shaping social memory of the 'Tan' and civil wars as public commemorations and idealized published accounts. As we shall see, interviews with friends, relatives and neighbours who knew the veterans suggest that there were contexts in which some of them gave surprisingly candid accounts of even the most controversial events. The boundaries between personal memory and folklore became particularly blurred in these quasi-folkloric hybrids, which tend to be less stylized than received tradition transmitted over many years. Details related by veterans became part of the listeners' own autobiographical memories. One of the most well-known Irish short-stories of the twentieth century, for instance, derived from an overheard conversation in an internment camp.

55 *Mayo News* (26 November 1921); Dominic Price, *The Flame and the Candle: War in Mayo 1919–1924* (Cork: The Collins Press, 2012), 187.

56 Email, Cronin to Author, 29 April 2016.

57 *Mayo News* (8 April 1922); *Connaught Telegraph* (4 June 1934); *Western People* (23 April 1966 and 17 April 1971).

## Guests of the Nation

In 1962, Frank O'Connor was interviewed about his famous short story, 'Guests of the Nation'. He explained that it had been inspired by a conversation between two IRA men he overheard while interned in Gormanstown Camp during the civil war. They were discussing the execution of two English soldiers who had been very well-liked by their captors. It was a sad story which reminded O'Connor of the 'voices of the English soldiers who lived around the barracks green when I was a boy'.[58] 'Guests of the Nation' may have been based on either of two actual incidents. One is the West Cork IRA's execution of two captured Essex Regiment soldiers, Percy Taylor and Thomas Watling, in December 1920.[59] The other took place in July 1921 in East Kerry, when the IRA shot John Steer (often referred to as 'Stay') and George Motley of the East Lancashire Regiment.[60] Both incidents were well known in the local areas.

There are differing accounts of the circumstances surrounding the executions of Watling, Taylor, Steer and Motley, not all sympathetic and some more convincing than others. Jeremiah Murphy inaccurately labelled Steer and Motley as 'typical, middle class English', 'gentlemen deserters' and spies.[61] In reality, all four were young, inexperienced and from very ordinary backgrounds. None of them were more than twenty-one years old. All the soldiers were kept prisoner for weeks before being were killed, and their guards got to know them well.

58 Frank O'Connor in 'Frank O'Connor – Interior Voices', RTE Television, broadcast (January 1962) <http://frankoconnor.ucc.ie/recording06.php?teanga> accessed 30 March 2016. I am grateful to Dr Robert P. Cremins for drawing my attention to this clip.

59 Essex Regiment, Regimental Enlistment Books, 1920 (Chelmsford, Essex Regiment Museum, Army Books 358). Thanks to archivist Ian Hook for his help and advice.

60 J. J. Ó Ríordáin, *Kiskeam Versus the Empire* (J. J. Ó Ríordáin, 1973), 84; John Scannell (MAI, BMH, WS 1114), 8; James J. Riordan (MAI, BMH, WS 1172), 15. See correspondence relating to their deaths and the subsequent discovery of their bodies in 'British Forces, Missing Personnel' (MAI, Collins Papers [hereafter CP], A/0909).

61 Jeremiah Murphy, *When Youth was Mine* (Mentor Press: Dublin, 1998), 157–61.

Watling and Taylor were executed because a scheme to contact and buy arms from one of the deserter's brothers (also a member of the Essex regiment stationed in the local barracks) went tragically wrong. Tom Barry accused them of deliberate treachery, but other accounts suggest that the letter one of the captured soldiers attempted to send to his brother had simply reached the wrong person, who then handed it in to the officer in command.[62] The consequences were dire. On 2 December 1920, John Galvin, Joseph Begley and Jeremiah O'Donoghue, three members of the local IRA, arrived in Laurel Walk – an isolated Bandon laneway – to make the deal. They walked straight into a British patrol lying in wait for them. All three were killed. Unlike the young deserters who paid the price for O'Donoghue's, Begley's and Galvin's deaths, the Essex Regiment officers on the patrol (and who testified at the court of enquiry) were all experienced Great War veterans.[63]

John J. Ó Ríordáin's account of Steer and Motley drew on interviews with some of the IRA men involved, including his father, James J. O'Riordan. Ó Ríordáin maintained the two soldiers were executed after an impending round-up by Crown Forces simply made it too dangerous to keep them. They knew too much about the local IRA to be released. James O'Riordan had tried without success to save Steer and Motley. In his witness statement, he said that they handed them over to the Kerry IRA for trial, but 'we felt that they were only deserters'.[64] O'Riordan was less reticent when speaking to his son: 'With tears in his eyes he still laments their death maintaining that they were completely innocent and that he himself would gladly have stood for the bullets which ended their lives'.[65]

62 Tom Barry, *Guerrilla Days in Ireland* (Dublin: Anvil Books, 1949; 1981), 55; *Southern Star*, 30 October 1971.

63 John Galvin, James Donoghue and Joseph Begley, Court of Inquiry (NAUK,WO 35/150); Medal cards of Sgt J. C. Benton, Captain R. E. G. Carolin, Major Richard Neave, Lieutenant Maxwell S. Solly, Corporal H. J. Staff, Lieutenant Frederick W. Treeby and Captain William Wreford Brown (NAUK, WO 372).

64 James J. Riordan (MAI, BMH, WS 1172), 15.

65 Ó Ríordáin, *Kiskeam Versus the Empire*, 84.

In both cases, the executions of these unfortunate soldiers had a considerable impact on some of the IRA men involved.

Watling and Taylor were held in the family home of John L. O'Sullivan, an IRA veteran who later became a Fine Gael TD. When the orders came to execute them, O'Sullivan himself refused to carry it out after his mother, who had grown very fond of both young men, tried to save their lives. Eventually the job was done by one of O'Sullivan's older brothers. In 1982, O'Sullivan spoke publicly about it for the first time to Kenneth Griffith: 'The revulsion of taking a human life goes very deep in a person'.[66] O'Sullivan had not kept silent though. The story was known locally. Bill O'Donoghue, the younger brother of Jeremiah O'Donoghue, recounted it to Father John Chisholm in 1969:

> John L [O'Sullivan] did tell the story that one of them was a perfect soldier [...] I think Dan Harte of Clonakilty was there, and Dan felt that you couldn't shoot a man like without first getting him to say some few prayers and that, and he told the lad like to say whatever few prayers he wanted to say, and the lad said he didn't wish to say any prayers, to get on with the job as quickly as they could. And finally Dan said to him [...] 'Blast you man, haven't you got a soul?' and your man just pointed at the sole of his boot, he said 'that's the only soul I've got ...'[67]

Certain events haunted (literally and metaphorically) the areas where they occurred and the individuals who witnessed or were involved in them. Scattered across a range of sources is evidence of a rich fantastical folklore, mostly ghost stories, relating to the independence struggle and civil war. It has, for the most part, received little scholarly attention. Perpetrators of atrocities were cursed.[68] Ghosts of IRA men, civilians, British soldiers, and policemen haunted ambush sites or the places where they died, or appeared as 'crisis apparitions' to the living at the moment of their deaths.[69]

66 Griffith and O'Grady, *Curious Journey*, 189.

67 Interview, Bill O'Donoghue, c. April 1969 (Chisholm interviews, private collection).

68 Kiely, *Drink to the Bird*, 26–9; Timothy Wilson, *Frontiers of Violence: Conflict and Identity in Ulster and Upper Silesia 1918–1922* (Oxford: Oxford University Press, 2010), 43.

69 Dan Breen, *My Fight for Irish Freedom* (Dublin: Anvil Books, 1981; 1st edn 1924), 154; Patrick F. Byrne, *The Second Book of Irish Ghost Stories* (Cork: Mercier Press, 1971),

O'Donoghue's account of the fate of Taylor and Watling is occasionally exaggerated, and some of the points of detail are wrong, but it contains a solid core of accurate information. Like so many local accounts of War of Independence stories, his also features a ghost:

> local people out around Carrigroe ... were very superstitious at that time in the country. They used to tell that very often in O'Sullivan's house that they could hear the footsteps coming up the stairs of this fellow, of whom Mrs O'Sullivan was very fond, like, coming back up the stairs at night.[70]

## Bridget Noble

IRA veterans were often reluctant to discuss some incidents, and circumspect when interviewed. Sometimes veterans lied, or they jointly agreed what to say beforehand, but attempts to silence or censor memory of certain events were never entirely successful. The internet is making it easier for historians and researchers to access local memory, and is fast becoming one of the most significant forums for its dissemination. The extent to which modern technology is accelerating this process became particularly clear to me when an open Google search led me to a remarkable piece of evidence about Bridget Noble, one of three women (along with Mary Lindsay and Kitty Carroll) known to have been deliberately killed by the IRA during the War of Independence.

Members of the Castletownbere Battalion, West Cork Brigade executed Noble in 1921. IRA General Orders forbid the execution of female 'spies',

---

45–51, 63–7; *Irish Independent*, 8 November 1928; Fitzgerald, *Knockraha 1977*, 88–90, 137; Tarquin Blake, *Haunted Ireland* (Collins Press: Cork, 2014), 222–5. A couple of academic studies have appeared in recent years, see Tait, 'Wandering Graveyards'; Mac Conmara, Tomas, 'Always in the Human Consciousness of the People': The Anglo-Irish War of Independence in County Clare, 1919–21, Oral History, Tradition and Social Memory' (unpublished thesis, University of Limerick, 2014).

70 Interview, Bill O'Donoghue, c.April 1969 (Chisholm interviews, private collection).

and it was a rare occurrence. Noble's case is not well-known outside the local area, and her body was never recovered. Patsy Dan Crowley, reputed to have executed her, was dead by the 1930s.[71] None of the Castletownbere Battalion veterans who gave witness statements mention Mrs Noble, not even Liam O'Dwyer, the Battalion OC who submitted an official IRA report on the incident.[72] In his report, O'Dwyer said that their local RIC contacts let them know that Noble had accused himself and Patsy Dan Crowley of killing William Lehane, a local 'landgrabber'.[73] Mrs Noble also named the seven local Volunteers who 'bobbed' her hair for associating with the police, for which one of them was later arrested.

There was very little additional information to be had about Bridget Noble, until I chanced upon the website of Skully, a musician and producer from Cork, and a piece of music he composed entitled 'Mrs Noble.' He sent me an excerpt from a recorded interview his father (Victor Sullivan) conducted with one of Mrs Noble's neighbours.[74] Eva Sullivan, the interviewee, was from Ardgroom Inner, the townland where Mrs Noble lived. Sullivan heard what happened from a local IRA man, and her remarkable story is worth quoting at length:

> Her name was Neill, and she married this cooper […] I don't know was he an English man or not. She married anyway. She married Noble. We never knew Noble, I don't know did he die or what, or did he run away from her, but she was there with her father all on her own […] at the time of the civil war, no, the time of the English and Irish war, she was telling tales of out of school anyway to the military, and she was tailed [?] up. Her hair was bobbed first. She got the warning, and then, one evening, after the hair being bobbed she knew what was coming. She set out for Castletown and I'm sure she was coming to Furious Pier to the British Army. I have a feeling

71 Ardgroom Company, 5th Battalion, 5th Cork Brigade (MAI, MSPC/RO/55).

72 Liam O'Dwyer (MAI, BMH, WS 1527); Comdt. 5th Bn, 5th Cork Brigade to Adj., 5th Cork Brigade, 21 October 1921 (MAI, CP, A/0649, Group IX). Castletownbere Battalion was 6th Battalion of the 3rd Cork Brigade during the War of Independence.

73 The report misdates the Lehane killing by two months, which occurred on 10 May 1920. *Cork County Eagle and Munster Advertiser* (15 May 1920).

74 Samples of Victor Sullivan's writing are available on the Cork Non-Fiction Writers Group blog. <http://www.corknonfictionwritersgroup.blogspot.ie/2016_04_01_archive.html>.

> maybe she was, but she set out walking. [...] Dan Holland was out that day with the bread and she asked Dan for a lift in [...] Dan had been warned, so he couldn't say no [...] 'be walking away' he said 'and I'll pick you up'. He had no intention of leaving Ardgroom. He stayed on and on and left her go away, and she was just going down the hill towards Eyries when they came along and captured her.[75]

Many of the details in Sullivan's story can be verified. Eva and her family were Protestants, by 1911, one of two in the townland.[76] Sullivan was about twelve years old in 1921, and she and her sister were members of Ardgroom Branch, Cumann na mBan.[77] Bridget Noble was in her forties. Alexander Noble, her husband, was a cooper from Scotland. He wrote from England in July and September 1921 to Eamon de Valera, asking for information: 'it is not clean work to take away my lone, defenceless wife'.[78] De Valera forwarded some of the correspondence to IRA GHQ. The Ministry for Defence did not confirm Bridget Noble's execution for 'espionage' until March 1922, and by then a court had already awarded Mr Noble £1,500 compensation for her death.[79] According to press reports, Mr Noble 'wept bitterly' after the local District Inspector confirmed he had 'certain authority for saying' that she was dead.[80]

Liam O'Dwyer states in his report that Mrs Noble was executed on 15 March 1921, eleven days after she was captured. In her recently released MSP application, Hannah Hanley née Neill (a local Cumann na mBan officer) says that she looked after her for a time.[81] Both O'Sullivan's and Hanley's accounts suggest that Noble was kept for far longer than eleven days. O'Sullivan says Noble was killed after a series of official reprisals in the area which took place in late May, 1921:

75 Interview, Eva Sullivan, c.1960s (Sullivan Family interviews, private collection).

76 <http://www.census.nationalarchives.ie/reels/nai001829919/>; <http://www.census.nationalarchives.ie/reels/nai001829929/>.

77 Ardgroom Company, 5th Bn, 5th Cork Brigade, Cumann na mBan (MAI, MSPC/CMB/21).

78 Noble to de Valera, 8 September 1921 (MAI, CP, A/0649, Group IX).

79 Sec. Minister for Defence to J. Travers Wolfe, 22 March 1922 (MAI, CP, A/0649, Group IX).

80 *Cork County Eagle and Munster Advertiser*, 28 January 1922.

81 Sworn statements by Hanley 30 July 1937 and 12 November 1942; Hanley, Particulars of service, 12 June 1942 (Hannah Hanley, MAI, MSPC/WMSP34REF27316).

> [...] She was taken to a shed in Ballycrovane, and she was gagged of course, and tied up there. I don't know how many days she was there – days and nights – but she was there the day that the army [Navy] sloop came in, landed the petrol for burning [...] They burned two that day. They burned Dwyer's and they burned Jer Connor's away in Inward Ardgroom, but they passed as close as I am to ye now, to that shed where she was tied up, with the tins of petrol to go burning the houses. And then she was taken from there, by boat, down the Kenmare river, and she was taken in at Collorus. I don't why they didn't do it where she was in Ballycrovane. They shot her at Collorus, and buried her there [...].

These house burnings were carried out after the local IRA killed three British soldiers.[82] Had Mrs Noble been kept prisoner for so long? Did the men involved deliberately withhold this information? Was the report an exercise in damage limitation? It is unlikely that there will ever be enough evidence to say for sure. Perhaps O'Dwyer's efforts in calling 'several meetings of the surviving officers in the area' to help the visiting Bureau investigator get 'as complete a record as possible' was partly an attempt to control the narrative.[83] But local people did not forget. Eva Sullivan and Skully's grandmother (who first told him the story) did not forget.

These quieter stories persisted and resisted attempts to smother them. The fact that some local IRA men contributed to their survival through telling their neighbours, relatives and friends about such incidents challenges polemical assumptions about what constitutes Irish revolutionary memory. For some veterans and some people it was and remains very simple. There was and is nothing to regret or feel remorseful about. For others, although there is no evidence that they stopped believing in the cause for which they fought, the fact that Taylor, Watling, Steer and Motley knew too much to be released without putting the IRA at risk, or being aware that Bridget Noble was giving information to the police, did not prevent lifelong regret at having to end those lives. It does not seem to have mattered whether or not the executed were guilty. Their victims' families, like those of the Loughnanes, of Michael Coen and Patrick Boland, still felt their loss every hour, every second of their lives.

82 *Cork Examiner* (28 May 1921 and 13 June 1921).

83 Liam O'Dwyer (MAI, BMH, WS 1527, investigators notes).

Attendees at even the most doctrinaire public commemoration may have been honouring more than what was said on the platform. There was no one response, or unified way of remembering. There was never just one Irish Revolution. There were thousands, as many pasts as there were historical actors, each one mediated by individual experience, culture and personality. Those who come after only ever have access to a few of them. If remembered histories are ghosts they haunt each other, and all written ones are plagued by them. Popular historical consciousness and 'folklore' of the Irish Revolution can be much more reflective and complex than is frequently assumed. In the oral tapestry of revolutionary memory, the death of civilian spies, British soldiers and IRA Volunteers alike often occupy the same haunted terrain, a weave of sadness rather than triumphalism. Regardless of whether the ghost of the dead young British soldier was 'real', his footsteps echoed between John L. O'Sullivan and his mother forever more. As the British paranormal investigator Peter Underwood once noted, the value of ghosts does not necessarily lie in their verticality:

> Oh, is there life after death, you ask? Well, to tell you the truth I don't know, but I do know that to live in the hearts and memories of those we leave behind is not to die.[84]

## Bibliography

Barry, Tom, *Guerilla Days in Ireland* (Dublin: Anvil Books, 1949; 1981).

Bebbington, D. W., *Victorian Religious Revivals: Culture and Piety in Local and Global Contexts* (Oxford: Oxford University Press, 2012).

Beiner, Guy, 'Mapping the "Year of the French": the Vernacular Landscape of Folk Memory', in C. E. J. Caldicott and Anne Fuchs, eds, *Cultural Memory: Essays on European Literature and History* (Oxford: Peter Lang, 2003), 191–208.

84 Underwood, Peter, *No Common Task: the Autobiography of a Ghost Hunter* (George G. Harrap & Co Ltd 1983), 228.

——, 'Recycling the Dustbin of Irish History: The Radical Challenge of "Folk Memory"', *History Ireland*, 14 (2006), 42–7.
——, *Remembering the Year of the French: Irish Folk History and Social Memory* (Madison, WI; London: University of Wisconsin Press, 2007).
Blake, Tarquin, *Haunted Ireland* (Cork: Collins Press, 2014).
Borgonovo, John, 'Separating Fact from Folklore: Review of "In the Name of the Republic"', *Irish Examiner*, 27 March 2013.
Bourke, Angela, *The Burning of Bridget Cleary: A True Story* (London: Pimlico, 1999).
Breen, Dan, *My Fight for Irish Freedom* (Dublin: Anvil Books, 1981; 1st edn 1924).
Briody, Micheál, *The Irish Folklore Commission: 1935–1970, History, Ideology, Methodology* (Helsinki: Finnish Literature Society, 2008).
Byrne, Patrick F., *The Second Book of Irish Ghost Stories* (Cork: Mercier Press, 1971).
Crowe, Catriona, ed., *Guide to the Military Service (1916–1923) Pensions Collection* (Department of Defence: Dublin, 2012).
Davies, Owen, *The Haunted: A Social History of Ghosts* (Basingstoke: Palgrave Macmillan, 2009).
Deasy, Liam, and Chisholm, John E., *Towards Ireland Free: the West Cork Brigade in the War of Independence 1917–1921* (Dublin: Mercier Press, 1973).
Devlin, Judith, *The Superstitious Mind: French Peasants and the Supernatural in the Nineteenth Century* (New Haven, CT; London: Yale University Press, 1987).
Duffy, Patrick J., *Exploring the History and Heritage of Irish Landscapes* (Dublin: Four Courts, 2007).
Dunne, Tom, *Rebellions: Memoir, Memory and 1798* (Dublin: Lilliput Press, 2004).
'Extracts from the Memoirs of Florence John Crowley', in *Dunmanway Doings*, Journal IV, Dunmanway Historical Society (2010), 16–29.
Fennell, Desmond, 'Against Revisionism', in Ciaran Brady, ed., *Interpreting Irish History: The Debate on Historical Revisionism 1938–1994* (Blackrock: Irish Academic Press, 1994).
Fitzgerald, James, *Knockraha 1977: Macra na Feirme Foras Feasa na Paroiste* (Knockraha, Cork, 1977).
Foster, R. F., 'History and the Irish Question', *Transactions of the Royal Historical Society*, 33 (1983) 169–92.
——, 'Remembering 1798', in McBride, *History and Memory in Modern Ireland* (Cambridge: Cambridge University Press, 2001), 67–94.
Gaughan, J. Anthony, *Austin Stack: Portrait of a Separatist* (Mount Merrion: Kingdom Books, 1977).
——, *Memoirs of Constable Jeremiah Mee, R. I. C.* (Dublin: Anvil Books, 1975).
Hazelgrove, Jenny, *Spiritualism and British Society Between the Wars* (Manchester: Manchester University Press, 2000).

Kennedy, Michael, and McMahon, Deirdre, *Reconstructing Ireland's Past: A History of the Irish Manuscripts Commission* (Dublin: Irish Manuscripts Commission, 2009).

Kiely, Benedict, *Drink to the Bird: A Memoir* (London: Methuen, 1991).

Kuijpers, Erika, Pollmann, Judith, Müller, Johannes, and van der Steen, Jasper, *Memory Before Modernity: Practices of Memory in Early Modern Europe* (Leiden: Brill, 2013).

Leerssen, Joseph, *Remembrance and Imagination: Patterns in the Historical and Literary Representation of Ireland in the Nineteenth Century* (Cork: Cork University Press in association with Field Day, 1996).

Leeson, David, *The Black and Tans: British Police and Auxiliaries in the Irish War of Independence, 1920–1921* (Oxford: Oxford University Press, 2012).

McAnallen, Dónal, 'Rev. Louis O'Kane (1905–1973) and his revolutionary recording of Irish Volunteers: Aspects of South and East Tyrone on Tape', *Dúiche Néill: Journal of the O'Neill Country Historical Society* 23 (2016) [proof copy].

Mac Conmara, Tomas, '"Always in the Human Consciousness of the People": The Anglo-Irish War of Independence in County Clare, 1919–21, Oral History, Tradition and Social Memory', (unpublished thesis, University of Limerick, 2014).

McCorristine, Shane, *Spectres of the Self: Thinking about Ghosts and Ghost-Seeing in England, 1750–1920* (Cambridge: Cambridge University Press, 2010).

McNally, Frank, 'Changed Utterly – An Irishman's Diary about Revisionism and 1916, Edward Kelly's Change of Mind', *Irish Times* (19 March 2015).

Malone, James, and Twohig, Patrick J., *Blood on the Flag: a Translation from the Irish of B'fhiú an braon fola* (Ballincollig, Co. Cork: Tower Books, 1996).

Markey, Anne, and O'Connor, Anne, *Folklore and Modern Irish Writing* (Sallins: Irish Academic Press, 2014).

Martin, F. X., and Byrne, F. J., *The Scholar Revolutionary: Eoin MacNeill, 1867–1945, and the Making of the New Ireland* (Dublin: Irish University Press, 1973).

Morrison, Eve, 'Bureau of Military History Witness Statements as Sources for the Irish Revolution' (MAI website, 2010).

——, 'Kilmichael Revisited: Tom Barry and the "False Surrender"', in David Fitzpatrick, ed., *Terror in Ireland: 1916–1923* (Dublin: Lilliput Press/Trinity History Workshop, 2012), 158–80.

——, 'Witnessing the Republic: The Ernie O'Malley Notebook Interviews and the Bureau of Military History Compared', in Cormac K. H. O'Malley and John P. Waters, eds, *Modern Ireland and Revolution: Ernie O'Malley in Context* (Dublin: Irish Academic Press, forthcoming 2016).

Murphy, Jeremiah, *When Youth was Mine* (Dublin: Mentor Press, 1998).

Ó Fathaigh, Pádraig, and McMahon, Timothy G., *Pádraig Ó Fathaigh's War of Independence: Recollections of a Galway Gaelic Leaguer* (Cork: Cork University Press, 2000).
Ó Giolláin, Diarmuid, *Locating Irish Folklore: Tradition, Modernity, Identity* (Cork: Cork University Press, 2000).
Ó Labhra, Colm, *Trodairí na Treas Briogáide* (Cló Uí Mheára: Aonach Urmhumhan, 1955).
Ó Ríordáin, J. J., *Kiskeam Versus the Empire* (J. J. Ó Ríordáin, 1973).
Ó Searcaigh, Cathal, *Light on Distant Hills* (London: Simon & Schuster, 2009).
Ó Tuathaigh, M. A. G., 'Irish Historical Revisionism', in Ciaran Brady, ed., *Interpreting Irish History: The Debate on Historical Revisionism 1938–1994* (Blackrock: Irish Academic Press, 1994).
O'Connor, Anne, *The Blessed and the Damned: Sinful Women and Unbaptised Children in Irish Folklore* (Oxford; New York: Peter Lang, 2005).
Ógáin, Ríonach uí, and Sherlock, Tom, *The Otherworld: Music & Song From Irish Tradition* (Dublin: Comhairle Bhealoideas Eireann, 2012).
Price, Dominic, *The Flame and the Candle: War in Mayo 1919–1924* (Cork: Collins Press, 2012).
Reynolds, John, 'The Templemore Miracles', *History Ireland* 1 (January/February 2009).
Richie, Daniel, 'The 1859 Revival and its Enemies: Opposition to Religious Revivalism within Ulster Presbyterianism', *Irish Historical Studies*, 40 (2016), 69–91.
Seymour, St John D., and Neligan, Harry, *True Irish Ghost Stories: Haunted Houses, Banshees, Poltergeists, and Other Supernatural Phenomena* (Mineola, NY; Dover; Newton Abbot: David & Charles, 2005; 1914).
Tait, Clodagh, 'Wandering Graveyards, Jumping Churches and Rogue Corpses: Tolerance and Intolerance in Irish Folklore', in James Kelly and Mary Ann Lyons, eds, *Death and Dying in Ireland, Britain and Europe: Historical Perspectives* (Sallins: Irish Academic Press, 2013), 277–96.
Thompson, David, *Woodbrook* (London: Vintage, 1994; 1st edn 1974).
Twohig, Patrick J., *Green Tears for Hecuba: Ireland's Fight for Freedom* (Ballincollig: Tower Books, 1994).
Underwood, Peter, *No Common Task: The Autobiography of a Ghost Hunter* (George G. Harrap & Co Ltd 1983).
Wilson, Timothy, *Frontiers of Violence: Conflict and Identity in Ulster and Upper Silesia 1918–1922* (Oxford: Oxford University Press, 2010).
Winter, J. M., *Sites of Memory, Sites of Mourning: The Great War in European Cultural History* (Cambridge: Cambridge University Press, 1996).

EAMON MAHER

# 5 Autobiography or Fiction?: Unravelling the Use of Memory in Francis Stuart and John McGahern

Cultural memory has come to the fore in a number of disciplines in recent decades, as researchers seek to read, in the words of Niamh Moore and Yvonne Whelan, 'the signs, symbols and sites of heritage that comprise the landscapes in which we live'. The book of essays which Moore and Whelan edited in 2007 set out to map 'the overlapping and oftentimes complex relationships between identity, memory, heritage and the cultural landscape'.[1] In a similar manner, poets, novelists and playwrights have invested much care in rendering a sense of place and time that will have the ring of authenticity – Joyce, Synge, Kavanagh, Kate O'Brien and Heaney spring to mind immediately in this context. However, a difficulty often arises for the writer; namely, how to recreate and share sensations that occurred in the past and which, when relived as memories, take on an altered hue. In order to be successful in this task, writers must be mindful of the danger of giving past experiences the inflection of contemporaneous perception – which is unavoidable to a certain degree, but which requires attention if one wishes to avoid an artificial rupture between past and present.

Few Irish writers have leaned on personal experience as a source of raw material for their fiction to the same extent as Francis Stuart (1902–2000) and John McGahern (1934–2006). Writing about Stuart's best-known novel, *Black List, Section H* (1971), Derek Hand notes that it is 'novelised autobiography',[2] a claim that could be made with equal validity about most of McGahern's fiction. This should not come as any great surprise, as

1 Niamh Moore and Yvonne Whelan, eds, *Heritage, Memory and the Politics of Identity: New Perspectives on the Cultural Landscape* (Aldershot: Ashgate, 2007), x.

2 Derek Hand, *A History of the Irish Novel* (Cambridge: Cambridge University Press, 2011), 204.

writers down through the centuries have noted that memory is at the heart of the image and the image is core to creative writing. The main concern of McGahern's credo, 'The Image' (1968), is how to reconcile image and style. He notes that the image cannot be separated from the rhythm and the vision, the latter being described as 'that still and private universe which each of us possess but which others cannot see'. Art, for McGahern, is 'an attempt to create a world in which we can live: if not for long or forever, it is still a word of the imagination over which we can reign', a 'Medusa's mirror, which allows us to celebrate even the totally intolerable'.[3] Memory plays a hugely important role in this process, as we shall see in our subsequent discussion of McGahern's work.

Images are fluid, subject to the mood invoked by a particular emotion at a particular time. Writing about events that happened years previously, no matter how clearly they are remembered, will never constitute an exact replica of past experience. So memory cannot be direct transposition: it is inevitably going to change in the telling, which is where the fictional element kicks in. Writing about *Black List, Section H* and the end of the Irish *Bildungsroman*, Richard Murphy quotes Declan Kiberd's comment that autobiography in Ireland became the autobiography *of* Ireland. He contrasted this with Stuart's claim that all literature is individual and concluded:

> Stuart's refusal to reconcile 'national' and 'individual', the very compromise that characterizes the canonical *Bildungsroman* form that his autobiographical novel assumes, offers a polemical challenge to the metonymic degradation that individuality undergoes in Irish life-writing.[4]

Both Stuart and McGahern were concerned with the individual graph of feeling that their characters are forced to undergo, and there are times when these feelings and the circumstances that trigger them are inspired by real events in the lives of the authors. However, unlike McGahern, who

3 John McGahern, 'The Image: Prologue to a Reading at the Rockefeller University', in Stanley van der Ziel, ed., *Love of the World* (London: Faber and Faber, 2009), 5.

4 Richard T. Murphy, 'A Minority of One: Francis Stuart's *Black List, Section H* and the End of the Irish *Bildungsroman*', *Irish University Review*, 34(2) (2004) 261–76, 261.

strongly contested the desirability of transferring his real-life experiences directly into his fictional output – he claimed that when events recounted in novels kept too close to the facts that the result was mere journalism and not literature[5] – Stuart was forthright about the self-referential tenor of his work. During a celebratory dinner to mark the Penguin edition of *Black List, Section H* in 1982, for example, he admitted to his biographer Kevin Kiely that much of his fiction was autobiographical. Kiely noted: 'He (Stuart) felt that imagined plots, circumstances and the invention of major characters made for bad fiction. The closer every element was to life, the better for his purposes as an artist'.[6] This is in stark contrast to McGahern's insistence that real life and fiction were incompatible and that merging the two resulted in bad art. Stanley van der Ziel notes that:

> Often in a novelist like McGahern, textual or literary truths are allowed to override historical or biographical ones. Good fiction is rarely a straightforward record of autobiographical fact, but some 'facts' in the memoirs of great novelists can easily be inspired not by what has happened to the subject but by their immersion in the works of literature.[7]

When first published in 2005, *Memoir* confirmed what many had always suspected, namely that there is a thin line between autobiography and fiction in McGahern's work. However, one would be well-advised not to see

5 In an interview with this author, McGahern made some revealing comments about the dangers associated with sticking too closely to facts. He maintained that a good work of fiction has to adapt reality in order to make it true, and believable: 'For some reason or other – there's an awful lot of confusion about this – a work of art has to conform to certain laws, and it's been my experience that I've made my worst artistic mistakes by keeping too close to what happened. For some reason, even when what happens by accident is close to the way you see it in work, it still needs to be reordered, reinvented in order to be true. In a very simple way, one of the differences between life and writing is that writing has always to be believable, whereas much of what goes on in life doesn't. Also, without the difficulty of having to reinvent, language loses tension'. 'Interview with John McGahern', in Eamon Maher, *John McGahern: From the Local to the Universal* (Dublin: The Liffey Press, 2003), 143–61, 147.

6 Kevin Kiely, *Francis Stuart: Artist and Outcast* (Dublin: The Liffey Press, 2007), 12.

7 Stanley van der Ziel, *John McGahern and the Imagination of Tradition* (Cork: Cork University Press, 2016), 14–15.

a complete merging of the two forms. In this regard, it is surely significant that certain passages from his novel *The Leavetaking* (1974–1984)[8] are reproduced word for word in *Memoir*. One possible reason for his doing this may well have been to alert readers that just as many of the events treated in the fiction were inspired by incidents in McGahern's life, so too *Memoir* can be read as a slightly fictionalized version of these same events. Whereas van der Ziel points out that some of the things that McGahern read about in books at times seemed more real than things that had actually happened, Dermot McCarthy is adamant that *Memoir* is just the last instalment of the McGahern life story, and especially the story of his relationship with the 'beloved', his mother:

> *Memoir* makes clear that 'the lost image', like 'the lost world', is the mother-image, the mother-world, and that the 'grave of the images of dead passions and their days' where McGahern sought them is memory and the unconscious, the personal labyrinth that he entered through the portal of writing and traversed through the art of fiction.[9]

The challenge for the writer of fiction is to present characters and situations that have, as Heaney once stated, the 'ring of a real hammer on a real anvil'.[10] There can be nothing fake in good art. If there is, the reader, the listener or the viewer will sense it and fail to be moved by what will

8 McGahern rewrote significant parts of the novel when working on the French translation by Alain Delahaye of *The Leavetaking*. In his Preface to the second edition [1984], McGahern wrote: 'The crudity I was attempting to portray, the irredeemable imprisonment of the beloved in reportage, had itself become blatant. I had been too close to the "Idea", and the work lacked that distance, that inner formality or calm, that all writing, no matter what it is attempting, must possess'. It is far from certain that the second version achieved 'that inner formality or calm' that McGahern was seeking either.

9 Dermot McCarthy, *John McGahern and the Art of Memory* (Oxford: Peter Lang, 2010), 10. This study of McGahern's work unveils incontrovertible evidence of how the writer borrowed heavily on his life experiences to fuel his art.

10 Seamus Heaney, 'Introduction', in John McDonagh and Stephen Newman, eds, *Remembering Michael Hartnett* (Dublin: Four Courts Press, 2006), 11. Heaney was actually quoting from Saul Bellow, who as a child lived close to a blacksmith's forge and heard the rhythmic sound of a hammer on an anvil every day. This summed up for Heaney what good poetry is all about: 'a ring of truth in the medium, the

appear to be mere artifice. How the objective of producing good fiction is achieved, through borrowing on real-life experiences or on one's imagination, makes little difference. What counts is the emotional charge, the sense of order and authenticity with which the writing is imbued. Both the authors who are discussed in this chapter achieve that objective, albeit in very different ways.

In Stuart's case, there is no ambivalence as to what he was attempting in *Black List, Section H* (1970). The events recounted bear a striking resemblance to his life: there are references to the tragic death of his father in Australia, to Stuart's painful experience of boarding-school education, his marriage to Iseult Gonne, the strained relationship with her mother, Maud, and suitor, Yeats, fighting on the Republican side in the Civil War, breeding horses and chickens, trips to Paris and London punctuated by serious drinking sprees, and, most controversially, the time spent in Hitler's Germany and the radio broadcasts he made from Berlin. Stuart was in a position to draw on a remarkably eventful and exciting life that offered much scope for exploitation. He was personally acquainted with some of the leading figures who fought in the War of Independence and the Civil War; similarly, he was part of the literary and cultural revolution that predated and outlasted these violent events, namely the Irish Revival. A Protestant who converted to Catholicism, a mystic whose sexual escapades were not always in tune with his religious beliefs, a soldier poet, a man whose jaundiced view of Irish society led to his leaving his wife and children for many years to live in war-torn Germany, there was undoubtedly much to ponder in the events that are laid bare in *Black List, Section H.* Before discussing that work in more detail, however, it is worth quoting Stuart's description of how the creative process worked for him:

> I am one of those writers who identify themselves closely with their fiction. This has a bearing on the fact that it has been at the start of a creative phase that I have written at my best; it has been out of a pressure of stored living and experiencing.[11]

sounding out of inner workings, the sense of being in the presence of a self-absorbed and undistracted endeavour'.

11 Cited by Kiely, *Francis Stuart*, 228.

There is a real convergence with McGahern in these lines: the 'pressure of stored living and experiencing' was what drove both men to write. It is this quality that draws readers to their work, since the subject matter in both writers is often far from uplifting, characterized as they are by an unhealthy preoccupation with Catholic guilt, repression and a general ambience that is nothing short of depressing. Hence, what attracts readers is the quality of the writing and the intensity of emotion one encounters in the characters. In fact, without the release valve of literature neither McGahern nor Stuart would have been able to put up with some of the things that happened to them in the course of their lives. There is a deeply personal, human quality in the work of both writers that gives their work a universal resonance: their close identification with the existential trauma of the world they depict leaves the reader uncomfortable, overwhelmed by the similarity with experiences they have encountered in their own lives. Good novelists do not therefore just tell their story; they aspire to making their story everyone else's story. Without this essential hook, the reader will find it difficult to identify with a work of fiction.

Stuart acknowledged the link between himself and the main character of *Black List, Section H*, once more to Kevin Kiely: 'I am H the central figure; his events are my events'.[12] There is no ambiguity here. Even someone with only a limited knowledge of Stuart's life will sense that the people evoked – Iseult and Maud Gonne, W. B. Yeats, AE, Liam O'Flaherty and a host of others – are presented as Stuart would have known them, without any attempt on the writer's part to change their names or character traits. The sexual difficulties he and Iseult endured are presented in a factual manner also, which would subsequently cause a lot of hurt to his wife and children. The following description is revealing of their problems:

> But inches were miles in the tricky geography of the body, where tenderness and loving trust would have guided him to the healing entrance, anguished desire, combined with Iseult's passivity in the face of his unhappy fumblings, prevented him. He became terrified of remaining outside with his hurt, unadmitted to the ravished sanctuary.[13]

12 Kiely, *Francis Stuart*, 12.

13 Francis Stuart, *Black List, Section H* (London: Penguin Books, 1982), 27.

Married at a very young age to a woman who was older and whom he portrayed as being frigid, Stuart never achieved a satisfying sexual relationship with Iseult. There were periods of heightened passion, but these were often followed by the withdrawal of sexual favours by his wife. Ideologically, they were at odds also. He did not like her easy assumption of the absolute rightness and moral purity of the nationalist cause. Equally, when it came to religion, the couple were at loggerheads, primarily as a result of Stuart's suspicion that Irish people used their religiosity as a means of self-aggrandisement:

> Under de Valera or Griffith, art, religion and politics would still be run by those who at best used them to give them power, prestige, and a good living, and, at worst, for this to H was more dangerous, as a means towards a sterile, high-toned conformism.[14]

Likewise, McGahern's characters who are veterans of the War of Independence (many of them modelled on his father) regularly bemoan what has emerged from their sacrifices during the struggle. An indigenous elite (composed of priests, politicians, the medical and legal professions) merely replaced their English or Anglo-Irish predecessors, as the idealistic fervour and strong sense of solidarity which characterized the struggle gave way to apathy and opportunism. It is this lethargy that causes the disillusioned IRA veteran Moran to declare in *Amongst Women* (1990): 'What did we get for it? A country, if you'd believe them. Some of our own johnnies in the top jobs instead of a few Englishmen. More than half of my own family work in England. What was it all for? The whole thing was a cod'.[15] To Moran and many like him, the struggle for independence and the idealism and enthusiasm that marked the early stages of taking control of the southern half of the country ended up being a huge disappointment, 'a cod'. His erstwhile friend and military comrade, McQuaid, does not share such a view, having fared extremely well as a cattle-dealer. On the last occasion he visits the Moran household on Monaghan Day (when the fair was held in Mohill), McQuaid refuses to bend to Moran's

14 Stuart, *Black List, Section H*, 78.

15 John McGahern, *Amongst Women* (London: Faber and Faber, 1990), 5.

will, when the latter states in relation to the current state of the country: 'Look where it has brought us. Look at the country now. Run by a crowd of small-minded gangsters out for their own good. It was better if it had never happened'.[16] Possibly irked by his former superior officer's compulsion to always have things on his terms, McQuaid utters the following comment as he gets into his Mercedes (most likely acquired as a result of his being one of the 'small-minded gangsters' to whom Moran referred) at the end of his last visit to his friend: 'Some people just cannot bear to come in second'.[17] On the social and material levels, Moran has indeed been overtaken, and not just by McQuaid, but by all those who were prepared to use their influence among the politicians or priests who, between them, were running the country.

The same suspicion of the conservative and inward-looking society that emerged in Ireland during the 1920s, 30s and 40s is demonstrated on numerous occasions in the writings of Stuart and McGahern. Both found it abhorrent that many of the laudable ideals of the leaders of 1916 were abandoned in the mad rush to establish a country dominated by bigotry and sectarianism. In an essay he wrote to mark the seventy-fifth anniversary of the Rising, McGahern marvelled at how many of the signatories of the Proclamation were writers and intellectuals: 'A more unlikely crowd to spark a nation to freedom would be hard to imagine [...]'. He felt that the best way of honouring 1916 would be to restore 'those rights and freedoms that were whittled away from the nation as a whole in favour of the dominant religion'. McGahern was not hopeful about Ireland's future: 'What we are likely to get, though, are more of the outward shows – maybe even a grant or two – while Wink goes out in search of Nod'.[18]

Stuart could not have married into a more politicized family, Iseult being the child of Maud Gonne and the step-daughter of Seán McBride.[19]

16 *Amongst Women*, 18.

17 *Amongst Women*, 22.

18 John McGahern, 'From a Glorious Dream to Wink and Nod', in Stanley Van der Ziel, ed., *John McGahern Love of the World: Essays* (London: Faber and Faber, 2009), 125–7, 127.

19 Her birth father was the French artist Lucien Millevoye.

In the background, Yeats was never far away, constantly offering advice and patronage. To be fair, Yeats did help Stuart at an early stage in his writing career by introducing him to many influential people and by speaking highly of his poetry. Stuart knew that he was a man of extremes, a trait that was pointed out to him by Iseult on one occasion: 'Whatever you don't react against violently, you come under the spell of. There's nothing between, no detachment, no balance, no perspective'.[20] Hence his obsession with the Gospels and the mystics, which he took to reading during a period of his life. On one occasion, he went to confession and admitted his awareness 'of being alone in the haunted room of my mind', something that made him identify with the suffering Christ:

> That's why the promise of Jesus to come and dwell with those who love Him has always had such an appeal. It fascinated me for a time and I made a study of the mystics to find out if the promise had ever actually been kept.[21]

As one reads through *Black List, Section H*, one begins to see Stuart's tortured soul: he was conscious of the harm he was causing to himself and to others, and yet could do nothing to prevent his descent into a shady underworld of sin and depravity. There is no real attempt on his part to win the good opinion of his readers. H is portrayed in a stark light: he is a broken vessel, a lowly man with little to commend him apart from the gift of writing, a gift that heals and comforts him. Richard Murphy wonders why Stuart's portrayed H in such a negative light and concludes that: 'Stuart invites moral condemnation as proof against his domestication into an Irish literary culture in which opposition to illiberal, anti-individualist Ireland has become the posture of the establishment'.[22] McGahern often lamented in a similar manner how post-Independence Ireland became something of a theocracy, with the Catholic Church assuming enormous power as a result of the new political establishment's deference to it on all important issues. By writing about events in his life that he was not proud of, by sharing his

20 McGahern, *Black List, Section H*, 114.

21 McGahern, *Black List, Section H*, 211.

22 Richard Murphy, 'A Minority of One', 267.

innermost thoughts and memories with his readers, Stuart experienced a type of catharsis that was very different from Joyce's famous epiphany in *Portrait of an Artist as a Young Man*, the sort of book that Stuart had absolutely no desire to reproduce. In *Black List* one encounters the following lines: 'No important book was ever produced by trying; it came about as the by-product of an all-obsessive and perilous inward journey' (156). It is hard to argue with this. Stuart knows that by surviving perilous situations, by giving in to his most base longings and instincts, by experiencing himself in all his brokenness, he in some way completed his apprenticeship as a writer. By finding words to express the events of his life and by naming his memories, Stuart discovered his role as a writer. He was not so naïve as to believe that what he wrote in *Black List* would win him any favours in the Ireland of the time. In fact, it placed him outside the common pale. The last lines of the novel reveal a somewhat contradictory tone of defiance and resignation:

> Whatever it was that was at the other end there was no way of telling. It might be a howl of final despair or the profound silence might be broken by certain words that he didn't yet know how to listen for.[23]

The vocabulary here is very similar to what one encounters at the end of McGahern's second novel, *The Dark* (1965), when Mahoney Senior and his son become reconciled in Galway after the young man makes the decision to leave university in order to take up a job with the ESB (Electricity Supply Board). Forgetting the sexual and physical abuse he has endured at his father's hands, young Mahoney feels strangely elated as the two of them stroll through the streets of the university town: 'You were walking through the rain of Galway with your father and you could laugh purely, without bitterness, for the first time, and it was the kind of happiness ...'[24] The surprising thing about Mahoney's reaction is that no significant turnabout has come about in his problematic relationship with his father. It is hard to explain his sudden exoneration of the man who abused him as a

23 Stuart, *Black List, Section H*, 351.

24 John McGahern, *The Dark* (London: Faber & Faber, 1965), 188.

child and tyrannized both him and his siblings. But it is right at the end of the novel that one comes across the lines that are very similar to the last quote given above from *Black List*. Sharing a bed once more with his father (who is now old and incapable of imposing his will on his son), there is a moment of accommodation as they prepare for sleep: 'It seemed that the whole world must turn over in the night and *howl* in its boredom, for the father and for the son and for the whole lot, but it did not'.[25] Many victims of abuse blame themselves for what happened to them at the hands of others. In the case of Mahoney, however, it would appear as though he chooses to blot out the painful memories of his father's rhythmic massaging of his stomach and genitalia under the guise of helping him to get rid of wind. McCarthy supplies the following explanation for the apparently inconsistent change of attitude in the young man towards his father:

> What emerges in the closing lines, rather, is a sense of calm based in a kind of emotional fatalism, a recognition that what has happened is now history, undeniable, unforgettable, horrible, but at last, behind him.[26]

Magnanimity is possible for Mahoney because he is certain at this point that his life with his father is over. In *Memoir* there is a description of McGahern's father abusing his son in an almost identical manner to what Mahoney Sr does to his first-born. Is this a specific transposition of personal memory in fiction? Or a fictionalized version of events in the memoir genre? Only the writer himself knows the answer to these questions. In a sense, it doesn't matter all that much, since the power of writing has a truth of its own. One should judge Stuart and McGahern, not on the extent to which real events made their way into their fiction, but rather by the way in which the reader is touched by what appears to be genuine, authentic emotion. Neither writer is found wanting in this essential area.

McGahern's admiration for, and indebtedness to, Marcel Proust were acknowledged in several of the interviews he gave, as well as in his critical writing. *A la recherche du temps perdu* (*In Search of Lost Time*) was one of

25 *The Dark*, 191. Emphasis added to highlight similar vocabulary to that used by Stuart.
26 McCarthy, *John McGahern and the Art of Memory*, 107.

McGahern's favourite texts. The French philosopher Gilles Deleuze has argued that Proust's magisterial novel is less about the past or about voluntary or involuntary memory than about the artistic development of Marcel, Proust's autobiographical avatar.[27] Deleuze's point in relation to Proust is equally valid for McGahern, a writer whose literary canvas involved a reworking or a reimagining of how the artist emerges from the slow distillation of memories and experiences accumulated over seventy-odd years. When, with the benefit of hindsight, one reads McGahern's final work of fiction published during his lifetime, *That They May Face the Rising Sun*, the signs are there that he would never write another novel. The tone is more pastoral than anything McGahern had written before; it has a gentler quality and the absence of a domineering father figure ensures that there is less tension in the narrative. Dermot McCarthy is correct in his assertion that the novel was 'the promised Mass he never said for his mother; it is a work that brings the rituals of confession, atonement, purification, and communion intrinsic in his fiction to a transcendent closure'.[28] Therefore, the announcement shortly after the publication of *That They May Face the Rising Sun* (2001) that McGahern was writing *Memoir* did not come as a huge surprise to many readers. His antipathy to the father and his unwavering devotion to the mother, who died from cancer when he was just ten years old, are still palpable in McGahern's final text. *Memoir* is a paean to Susan McGahern, the kindly teacher, loving mother and devout Catholic, as much as it is an ill-disguised assault on Frank McGahern, IRA veteran of the War of Independence, Garda sergeant, violent, self-obsessed bully who regularly beat his children and made them feel as though they were the bane of his life. That the writer could still hold on to such antipathy for his father late in life is indicative of how poisoned the relationship between them was.

McGahern's evocation of the time spent with his mother is the most moving thing about *Memoir*. The young boy describes his unbounded

27 Cf. Ronald Bogue, *Deleuze and Guattari* (London: Routledge and Kegan Paul, 1989), 36. I would like to acknowledge my indebtedness to Mary Pierse for introducing me to this reference.

28 Dermot McCarthy, *John McGahern and the Art of Memory*, 8.

joy when the two of them are reunited after her first bout of cancer. He jealously sought to be alone with his beloved, whom he associated with immense happiness:

> The happiness of that walk and night under the pale moon was so intense that it brought on a light-headedness. It was as if the whole night, the dark trees, the moon in the small lake, moonlight making pale the gravel of the road we walked, my mother restored to me and giving me her free hand, which I swung heedlessly, were all filled with healing and the certainty that we'd never die.[29]

But the sad thing is that we do die – in the case of Susan McGahern, all too early – and afterwards those who are left behind have to deal with the legacy of grief and bewilderment. The loss of his mother was central to the discovery of a literary vocation by McGahern. *Memoir* opens with a detailed description of the landscape that the writer contemplated throughout his life, a landscape that he associated with the time spent with his mother. This may explain the capacity the Leitrim countryside has to transport the writer back to when he used to walk through these fields and along those lanes with his beloved mother. The lanes were the most conducive when it came to triggering what Proust would refer to as 'involuntary memory':

> I must have been extraordinarily happy walking that lane to school. There are many such lanes around where I live, and in certain rare moments over the years while walking in these lanes I have come into an extraordinary sense of security, a deep peace, in which I feel I can live for ever. I suspect that it is no more than the actual lane and the lost lane becoming one for a moment in an intensity of feeling, but without the usual attendants of pain and loss. These moments disappear as soon as they come, and long before they can be recognized and placed.[30]

Proust embraced Bergson's belief that all human memories are preserved, that what has been 'lost' can be found. McGahern shared Proust's view that one cannot will the past into being, which makes the distinction Proust spelled out between voluntary and involuntary memory all the more important:

29 John McGahern, *Memoir* (London: Faber & Faber, 2005), 64.
30 McGahern, *Memoir*, 4.

> For me, voluntary memory, which is above all a memory of the intellect and of the eyes, gives us only facets of the past that have no truth; but should a smell or a taste, met with again in quite different circumstances, reawaken the past in us, in spite of ourselves, we sense how different that past was from what we thought we had remembered, our voluntary memory having painted it, like a bad painter, in false colours.[31]

When McGahern speaks about the danger of sticking too closely to the facts, of recounting events exactly as they happened through the medium of fiction, it is Proust's description of voluntary memory that he has in his sights. It was involuntary memory that inspired the passage from *Memoir* quoted above, the type of memory that cannot be summoned at will and which happens of its own accord. Van der Ziel observes that the narratives of McGahern's novels are rarely arranged chronologically and offers one possible explanation for this in the writer's awareness of how the strict chronology of the traditional realist novel, which relies on linear time, does not conform to the way time is perceived by human beings.[32] In order to illustrate a similar evocation of lost time in McGahern and Proust, van der Ziel quotes the famous passage from *In Search of Lost Time* in which the French writer makes the following observation: 'It is a waste of effort for us to try to summon [the past], all the exertions of our intelligence are useless. The past is hidden outside the realm of our intelligence and beyond its reach'.[33] Consequently, it makes no sense for creative writers to consciously seek to recreate or relive events from their own lives. There are moments – such as what happens to McGahern while walking along the lanes of his youth – when the past returns, unbidden, and transports the subject to former times, but these moments cannot be summoned into being. They must come spontaneously, or not at all.

We may not sure what McGahern's opinion of Stuart's *Black List, Section H* was, but it appears likely that he would have appreciated the powerful way in which Stuart describes the birth of an artistic vocation

31 Marcel Proust, *Against Sainte-Beuve and Other Essays*, trans. John Sturrock (London: Penguin, 1980), 235–6. Once more, I'd like to thank Mary Pierse for making me aware of this quote.

32 Van der Ziel, *John McGahern and the Imagination of Tradition*, 166.

33 Cited by van der Ziel, 171.

and the trials and tribulations he underwent on the path to that goal. Joyce's *Portrait* did meet with McGahern's approval, especially because of the 'scrupulous meanness' of its style and the Flaubertian objectivity in which everything is couched. McGahern loved to quote the famous line from Flaubert: 'The author is like God in nature, present everywhere but nowhere visible', which for him was the ultimate achievement in literature. All great writers draw closely on their own experiences when fashioning their art – is there a more autobiographical writer than Joyce? However, art demands that emotion be controlled, that memory be distilled, that the act of creation open out onto something that is purer and more uplifting than simple transposition. It is in this crucial respect that Stuart and McGahern, two highly autobiographical writers who drew closely on memory, succeeded in creating good literature. Their works capture something that is central to human experience, with the personal assuming a universal resonance. So it was that with the approach of death which he knew was imminent after he was diagnosed with cancer during the composition of *Memoir*, McGahern could write:

> We grow into an understanding of the world gradually. Much of what we come to know is far from comforting, that each day brings us closer to the inevitable hour when all will be darkness again, but even that knowledge is power and all understanding is joy, even in the face of dread, and cannot be taken from us until everything is. We grow into a love of the world, a love that is all the more precious and poignant because the great glory of which we are but a particle is lost almost as soon as it is gathered.[34]

These lines offer an excellent summation of the fragile hold we have on life. It is through art that time is abolished and that one has a premonition of the state commonly referred to as eternity.

In a highly revealing article on Joyce and Stuart, S. J. Catterson draws the following comparison between *Black List* and *Portrait*: 'Written when he was in his sixties, *Black List* is a portrait of the artist as an old man, retro- and well as introspective. *A Portrait* is the first stage in a quest, *Black List* is

34 McGahern, *Memoir*, 36.

the legacy of one'.[35] In a similar way to Stuart, McGahern's literary 'quest' was a constant refining and re-sculpting of a life that brought with it pain and joy in equal measure. Both writers knew that the looking back, the re-creation through language of their past, was never going to be picture perfect. One has to allow for the fallibility of the memory, the manner in which events become distorted, unclear, and different, when recreated at a remove of a few decades. For Stuart and McGahern, it was the intensity of feeling and the well of imagination that brought about the conditions that allowed for the artistic process to be kick-started. Stuart once noted, 'whatever is imagined intensely enough becomes reality',[36] a comment that is eerily similar to the one we already quoted from 'The Image', where McGahern speaks of the 'Medusa's mirror' which allows us to 'see and to celebrate even the totally intolerable'. Because they feel deeply, 'purely', the imagination is the vehicle employed by the two writers to turn intense memories into art. Their work bears witness to the success of their respective quests and to the resilience of their character.

## Bibliography

Bogue, Ronald, *Deleuze and Guattari* (London: Routledge and Kegan Paul, 1989).

Catterson, S. J., 'Joyce, the "*Künstlerroman*" and Minor Literature: Francis Stuart's *Black List, Section H*', *Irish University Review*, 27(1) (1997), 87–97.

Hand Derek, *A History of the Irish Novel* (Cambridge: Cambridge University Press, 2011).

Heaney, Seamus, 'Introduction', in John McDonagh and Stephen Newman, eds, *Remembering Michael Hartnett* (Dublin: Four Courts Press, 2006).

Kiely, Kevin, *Francis Stuart: Artist and Outcast* (Dublin: The Liffey Press, 2007).

McCarthy, Dermot, *John McGahern and the Art of Memory* (Oxford: Peter Lang, 2010).

McGahern, John, *Amongst Women* (London: Faber and Faber, 1990).

35 S. J. Catterson, 'Joyce, the "*Künstlerroman*" and Minor Literature: Francis Stuart's *Black List, Section H*', *Irish University Review*, 27(1) (summer 1997), 87–97, 93.

36 Cited by Caterson, 94.

——, *The Dark* (London: Faber & Faber, 1965).
——, 'From a Glorious Dream to Wink and Nod', in Stanley Van der Ziel, ed., *John McGahern Love of the World: Essays* (London: Faber and Faber, 2009).
——, 'The Image: Prologue to a Reading at the Rockefeller University', in Stanley van der Ziel, ed., *Love of the World* (London: Faber and Faber, 2009).
——, *Memoir* (London: Faber & Faber, 2005).
Maher, Eamon, 'Interview with John McGahern', in *John McGahern: From the Local to the Universal* (Dublin: The Liffey Press, 2003).
Moore, Niamh, and Whelan, Yvonne, eds, *Heritage, Memory and the Politics of Identity: New Perspectives on the Cultural Landscape* (Aldershot: Ashgate, 2007).
Murphy, Richard T., 'A Minority of One: Francis Stuart's *Black List, Section H* and the End of the Irish *Bildungsroman*', *Irish University Review* 34(2) (2004), 261–76.
Proust, Marcel, *Against Sainte-Beuve and Other Essays*, trans. John Sturrock (London: Penguin, 1980).
Stuart, Francis, *Black List, Section H* (London: Penguin Books, 1982).
Ziel, Stanley van der, *John McGahern and the Imagination of Tradition* (Cork: Cork University Press, 2016).

SARA DYBRIS MCQUAID

# 6 Notes on Studying Public Policies of Memory: The Parades Commission in Northern Ireland and the Institutionalization of Memory Practices

This chapter seeks to contribute to the discussion of how political science and memory studies might engage each other more directly at a number of junctions, and most importantly, to begin a discussion on what kinds of inquiries such engagement might inspire. The impetus behind these reflections derives from an initial observation that there seems to be some missing contact points between political science and memory studies. Indeed, political scientists and historians have written widely about the politics of memory and the politics of history, particularly in the top down construction of authentic authority and legitimate tradition of rule – in establishing loyalty to, and membership of national and communal collectives.[1] However, it often remains undertheorized what kind of a thing or a process collective memory is in such procedures – put in the political science lingo, collective memory is rarely treated as an independent variable, or a focal point in these studies.[2] On the other hand, memory studies (with no lack of theoretical sophistication on the conceptualization of different forms of collective memory) often engages politics solely in the arena of

1 This approach taps into the school of history where nationalism is considered to be a process of presentist, instrumentalist social engineering. This is manifested particularly in the work of Hobsbawm and Ranger, in *The Invention of Tradition* (Cambridge: Cambridge University Press, 1983).

2 There is, however, a lot of work done on the role of symbols and traditions in formations of nationalism, not least in Anthony D. Smith's 'Ethnosymbolism', in *Ethno-symbolism and Nationalism* (London: Routledge, 2009) and 'Mythomoteurs', in *The Ethnic Origins of Nations* (Oxford: Blackwell, 1986).

representation, media and communication, where competing memory discourses are formed and memory actors vie for recognition and domination. This engagement rarely stretches to any detailed interrogation of the political institutional structures, or bureaucratic and policy processes through which they are filtered.

Transitional justice is an arena where the two areas of study have met more frequently, especially since transitional justice has come to cover not just processes of dealing with the past in terms of accountability and justice but also processes of reconciliation[3] – including artistic expressions and correlating accounts of conflict.[4] However, in this field (some call it a field, others a cloak),[5] the focus is often on legal frameworks for dealing with the past, which are overtly about memory and justice. Still, policy processes remain somewhat overlooked. Furthermore, while memory studies and transitional justice are said to have entered into a critical dialogue on the codified (tribunals, inquiries) and less codified (memorials and commemorations) forms and modes of remembrance,[6] policy and funding processes are rarely considered in any dynamic way.

There are of course always exceptions, but these broad characterizations of the field should be recognizable. Certainly, a number of scholars have encouraged the closer study of the politics of memory explicitly as policy processes,[7] whilst others have argued that there is little work within the academic study of cultural memory on the ways in which memory or the past is invoked discursively in policy or law.[8] This chapter, then, is an

3 Ruti Teitel, 'Editorial Note-Transitional Justice Globalized', *International Journal of Transitional Justice*, 2(1) (2008), 1–4.

4 *The Rule of Law and Transitional Justice in Conflict and Post-Conflict Societies: Report of the Secretary-General*. UN, 2004.

5 Christine Bell, 'Transitional Justice, Interdisciplinarity and the State of the "Field" or "Non-Field"', *International Journal of Transitional Justice*, 3(1) (2009), 5–27.

6 Ann Rigney, 'Reconciliation and Remembering: (How) Does it Work?', *Memory Studies*, 5 (2012), 251–8.

7 See for example E. Meyer, 'Memory and Politics', in Astrid Erll and Ansgar Nünning, eds, *Cultural Memory Studies* (New York: Walter de Gruyter, 2008), 173–80.

8 See Anna Reading, 'Identity, Memory and Cosmopolitanism: The Otherness of the Past and a Right to Memory', *European Journal of Cultural Studies*, 14(4) (2011), 379–94.

attempt at pointing a way forward. The first section of this chapter sketches out a methodology for studying formations of collective memories and public policies together, (and moreover for incorporating institutions seriously into the analysis) – as a way to analyse how policy becomes practice, is produced by practices, or co-produces practices in relation to the complex dynamics of collective memory. In the second section, I will offer the empirical case of the Northern Ireland Parades Commission to illustrate how a coupling of policy studies and memory studies might alter or add to our research agenda.

## Policy Studies and Memory Studies

There is a critical overlap between policy studies and memory studies. Constructivist approaches to policy studies like frame analysis and narrative analysis obviously take symbols and language seriously in terms of the sense-making and storytelling that is part of any political agenda setting.[9] In this school of policy analysis, meaning is understood to be contextual, contingent and historical. In memory studies, the object is likewise 'to analyse the narratives and discourses that underlie agendas, the processes that bring these narratives into the public domain and the political mechanisms at work in their delivery'.[10] Both fields of studies are interested in relations of power and the exercise of power in policy and memory and memory struggles. However, I would like to contend that particularly the study of political, institutional, and indeed bureaucratic mechanisms at work in delivering policies on the past can be boosted in memory studies.

Put differently, policy studies asks who acts where? Doing what? On which scale? In an argumentative turn of policy analysis we might add: using what language, invoking which loyalties and calling what threats and

9 Frank Fischer, *Reframing Public Policy: Discursive Politics and Deliberative Practices* (Oxford: Oxford University Press 2003).

10 Rigney, 'Reconciliation and Remembering'.

dangers into being to inspire fear and othering?[11] Memory studies similarly asks how and by whom, through which means, with which intention and which effect, are past experiences brought into play and made relevant?[12]

This all pertains to the agenda-setting part of policy, but beyond shifting frames, I am further interested in the delivery and the evaluation of policy and how those processes are constituted by and constitutive of collective memory concerns. Here we should include the technocratic conduits and conveyors, between governors and governed, by which I mean, for example, steering mechanisms, funding bodies, and hosting procedures. That is, we should examine the specific sites of policy implementation, which are shaped by and reshape agendas, and where instigators and recipients are also bureaucratic elites in the public, private and community sector. Such an endeavour might be thought of as a *nouveau mariage* of new institutionalism and the anthropology of policy, perhaps. Before turning to the empirical example, we should briefly consider the question of context.

If we are to take seriously the importance of policy context, I would suggest that we incorporate the concept of political culture. This means that we now have two fields of study (policy studies and memory studies) that (in some variants) share the same approach but not (yet) the same object, and then we have two central concepts that need to be theorized: collective memory plus political culture. The two central concepts in some respects share an ontological split between the individual and the collective. In Olick's seminal text 'Collective Memory: The Two Cultures' he sets out a baseline distinction between collected and collective memory.[13] What he refers to as *collected memory* is simply an aggregate of individual memories, whereas *collective memory* is not reducible to the individual but concentrates on the social frameworks that provide terms and definitions (including language, symbols and systems) for cognitive processes.[14]

11 Frank Fischer and John Forester, eds, *The Argumentative Turn in Policy Analysis and Planning* (Durham, NC; London: Duke University Press, 1996).

12 Graham Dawson, *Making Peace with the Past?* (Manchester: Manchester University Press, 2007).

13 Jeffrey K. Olick, 'Collective Memory: The Two Cultures', *Sociological Theory*, 17(3) (1999), 333–48.

14 Ibid., 336

Beyond distinguishing between collected and collective memory, others, like Aleida Assmann, have sought to differentiate forms of collective memory. Analytical distinctions in this respect centre around the collective that is referred to; the distance from which we remember; the modes of recall; and the drivers of memory.[15]

Conceptualizations of political culture share this ontological split between the individual and the collective. At one end of the spectrum we have the classic comparative distinctions of political culture. Here the term refers to aggregate patterns of subjective orientations toward political outcome, that is parochial, participant, or subject,[16] or distinctions between mass/elite, coalitional/contradictive, or homogenous/heterogeneous systems and society.[17] At the other end, we have a more inter-subjective conceptualization where political culture is analysed by mapping the discourses and symbolic practices by means of which both individuals and groups articulate their relationship to power, elaborate their political demands and put them at stake.[18] Political cultures, as such, structure the way people launch claims about what should be private and what should be public.

In transitional societies (emerging from violent conflict) political culture is often reframed and remade in a negotiation (or contestation) of meaning and an affirmation or renewal of patterns of values and behaviour. Such a negotiation includes what should be publicly considered unjust and what contextually unremarkable; what should be remembered and what forgotten. Here, we might think about the political economy of memory in its cultural constitution. Societies emerging from violent conflicts face difficult decisions in terms of dealing with the past. These decisions are conditioned by a number of factors that include first, how violence was ended (i.e. victory, defeat, compromise); second, the type of transition

15 Aleida Assmann, 'Memory, Individual and Collective', in *The Oxford Handbook of Contextual Political Analysis* (Oxford: Oxford University Press, 2006), 210–24.

16 Gabriel A. Almond and Sidney Verba, *The Civic Culture* (Boston, MA: Little, Brown and Company, 1965).

17 Arend Lijphart, *Patterns of Democracy* (New Haven, CT: Yale University Press, 1999).

18 See, for example, Jeffrey K. Olick and Daniel Levy, 'Collective Memory and Cultural Constraint: Holocaust Myth and Rationality in German Politics', *American Sociological Review*, 62(6) (1997), 921–36.

(i.e. from authoritarianism to democracy, from contested power to legitimate order); third, the existing political culture (i.e. the cultural constitution of political identities and activities); and fourth, the modes and materials for circulating memories between horizontal and vertical planes (i.e. between societal groups and the state).

Obviously, policy processes are framed and play out differently in different political cultures and different contexts of transition. They are also more than discursive struggles. They have concrete actors, systems and practices. To move beyond a mere representational analysis and thoroughly investigate the public policy of memory, we should consider sites of responsibility, decision-making, resources and not least the institutions that regulate and create specific templates for memories, for such institutions are embedded in local contexts and are constituted by as well as help constitute concrete cultural practices. Recent work in pragmatic sociology[19] situates political culture in contexts of institutional policies, political geographies and collective memory. Considering changes and particular continuities in these sites will keep any analysis of the policy of memory contextually responsive (and this overlaps in part with the cultural political economy school of policy analysis that takes material and social life into consideration).

Continuities and changes may take place at a number of levels, and so an analysis should reflect upon the internal institutional or organizational context of personnel, sponsors and funders being in place and being replaced. Furthermore, the proximate context of the policy environment and the macro context of changing directions of policy and institutions designed to implement policy, including economic fluctuations and changing sites of policy debates should be taken into consideration. A significant question is therefore: how are resources and discourses shepherded around in these scenarios?

Like the historian and sociologist Margaret Somers, I am quite conscious that we should not let culture disappear behind social structures and

19 See Paul Lichterman and Daniel Cefaï, 'The Idea of Political Culture', in *The Oxford Handbook of Contextual Political Analysis* (Oxford: Oxford University Press, 2008), 392–414.

psychological orientations. In order to locate another level of the public policy of memory, I would contend that we have to look 'below' the grand designs of 'repressive forgetting' or 'truth and reconciliation commissions' in constitutional settlements to more piecemeal or lower entry policy delivery. This is where I suggest we look at cultural policy and cultural politics and concretely at the institutions or mechanisms that manage and deliver these, thus regulating certain fields of culture (without losing sight of the economy and politics). This would also give us an opportunity to look beyond the usual suspects in protracted conflicts (in which cultural conflicts and identity politics are actually never far from sight). Rather than looking at community leaders, politicians, media and governments, we would be looking at leaders in the community sector, boards of trustees, funding bodies, teams of curators and academic research projects. This would allow us to appreciate competing forms and frames beyond the sectarian lenses that often dominate conflicted societies.

## The Parades Commission in Northern Ireland as a Case in Point

Turning now to the Parades Commission in Northern Ireland as a test case for these thoughts, I must begin by emphasizing that it is not possible to explore all the dimensions of the methodological framework set out above in this short space. Nevertheless, the Parades Commission serves as a good case for examining the public policy of memory in a number of ways. Firstly, it is a bureaucratic conduit to regulate contested ways of commemorating the past, through various technologies. Secondly, it was conceived as part of the transition from violent conflict to political compromise in Northern Ireland in which parade-related violence has persisted. Thirdly, transition itself is perched on a political elite compromise between the forces of constitutional unionism and nationalism (that is, parties either seeking to maintain the union between Northern Ireland and the rest of the United Kingdom, or seeking unification of Ireland north and south), in which the Parades Commission is also negotiating cultural demands beyond constitutional demands. Fourthly, the political culture in Northern Ireland is fraught with tensions not just between communities, but between the official political compromise and the vernacular experiences of the peace process, in which parades are used to articulate political demands. Finally, the Parades Commission is a hugely contested body, with changing members, making continuous rulings in a policy framework that remains under development as politicians disagree on policies to deal with the past.

In this context, Meyer's suggestion that we should treat the politics of history (how communities assign significance to the past and instrumentalize the past as leverage in present disputes) as policies for the past (usually thought of as a closed off period due to regime change with a specific judicial dimension), is useful.[20] As such, we would identify a form of politics of history (parading) instead as a policy area, where public policy is developed to regulate this particular form of politics of history, and where such regulation at the same time gives shape to new formations of collective memory. In very simple terms, policy analysis either examines existing policies in a field or develops new policies in a field. A policy analysis of the regulation of parades in Northern Ireland would look at agenda-setting; policy formation; delivery; and evaluation. Such an analysis would engage the main stakeholders, advocacy networks and adjacent policy areas (In this case for example public order legislation and human rights). It might also explore specific policy instrumentations (the tools and modes of carrying out policy) as a form of theorization of the relationship between the governing and the governed and a distinct vehicle for producing specific effects.[21] Memory studies, on the other hand would explore parades as commemorative practices, looking at the social, cultural, political, material, spatial and temporal dimensions of the phenomenon in terms of being shaped by, and shifting shapes of collective memory. If we combine the two,

20 Meyer, 'Memory and Politics'.

21 P. Lascoumes and P. Le Gales, 'Introduction: Understanding Public Policy through Its Instruments – From the Nature of Instruments to the Sociology of Public Policy Instrumentation', *Governance: An International Journal of Policy, Administrations and Institutions*, 20(1) (2007), 1–21.

we may examine the Parades Commission as a site where commemorations are hyper-codified in policy instrumentation as well as cultural terms. A particular investigation could engage the Parades Commissions website, which functions as a portal to describe and notify commemorative parades whilst also being a dynamic intervention in political culture and particular formations of collective memories.

Before developing the analysis further, it may be helpful to give some brief background to the issue of parading and the body of the Parades Commission. Commemorative parading is prolific in Northern Ireland, where upwards of 5000 parades take place every year (this number also includes all other forms of parades, charity runs etc.).[22] It can also be a much contested practice as many parades are exclusive stagings, which rehearse a master commemorative narrative composed of a sectarian selection of historical events, reminding an in-group of its distinct social identity and historical development in a ritual performance.[23] In this case the groups are roughly defined as Protestant/Unionist/Loyalist and Catholic/Nationalist/Republican. It may even be said that communities are remembering against each other, and often in an antagonistic mode.[24] Both in historical and contemporary terms, tensions arise when such exclusive narratives are taken to the streets and asserted across territories where they do not, or no longer, resonate and are met by organized resistance. Every year, well before and well beyond the political peace agreement in 1998, the marching season has been a trigger for confrontations and riots.[25]

22 *Parades Commission for Northern Ireland: Annual Report and Financial Statements for the year ended 31 March 2015* (Belfast: Parades Commission, 2015).

23 Eviatar Zerubavel, *Recovered Roots: Collective Memory and the Making of Israeli National Tradition* (Chicago: University of Chicago Press, 1995); Paul Connerton, *How Societies Remember* (Cambridge: Cambridge University Press, 1989).

24 For an exploration of different modes of memory see A. C. Bull and H. Lauge Hansen, 'On Agonistic Memory', *Memory Studies* (27 November 2015) <http://mss.sagepub.com/content/early/2015/11/26/1750698015615935>. Abstract. Accessed 13 June 2016.

25 It should be stressed that only a couple of hundred of these parades are contested to an extent that results in a determination by the Parades Commission. In 2014, 200 Unionist/Loyalist parades and ten Nationalist/Republican parades were considered.

Importantly, the parading issue is not just about confrontations between communities, but also between communities and the agents of the state.

A lot of fascinating work has been done on mapping the landscape of parades:

Neil Jarman's seminal book 'Material Conflicts' (1997)[26] introduced the main themes surrounding parades in Northern Ireland, charted parades as a historical phenomenon, explored how they function as vehicles that constitute the memory and collective identity of communities and analysed the dynamics of forging iconography, rituals and material culture. Dominic Bryan has been instrumental in developing a research agenda on the history of disputes over parades,[27] conflict over routes,[28] and the monitoring of parades.[29] A key point in this work is that analysing parades should amount to more than a cultural reading. Indeed, parades inherently combine a cultural tradition with the politics of the streets and it has been forcefully argued by Bryan that parades should be understood as highly political interventions with a scope far beyond reiterating heritage.[30] More recently, analyses have been developed that specifically relate the conceptual ordering of collective memory studies to the manifold layering of memory in the politico-cultural practice of parading in order to capture tensions between official and vernacular patterns of competitive commemoration in a 'post-conflict' scenario in Northern Ireland.[31]

26 N. Jarman, *Material Conflicts* (London: Berg, 1997).

27 Dominic Bryan, *Orange Parades: The Politics of Ritual, Tradition and Control* (London: Pluto Press, 2000).

28 D. Bryan, T. Fraser and S. Dunn, *Political Rituals: 'Loyalist' Parades in Portadown* (Ulster: University of Ulster, 1995).

29 D. Bryan and N. Jarman, *Independent Intervention: Monitoring the Police, Parades and Public Order* (Belfast: Democratic Dialogue, 1999); Dominic Bryan, 'The Anthropology of Ritual: Monitoring and Stewarding Demonstrations in Northern Ireland', in *Anthropology in Action*, 13(1) (2006), 22–31.

30 Dominic Bryan, 'Ireland's Very Own Jurassic Park: The Mass Media, Orange Parades and the Discourse on Tradition', in I. A. D. Buckley, ed., *Symbols in Northern Ireland* (Belfast: The Institute of Irish Studies, 1998), 23–42.

31 See also Sara McQuaid, 'Parading Memory and Re-member-ing Conflict: Collective Memory in Transition in Northern Ireland', *International Journal of Politics, Culture and Society* (2015). <http://link.springer.com/article/10.1007/s10767-015-9210-6> accessed 13 June 2016.

While there is, then, a literature on the politics of parades as politics of memory and history, as well as the policy and practices surrounding parades, there is less direct engagement with the Parades Commission as a bureaucratic technology through which these issues are caught up, filtered and reshaped.[32] The Parades Commission is the main body that gives expression to the law in terms of governing this local cultural practice – and the cultural practice has itself of course occasioned the creation of the regulatory body. The Parades Commission thus offers a compelling entry point for exploring the contact points between memory studies and policy studies through the framework set out above – that is, as a site for examining parading as a public policy of memory. The Parades Commission was established on the basis of the North Report (1997) which reviewed the contentious issues to do with parades and protests, a year before the Belfast/Good Friday Agreement in 1998. It now operates under the Public Processions (NI) Act 1998. The Parades Commission was conceived to be an independent body, which could make determinations on parades and related protests, considering a wide range of implicit concerns beyond obvious public order issues.

Crucially, the Parades Commission would take over this responsibility from the police, which was regarded with increasing distrust by all sides of the community in regards to making decisions and determinations on parades. Members of the Parades Commission (a chairman and six members) are all appointed by the Secretary of State for Northern Ireland. All parades and related protests in Northern Ireland now have to be notified to the Parades Commission twenty-eight days prior to the event (fourteen days prior for related protests). This notification process can be done through the Parades Commission website, which also logs past parades and allows for advanced searches of parades by time, place and organization. Based on this notification, the police flag up any 'sensitive' parades, and this is where the crucial process of determination begins. Although the Parades Commission remains the principal body tasked with responding to disputes over parades, a variety of other interested parties have taken

32 A notable exception would be M. Hamilton and D. Bryan, 'Mediation and the Law: The Parades Commission in Northern Ireland', *Ohio State University Journal of Peace Studies*, 22(1) (2006), 133–87.

a role in trying to mediate disputes.[33] Indeed, a number of organizations in (London)derry have adopted codes of conduct, which they instead call 'voluntary parades guidance' to dissuade determinations by the Parades Commission. In 'The Maiden City Accord' a number of parade organizations have agreed to set out formal guidance on the practices of parading. Among these is the active discouragement of alcohol consumption and a stipulation that all places of worship along the routes should be contacted to avoid disruption of religious services.[34]

Before turning to the substance of determinations, I want to dwell for a moment on the Parades Commission website as a bureaucratic technology of policy instrumentation that at once governs and shapes commemoration but also archives the determinations of those commemorative practices. On the website a parade is notified and searchable by date, time, district, organization, purpose and level of sensitivity. That is, an organization might self-identify under the rubric of 'Nationalist/Republican, National Graves, Commemoration Committee, H-Block, Sinn Féin' or 'Flag protest/Parade' (to mention two of seventeen possible designations). Similarly a purpose might be identified as 'Dedication/Celebration' or 'Remembrance/Commemoration' (to mention two out of fourteen possible purposes). Part of the notification process is to specify what 'uniform regalia' will be worn and what 'Banners/Flags will be carried'. Clearly, the options given are a mix between generic descriptors and particular local commemorative symbols and practices. The hallmark of the political culture in Northern Ireland, as separated into unionist and nationalist traditions, is recognized and recognizable through the collapse of historical and contemporary organizations or collectives into the same category (i.e. 'Nationalist/Republican, National Graves, Commemoration Committee, H-Block, Sinn Féin'), but also complicated by the emergence of new forms of parades and protests. It is notable that 'Flag protest/Parade' is given as an option both when selecting a 'type of organization' and a 'purpose'. This particular form of parade/protest has emerged since the decision in 2012 to fly the Union

33 N. Jarman, M-K Raillings and J. Bell, *Local Accommodation: Effective Practice in Responding to Disputes Over Parades* (Belfast: Institute for Conflict Research, 2009).

34 S. McBride, 'Voluntary Parade Guidance Adopted', *News Letter*, 6 August 2014.

Jack from Belfast City Hall only on designated days. It is clear, then, that the website is ordering commemoration and perpetuating typologies of memory collectives, allegiance and purpose. It acts as a social framework that provides terms and definitions but in a way that is also somewhat open to let users name new dynamics and relocate remembrance at least on an intra- if not an inter-community basis.

On the website it is also possible to see whether a parade has been notified in due time, whether it is deemed sensitive, whether there are any related protests and whether any restrictions have been imposed. Hamilton and Bryan have argued that 'the Parades Commission represents an attempt to address longstanding structural inequalities in the public sphere and illustrates the potential capacity of an institution to address entrenched identity-based conflict and deepen the roots of democratic governance'.[35] However, on the one hand it is clear that the Parades Commission is a response to a number of anxieties arising simultaneously from the communities that wish to parade, and the communities in which they wish to parade. In this context, the website also functions as a transparent participatory tool in understanding the procedures governing determinations. On the other hand, the State is literally trying to embody this anxiety in structures like the Parades Commission by expanding and delegating the procedures to accredit and perform commemorative parades. This is underscored by the fact that parades must have trained stewards as well as adhere to a 'code of conduct' set out by the Parades Commission. In fact, we may argue that the bureaucratic solution mirrors the political economy of the peace process itself. If there cannot be substantive agreement (on the right to parade or the future constitutional position of Northern Ireland), there can at least be procedural agreement (that is, how to work towards disparate destinations in both cases). The question is whether this way of dealing with anxiety provokes more or less ontological security for groups that are remembering collectively to ensure the continuation of identity in shifting frames of political compromise.

35 Hamilton and Bryan, 'Mediation and the Law'.

There are a number of conditions that can be imposed on a contentious parade, ranging from regulating songs to be sung and symbols to be displayed to restrictions on routes. These might also include instructions on behaviour when passing places of worship or other locations of symbolic significance. Notably, any determination will include a section on 'Background'. In this section the pedigree of the parade is assessed: whether it has previously occasioned violence; whether parade-related protests have been announced; and if that is the case, whether there has been dialogue between the organizers and the protesters. Furthermore, the background section of a determination considers the immediate state of community relations in the psycho-social context of the parade. In this, the Parades Commission represents an official peace process discourse, which tries to regulate vernacular discourses and traditions. Again, these acts of regulation at once shape the contemporary parading tradition and provoke resentment and protest.

## Determining the Past and the Present

In the following, I would like to introduce a brief example of a Parades Commission determination. The example is drawn from North Belfast, which in recent years has been a hot spot of competitive commemoration between so-called dissident republicans and an array of Protestant marching institutions. The competition for access to the public sphere in North Belfast has a long history[36] and is compounded by the continuing presence of organizations that espouse violence, and particularly by the relationship between these organizations and the parading culture. In the context of 'post-conflict' Northern Ireland, parades in this area are used as instruments to vie for domination of fluid space (political as well as

36 F. Burton, *The Politics of Legitimacy: Struggles in a Belfast Community* (London: Routledge, 1978); P. Shirlow and B. Murtagh, *Belfast: Segregation, Violence and the City* (London: Pluto Press, 2006).

territorial), and are to some extent regulated as such, more so than commemorative practices.[37]

The case involves a parade billed by its organizers, the Republican Network for Unity, as a memorial march for Henry Joy McCracken, who was a Protestant Irish nationalist and a founding member of the Society of United Irishmen. He was hanged in 1798 as one of the leaders of the failed rebellion against British rule in Ireland. Since 2011, the Republican Network for Unity has organized a memorial march in his name, terminating at McCracken's graveside memorial in Clifton Street Cemetery, North Belfast. Since its inception, the march has stirred up tensions, partly because it is organized from an anti-peace-agreement platform; partly because it proposes to go along a contentious route, parts of which run along a particularly fractious interface; and partly because chapters of the Royal Black Institutions (another protestant fraternal order) march at the same time in the same area.[38] In 2013, the Parades Commission argued that 'Should the parade [the Henry Joy McCracken Memorial March] process without restriction there will be an adverse effect on already fragile community relations, disruption to community life and potential for public disorder'.[39] The final determination for the Henry Joy McCracken Memorial March prescribed a new route, called for trained stewards, laid down outer perimeters for the congregation and the dissolution of the parade as well as emphasized the imperative for 'respectful behaviour' (i.e. no singing, chanting, loud drumming, or display of paramilitary clothing or insignia) in the vicinity of interface areas. At the same time the organizers of the parade are responsible for all participants, bands and supporters being made aware of the code of conduct and the determination. This determination provoked a very angry response from the acting chair of the Republican Network for Unity, who dispersed the parade at the point of the re-route

37 McQuaid, 'Parading Memory and Re-member-ing Conflict'.

38 Ibid.

39 Parades Commission: Determination #13 25.8.13.

and encouraged the marchers to instead go home and 're-double efforts to build a United Republican Movement […]'.[40]

At least three things spring to mind from this case. First, the elaborate bureaucratic procedures clearly shape cultural traditions as mapping and remapping take place in an attempt to fixate or supplant memory and geography. Demographic and territorial changes over time are overridden by 'traditional routes', which are again overridden by Parades Commission rulings either restricting, rerouting, or observing another geography in the shape of interfaces, churches or war memorials. Here policy bodies quite literally attempt to steer the expression and (re)location of memory. Second, recent memory already plays into determinations, as the historical trajectory of any given parade is taken into consideration.[41] Thus, recent memory then guides the policy response of the Parades Commission. Third, determinations are themselves contested, and give rise to what we might call 'determination memory', where the perceived injustice of specific, or sequential regulations lead to simmering anger and more commemorative violence.[42] Here new forms of memory and contestation develop directly from policy instruments and are archived by governing bodies for retrieval.

## Conclusion

This chapter has sketched a methodological framework for developing contact points between policy analysis and memory studies. For this purpose, it suggests exploring the linkages and dynamics between processes

40 Republican Network for Unity, 'RNU self disperse Henry Joy march. No back door for Republicanism' (2013), <http://www.republicanunity.org/rnu-self-disperse-henry-joy-march-no-back-door-for-republicanism/> accessed 13 June 2016.

41 McQuaid, 'Parading Memory and Re-member-ing Conflict', 8.

42 J. Byrne, S. McDowell and M. Braniff, 'Violence, space and memory in the new Northern Ireland' (2012). <https://www.opendemocracy.net/opensecurity/maire-braniff-sara-mcdowell-jonny-byrne/violence-space-and-memory-in-new-northern-irela> accessed 13 June 2016. See also McQuaid, 'Parading Memory and Re-member-ing Conflict'.

of collective memory formations and policy formations. In the subsequent brief empirical sketch of the Parades Commission in Northern Ireland, I tried to open up an analysis of the public policy of memory at the intersection of bureaucratic frames and commemorative forms. I further argued that the synergy afforded by coupling the fields of memory and policy studies not only points to new sites of investigation, but also to new research agendas. In this, a firmer understanding of the technocratic shaping of memory is gained in memory studies, including how political languages form and are reformed in the policymaking process by institutional and bureaucratic practices. In policy analysis, power struggles can sometimes disappear behind technical discourse, but a firmer appreciation and conceptualization of structures and carriers of collective memory, in particular political cultures, may lead to more reflexive policy instrumentation in terms of engaging these dynamics.

## Bibliography

Almond, Gabriel A., and Verba, Sidney, *The Civic Culture* (Boston, MA: Little, Brown and Company, 1965).

Assmann, Aleida, 'Memory, Individual and Collective', in Robert E. Goodin and Charles Tilly, eds, *The Oxford Handbook of Contextual Political Analysis* (Oxford: Oxford University Press, 2006).

Bell, Christine, 'Transitional Justice, Interdisciplinarity and the State of the "Field" or "Non Field"', *International Journal of Transitional Justice*, 3(1) (2009), 5–27.

Bryan, Dominic, 'The Anthropology of Ritual: Monitoring and Stewarding Demonstrations in Northern Ireland', in *Anthropology in Action*, 13(1) (2006), 22–31.

——, 'Ireland's Very Own Jurassic Park: The Mass Media, Orange Parades and the Discourse on Tradition', in I. A. D. Buckley, ed., *Symbols in Northern Ireland* (Belfast: The Institute of Irish Studies, 1998), 23–42.

——, *Orange Parades: The Politics of Ritual, Tradition and Control* (London: Pluto Press, 2000).

——, Fraser, T., and Dunn, S., *Political Rituals: 'Loyalist' Parades in Portadown* (Ulster: University of Ulster, 1995).

——, and Jarman, N., *Independent Intervention: Monitoring the Police, Parades and Public Order* (Belfast: Democratic Dialogue, 1999).

Bull, A. C., and Lauge Hansen, H., 'On Agonistic Memory', *Memory Studies*, 27 November 2015, <http://mss.sagepub.com/content/early/2015/11/26/175069 8015615935> abstract accessed 13 June 2016.

Burton, F., *The Politics of Legitimacy: Struggles in a Belfast Community* (London: Routledge, 1978)

Byrne, J. J., McDowell, S., and Braniff, M., 'Violence, Space and Memory in the new Northern Ireland', (2012) <https://www.opendemocracy.net/opensecurity/maire-braniff-sara-mcdowell-jonny-byrne/violence-space-and-memory-in-new-northern-irela> accessed 13 June 2016.

Connerton, Paul, *How Societies Remember* (Cambridge: Cambridge University Press, 1989).

Dawson, Graham, *Making Peace with the Past?* (Manchester: Manchester University Press, 2007).

Fischer, Frank, and Forester, John, eds, *The Argumentative Turn in Policy Analysis and Planning* (Durham, NC; London: Duke University Press, 1996).

——, *Reframing Public Policy: Discursive Politics and Deliberative Practices* (Oxford: Oxford University Press 2003).

Hamilton M., and Bryan, D., 'Mediation and the Law: The Parades Commission in Northern Ireland', *Ohio State University Journal of Peace Studies*, 22(1) (2006), 133–87.

Hobsbawm, E. J., and Ranger, T. O., *The Invention of Tradition* (Cambridge: Cambridge University Press, 1983).

Jarman, N., *Material Conflicts* (London: Berg, 1997).

——, Raillings, M.-K., and Bell, J., *Local Accommodation: Effective Practice in Responding to Disputes Over Parades* (Belfast: Institute for Conflict Research, 2009).

Lascoumes, P., and Le Gales, P., 'Introduction: Understanding Public Policy through Its Instruments – From the Nature of Instruments to the Sociology of Public Policy Instrumentation', *Governance: An International Journal of Policy, Administrations and Institutions*, 20(1) (2007).

Lichterman, Paul, and Cefaï, Daniel, 'The Idea of Political Culture', in *The Oxford Handbook of Contextual Political Analysis* (Oxford: Oxford University Press, 2008).

Lijphart, Arend, *Patterns of Democracy* (New Haven, CT: Yale University Press, 1999).

McBride, S., 'Voluntary Parade Guidance Adopted', *News Letter* (6 August 2014).

McQuaid, Sara, 'Parading Memory and Re-member-ing Conflict: Collective Memory in Transition in Northern Ireland', *International Journal of Politics, Culture and Society*, (2015) <http://link.springer.com/article/10.1007/s10767-015-9210-6> accessed 13 June 2016.

Meyer, E., 'Memory and Politics', in Astrid Erll and Ansgar Nünning, eds, *Cultural Memory Studies* (New York: Walter de Gruyter, 2008), 173–80.
Olick, Jeffrey K., and Levy, Daniel, 'Collective Memory and Cultural Constraint: Holocaust Myth and Rationality in German Politics', *American Sociological Review*, 62(6) (1997), 921–36.
——, 'Collective Memory: The Two Cultures', *Sociological Theory*, 17(3) (1999), 333–48.
*Parades Commission for Northern Ireland: Annual Report and Financial Statements for the Year ended 31 March 2015* (Belfast: Parades Commission, 2015).
Reading, Anna, 'Identity, Memory and Cosmopolitanism: The Otherness of the Past and a Right to Memory', *European Journal of Cultural Studies*, 14(4) (2011), 379–94.
Republican Network for Unity, 'RNU self disperse Henry Joy march. No back door for Republicanism', (2013) <http://www.republicanunity.org/rnu-self-disperse-henry-joy-march-no-back-door-for-republicanism/> accessed 13 June 2016.
Rigney, Ann, 'Reconciliation and Remembering: (How) Does it Work?', *Memory Studies*, 5 (2012), 251–8.
*The Rule of Law and Transitional Justice in Conflict and Post-Conflict Societies: Report of the Secretary-General*, UN, 2004.
Shirlow P., and Murtagh, B., *Belfast: Segregation, Violence and the City* (London: Pluto Press, 2006).
Smith, Anthony, 'Ethnosymbolism', in *Ethno-symbolism and Nationalism* (London: Routledge, 2009)
——, 'Mythomoteurs', in *The Ethnic Origins of Nations* (Oxford: Blackwell, 1986).
Teitel, Ruti, 'Editorial Note-Transitional Justice Globalized', *International Journal of Transitional Justice*, 2(1) (2008), 1–4.
Zerubavel, Eviatar, *Recovered Roots: Collective Memory and the Making of Israeli National Tradition* (Chicago: University of Chicago Press, 1995).

STEPHEN HOPKINS

# 7 The Irish Republican Movement and the Contested Past: 'Official Memory' and the Politics of Dissent

This chapter will analyse the politics of contested memory in relation to the recent post-Good Friday Agreement (GFA) trajectory of the Irish Provisional Republican movement, consisting of Sinn Féin (SF) and the Irish Republican Army (IRA). This will involve an examination of attitudes to critical aspects of the movement's past, as exhibited by a range of emblematic individuals in leadership positions within the mainstream movement, including SF's President, Gerry Adams, and Northern Ireland's Deputy First Minister, Martin McGuinness. It will be argued that a self-perpetuating leadership group based around Adams and McGuinness has effectively sought to construct what Elizabeth Jelin has termed an 'official memory' or a 'master narrative' for the movement.[1] This analysis will also examine the efforts to challenge this 'official memory' from within the milieu of so-called 'dissident' republicanism. A small but growing number of prominent ex-Provisionals (such as Anthony McIntyre,[2] Brendan Hughes,[3]

1 E. Jelin, *State Repression and the Struggles for Memory* (London: Latin American Bureau, 2003).

2 For a collection of articles and interviews, see A. McIntyre, *Good Friday: The Death of Irish Republicanism* (New York: Ausubo Press, 2008).

3 For an extensive oral history interview with Brendan Hughes, see E. Moloney, ed., *Voices from the Grave: Two Men's War in Ireland* (London: Faber and Faber, 2010).

Richard O'Rawe[4] and Tommy McKearney[5]) have challenged the leadership's account of the past, particularly with regard to crucial events such as the 1980–1981 hunger strikes. It is also the case that some relatively obscure former foot-soldiers, such as Gerry Bradley[6] and Eamonn Collins,[7] have also engaged in memoir-writing that is profoundly critical of the leadership's version of the past.

The chapter will utilize a range of sources, including newspapers, party journals and websites, speeches, but particularly life-writing by key individuals.[8] These are important resources for researchers interested in the contemporary debates around conflicting memories of the Troubles, and the growing fragmentation *within* the 'republican family', which had been relatively united during the conflict itself. The chapter investigates the role of the republican movement's organizational culture in the contemporary struggle over memory within Irish republicanism. The Provisional movement, in keeping with other avowedly revolutionary organizations such as the international communist movement, placed a very significant emphasis upon 'unity in action' or 'democratic centralism'. As the Provisionals have adjusted to a non-revolutionary era in the wake of the compromise enshrined in the 1998 GFA, it has become of paramount importance for the 'leadership group' to seek to control the movement's collective memory of the past as well as its narrative reconstruction of the struggle. Analysis of these issues, and their reception within the broader Catholic nationalist and republican community, is a key element of a more nuanced understanding

4 For Richard O'Rawe's memoir, recollecting the 1981 hunger strike by republican prisoners, see *Blanketmen: An Untold Story of the H-Block Hunger Strike* (Dublin: New Island, 2005). For further details of the dispute unleashed by O'Rawe's allegations, see his *Afterlives* (Dublin: Lilliput Press, 2010).

5 See T. McKearney, *The Provisional IRA: From Insurrection to Parliament* (London: Pluto Press, 2011).

6 G. Bradley with B. Feeney, *Insider: Gerry Bradley's Life in the IRA* (Dublin: O'Brien Press, 2009).

7 E. Collins with M. McGovern, *Killing Rage* (London: Granta, 1997).

8 For a fuller discussion of the significance of memoir-writing as a source for interpreting the Troubles, see S. Hopkins, *The Politics of Memoir and the Northern Ireland Conflict* (Liverpool: Liverpool University Press, 2013).

of the 'memory struggles' which characterize the legacy of the Troubles in Northern Ireland.[9]

## Collective Memory and the Irish Republican Past

It has become a commonplace that the Irish republican movement has always placed enormous store by the sacrifices of earlier generations. The Provisional movement has claimed a 'deep identification' with the heroes of Easter 1916; indeed, the rationale of the movement could, at least in part, be understood in terms of 'the mythic resurrection of some sacred national "tradition"'.[10] As Richard O'Rawe, the public relations officer of the IRA prisoners at the time of the 1981 hunger strike, put it: 'we existed in an idealistic time warp, wallowing in the vision of historical Irish heroes struggling for freedom. There was an almost biblical reverence for the 1916 Proclamation [...] sometimes elevating the sacrifice of the signatories to a hallowed act'.[11] In this sense, the collective memory of the movement may be interpreted as what the poet Seamus Heaney described, with specific reference to the hunger strike, as a 'sacred drama', in which the moral authority of the dead has often outweighed the demands and interests of the living generations of Irish people.[12] In citing this comment, David Rieff makes the important point that 'as Heaney knew well, sacred drama is the antithesis of any decent politics. For once the sacred has been invoked, there can be no compromise with one's adversaries, only their unconditional

9 For a recent analysis of SF as a party adjusting to a non-revolutionary era, see S. Whiting, 'Mainstream Revolutionaries: Sinn Féin as a "Normal" Political Party', *Terrorism and Political Violence*, 28(3) (2016), 541–60.

10 R. Kearney, 'Myth and Martyrdom: Foundational Symbols in Irish Republicanism', in R. Kearney, ed., *Navigations: Collected Irish Essays, 1976–2006* (Dublin: Lilliput Press, 2006), 34.

11 R. O'Rawe, *Blanketmen*, 79–80.

12 S. Heaney cited in D. O'Driscoll, *Stepping Stones: Interviews with Seamus Heaney* (New York: Farrar, Strauss and Giroux, 2008), 259.

surrender. To the extent that this can still be called politics, it is a politics of totalitarianism'.[13] However, Heaney's complex reactions to the hunger strike, both at the time and subsequently, capture something of the deep ambivalence that many Irish Catholic nationalists felt with regards not only to these specific events, but also perhaps to republicanism more generally. Heaney answered affirmatively when asked whether he had sympathy for the hunger strikers as men, even if he had none for the political ideology which inspired them; he even attended the wake of Thomas McElwee (the ninth of the strikers to die), and talked of the 'cruelty of the predicament' which nationalists who refused to support the IRA campaign felt. But, at the same time, Heaney was 'wary of ennobling their sacrifice beyond its specific historic and political context'.[14]

In a movement bound up with physical force and violent conflict, these memories are marked by sacrifice, victimhood, and suffering of wounds endured, on the one hand; but also by questions of legitimacy, responsibility and recognition of hurts inflicted, on the other. The Provisional movement has invested a huge amount of effort to claim that its violence was the work of an oppressed people, and that the Catholic nationalist population was the primary community of victims in the Northern Irish conflict. But, as Rieff argues, 'there are few phenomena more uncontrollable socially and, hence, more dangerous politically than a people or a social group that believes itself to be a victim'.[15] In Ireland, the 'coveted dual role of leader and victim' has been sought after not simply by Provisionals like Gerry Adams,[16] but also by mainstream leaders such as De Valera, MacBride and Haughey.[17] Seán

13 D. Rieff, *In Praise of Forgetting: Historical Memory and its Ironies* (New Haven, CT; London: Yale University Press, 2016), 103.

14 S. Heaney in D. O'Driscoll, *Stepping Stones*, 260.

15 D. Rieff, *In Praise of Forgetting*, 117.

16 The most recent, and egregious, example of Adams' adoption of a victim status came with his arrest and questioning in May 2014 in relation to the IRA's 1972 murder of mother-of-ten Jean McConville. Supporters of Adams produced a large wall mural of their leader, and alleged that he had been the victim of 'securocrats' in the police service.

17 R. Kearney, 'The Triumph of Failure: The Irish Prison Tradition', in R. Kearney, ed., *Navigations*, 411, 52. De Valera's political status was enhanced by his record of

MacBride, in his introduction to the prison diary of Bobby Sands, made the point that Catholic nationalists' sympathy towards the hunger strikers was due precisely to the interpretation of their actions within the context of a collective memory of British colonial misrule during, for example, the Great Famine, and the sense of victimhood which accompanied it.[18] Kearney summarized MacBride's argument as follows: the hunger strike 'was not some isolated political happening of our time but a deep symptom of a historically recurring persecution'. MacBride had identified 'the basic rationale of the republican prison campaign in terms of the memory of the nationalist people's historical suffering, a memory of heroic martyrdom [...]'.[19] Of course, the danger identified by Rieff is clear here: the Provisional movement was only too willing to exploit this sympathy to recruit new volunteers, and to legitimize its violent 'resistance' to perceived oppression.

In the first volume of his memoirs, Gerry Adams made the claim that 'out of the prison struggle came a revitalized SF, conscious perhaps for the first time of our ability to galvanize public support and to marshal support through elections. The hunger strikes were the beginning of the end of spectatorism in Republican politics'.[20] The problem for Adams is that this end to spectatorism has increasingly cut both ways in republicanism: in the post-GFA era, both internal and external critics of the contemporary political trajectory of the movement have argued that the sacrifices of the hunger strikers (and those of many other republicans who died in the Troubles) cannot and should not be used to justify the ideological about-turns of a leadership which has compromized its republican 'soul', and betrayed its sacred duty. McIntyre encapsulated this sentiment in 2004: 'when I visit the Republican plot in Belfast the thing furthest from

imprisonment on behalf of the republican cause in the wake of the Easter Rising. In a more ambivalent fashion, Haughey's position as a 'scapegoat' was arguably one of the factors behind his electoral success.

18 S. MacBride, 'Introduction', in B. Sands, ed., *One Day in My Life* (Cork and Dublin: Mercier Press, 1983), 7–21.

19 R. Kearney, 'The Triumph of Failure: The Irish Prison Tradition', 50–1.

20 G. Adams, *Before the Dawn: An Autobiography* (Dingle: Brandon, 1996), 49.

my mind is the politics of Sinn Féin, which sit like an ugly scar defiling the very sacrifice inscribed in each Republican grave'.[21]

Paradoxically, it is even more important to the movement now, in the *aftermath* of the conflict, since the guns have fallen silent, to justify the *prior* necessity and legitimacy of the IRA's violence, even as this phase of the struggle has been superseded. As English has convincingly argued, 'Provo politics were centrally defined by the necessity for political violence. Force was essential to the achievement of republican goals'.[22] Of the four key justifications for violence put forward by the movement (namely, communal defence; the inequity of the Northern Ireland state and the collective victimhood of the Irish Catholic people who live there; the illegitimacy of British rule; and, the irreformable character of this state), it is arguably only the first and last of these issues that have genuinely been transformed in the era of the peace process. To establish the essential continuity of the movement, SF leaders have attempted to promote the subtle argument that the Provisional IRA's 'armed struggle' was both legitimate and necessary in the period up until 1994, but this commitment to violent methods has since become unnecessary, as the Catholic people no longer require physical defence. Loyalist paramilitaries have instituted their own ceasefire, the British Army has withdrawn, and the Northern Ireland state has been, and will continue to be, reformed by non-violent means, according to SF's interpretation of the transitional character of the 'peace process'. For dissenters, this rationalization is, of course, sophistry; the reform of the state *within* the confines of UK sovereignty had been available since the early 1970s, but as revolutionary nationalists or socialists (or both), this goal was simply not what the Provisionals had been fighting for, nor an objective that they should still be pursuing. As Bernadette Sands-McKevitt, the sister of Bobby Sands, has put it: 'Bobby did not die for cross-border bodies with executive powers. He did not die for Nationalists to be equal British citizens within the Northern Ireland state'.[23] This is the context

21 A. McIntyre, *Good Friday*, 15.

22 R. English, *Armed Struggle* (London: Macmillan, 2003), 342.

23 B. Sands-McKevitt cited in D. O'Hearn, *Bobby Sands: Nothing but an Unfinished Song* (London: Pluto Press, 2006), 283.

within which the collective memory of Provisional republicans has become fragmented and fractured, with dissenters challenging the 'ownership' of the 'memoryscape' by the leadership group. The recent centenary commemorations of the 1916 Easter Rising confirmed the critical significance of the 'republican dead' for contemporary political manoeuvring amongst different factions within the broad 'republican family'.[24] The SF leadership attempted (without great success) to wrest control of this centenary from the parties of government in the Republic of Ireland, but the speeches of SF leaders were instructive in their emphasis upon the essential continuity of their movement with the bloody insurrection and heroic sacrifice of the executed martyrs of Easter. As Seán Lynch, an ex-IRA prisoner and current SF Member of the Stormont devolved assembly, argued in a speech commemorating the Easter Rising in Pettigo, Co. Fermanagh: 'There are those who would have us believe that these men and women [the volunteers of the Provisional IRA] cannot be equated with those of 1916. They are hypocrites. Bobby Sands was a revolutionary and visionary in the same vein as James Connolly and Pádraig Pearse'.[25] Yet, this determined attempt to lay claim to the historical mantle of republican legitimacy, and to reassure SF members that the movement remains true to its self-proclaimed revolutionary vocation, is a distinctly double-edged sword.

David Rieff has recently argued persuasively (borrowing from W. B. Yeats' poem 'September 1913') that this 'romantic' nationalist vision of the past outlived the Fenian movement of the nineteenth century, and was quite consciously replicated, first by the Easter rebels themselves, and then by successive generations of republican 'diehards' during the twentieth century. According to Rieff, it was only the signing of the GFA that 'finally put romantic Ireland in its grave'.[26] However, the SF approach to the centenary of the Rising, and to establishing and controlling the collective memories and mythological narratives of the republican past might give us

24 See S. Hopkins, '"Our Whole History has been Ruined!" The 1981 Hunger Strike and the Politics of Commemoration and Memory', *Irish Political Studies*, 31(1) (2016), 44–62.

25 *Irish Times* (29 March 2016).

26 D. Rieff, *In Praise of Forgetting*, 104.

pause for thought in making such a judgment. In the last two decades, as SF has moved to centre stage in the political systems of Northern Ireland, and increasingly in the Republic also, the movement has become divided over both its future strategic direction, but also the 'contested authorship' of its own history and memory.[27]

## The 'Leading Group' and Republican Organizational Culture

The current leadership of SF and the broader republican movement has been in place since the late 1970s, based around Gerry Adams, who has been SF's President since 1983, and Martin McGuinness, its Deputy First Minister in the Northern Ireland Executive.[28] During the 1980s, internal challenges to the power base of this 'leading group' were absorbed or marginalized. For instance, once the leadership had endorsed a departure from traditional republican orthodoxy in 1986 by ending the policy of abstentionism to the Republic of Ireland's parliament, a significant number of Provisionals (gathered around the previous SF President, Ruairí Ó Brádaigh) left the movement, and set up a rival organization, *Republican* SF. Other rivals to the Adams/McGuinness leadership were demoted, such as Ivor Bell. This leadership proved adept at obstructing and diverting any new challenge to their control of SF and the IRA, and whilst the 'war' against the British state was being waged they were aided in this endeavour by a natural predisposition to 'unity in action', as manifested by military discipline and political loyalty.

The ordinary membership of the movement (whether in the military 'wing' or in SF, or both) was largely willing to accept the necessity of this 'top-down' organizational culture, as the military elitism of the 'armed

27 A. McIntyre, *Good Friday*, 104.

28 For biographical studies of Adams, see C. Keena, *Gerry Adams: A Biography* (Cork: Mercier, 1990); and D. Sharrock and M. Devenport, *Man of War, Man of Peace? The Unauthorised Biography of Gerry Adams* (London: Macmillan, 1997).

struggle' took precedence. This culture was marked by a strong hierarchy of conformity and a perceived pressure to uphold the unity of the republican 'resistance community'. As key Belfast IRA leader Brendan Hughes recognized, an important factor in this prevailing culture was the sense that any dispute within the movement's ranks risked giving succour to the 'enemy'.[29] The IRA had a formal decision-making process, with the General Army Convention (GAC) at the apex; this body of delegates would elect a twelve-man Army Executive, which in turn elected a seven-man Army Council, responsible for the day-to-day running of the organization, and for appointing the IRA's Chief of Staff. In practice, however, Army conventions were very rare during the violent conflict; there was no meeting at all of the Provisional IRA GAC from its initial meeting in September 1970 until 1986.[30] Within SF, the public face of the republican movement, there was a greater degree of apparent transparency, with an *ard fheis* taking place on an annual basis. However, during the Troubles it was clear to all republicans that SF was effectively a 'creature' of the IRA, with only limited scope for autonomous action; the relationship was far from being one of equals.[31]

In fact, though, the reality of power within the Provisional movement was even less transparent than this outline would suggest. During recent controversies surrounding the movement's decision-making in the fraught environment of the hunger strikes from 1980 to 1981, it has been alleged that informal yet authentic power rested with a secretive 'think tank' or 'kitchen cabinet', established outside the public channels of internal party accountability (in the case of SF), and outside the conspiratorial, but supposedly democratic, control of the Army Council of the IRA.[32] To some extent this self-selected 'leading group' (which was based around Adams and

29 B. Hughes cited in A. McIntyre, 'Provisional Republicanism: Internal Politics, Inequities and Modes of Repression', in F. McGarry, ed., *Republicanism in Modern Ireland* (Dublin: University College Dublin Press, 2003), 189.

30 R. English, *Armed Struggle*, 43, 114.

31 Ibid., 115.

32 R. O'Rawe, *Afterlives*, 5. For a detailed treatment of the controversy generated by O'Rawe's claims, see S. Hopkins, 'The Chronicles of Long Kesh: Provisional Irish Republican Memoirs and the Contested Memory of the Hunger Strikes', *Memory Studies*, 7(4) (2014), 425–39.

McGuinness, but included figures such as Danny Morrison, Ted Howell, Jim Gibney, Tom Hartley and Gerry Kelly) acted as an army within the army, and as a party within the party.[33] As with other revolutionary movements (e.g. the communist parties' adoption of the organizational principle of 'democratic centralism'), there was both a formal and informal structure of internal power within the republican movement.[34] Whilst it is significant that 'democratic centralism' was introduced during the wartime conditions of the early Bolshevik period, this organizational culture outlived the emergency conditions which gave rise to it, and communist parties tended to adopt and retain a model which entailed the 'permanent regimentation of martial order'.[35] In my view, there are clear parallels with the Provisionals' maintenance of a highly centralized structure even in the 'post-war' era. The significance of this organizational culture for the memory work of the movement will be analysed in the next section.

## Orthodoxy, Dissent and Control of the Past

Despite a number of potentially bewildering strategic and ideological shifts since the early 1990s, which were designed to permit the Provisional movement to break out of its political isolation, the leadership have largely managed to contain the growth of open dissent. As Whiting has argued, when SF has voted to support such shifts, 'rather than being an exercise in intra-party democracy and grassroots consultation, such processes have been described as the management of opposition as opposed to

33 E. Moloney, *A Secret History of the IRA* (London: Allen Lane Penguin, 2002), 401–5.

34 For more detail on the operation of democratic centralism within communist parties, see R. Tiersky, *Ordinary Stalinism: Democratic Centralism and the Question of Communist Political Development* (London: Allen & Unwin, 1985); and M. Waller, *Democratic Centralism: An Historical Commentary* (Manchester: Manchester University Press, 1981).

35 R. Tiersky, *Ordinary Stalinism*, 41.

encouraging debate'.[36] Between 1994 and 2007, the IRA's volunteers and SF's activists were asked to endorse the ending of the IRA campaign of violence; the decommissioning of its arsenal of weapons and explosives; the recognition of the Northern Ireland Assembly at Stormont (under the auspices of the Westminster parliament and continuing British sovereignty); the sharing of power in the Northern Ireland Executive with hard-line unionists from the Democratic Unionist Party; an acknowledgement (implicit perhaps) that any change to the constitutional position of Northern Ireland within the UK effectively requires majority consent within this jurisdiction; and, a recognition of the legitimacy of the reformed Police Service of Northern Ireland. In the face of such a deep-rooted overturning of previous commitments, the IRA has largely remained quiescent (even if it has not disbanded), and SF has moved from the margins to centre stage.[37]

In handling such circumstances, the leadership have relied upon shaping and controlling the narrative of the movement's recent past, in order to provide reassurance to its members and activists, and retain the contemporary cohesion of the movement. Memory work and commemoration have served to promote cohesion within the Provisional movement in two ways: firstly, via the 'importance of the transmission of memory as a means of underlining historical continuity, and, secondly, the building of a sense of solidarity, of fictive kinship, presenting the community as a united political clan sharing the same values and memory'.[38] Since 293 IRA volunteers were killed during the Troubles,[39] it is not surprising that the leadership has used the memory of the dead in an attempt to sanctify contemporary

36 S. Whiting, 'Mainstream Revolutionaries', 550–1.

37 See K. Bean, *The New Politics of Sinn Féin* (Liverpool: Liverpool University Press, 2007); and S. McDowell, 'Armalite, the Ballot Box and Memorialization: Sinn Féin and the State in Post-conflict Northern Ireland', *The Round Table*, 96(393) (2007), 727.

38 K. Brown and E. Viggiani, 'Performing Provisionalism: Republican Commemorative Practice as Political Performance', in L. FitzPatrick, ed., *Performing Violence in Contemporary Ireland* (Dublin: Carysfort Press, 2009), 229.

39 D. McKittrick et al., *Lost Lives* (Edinburgh: Mainstream, 1999), 1479.

strategies.[40] The argument promoted by the leadership is that SF remains a party with a fundamental fidelity to traditional republican goals, even if the means to achieve these have been transformed. Whiting cites SF National Chairperson Mitchel McLaughlin, speaking to the *ard fheis* in 1997, to the effect that 'SF will enter peace talks and finish the political task of the IRA's dead'.[41]

There has been some unease amongst the rank-and-file, particularly with regard to issues such as decommissioning, which the leadership had expressly ruled out not long before it undertook a complete volte-face. In the main, though, defections to rival republican organizations have been piecemeal, and the so-called 'dissidents' remain fragmented, and unable to mount an effective challenge either to the British state in Northern Ireland, or to the hegemony of the SF leadership within the wider 'republican family'. Insofar as there were discrete waves of defections, these actually served to weaken the internal challenge to the leadership. SF's electoral growth in both jurisdictions, combined with the credibility and longevity of the Adams/McGuinness leadership, and the organizational culture of conformity inherited from the 'war' years, all helped to contain the prospects for internal dissent. Nonetheless, several important ex-IRA figures, including individuals with impeccable credentials in terms of serving lengthy jail sentences for the movement, have not been persuaded by the leadership group's efforts to shape the historical memory of the Provisionals. In online forums like *The Blanket* (and latterly *The Pensive Quill*) or journals such as *Fourthwrite*, dissenting activists such as Anthony McIntyre and Tommy McKearney, have contested the leadership's version of key events in the movement's past, in order to argue vehemently against the claims of a continuity of purpose spanning the entire period since the Provisionals' foundation in 1970.

40 H. Patterson, 'Beyond the "Micro Group": The Dissident Republican Challenge', in P. M. Currie and M. Taylor, eds, *Dissident Irish Republicanism* (London and New York: Continuum, 2011), 81; see also S. McDowell, 'Armalite, the Ballot Box and Memorialization', 730.

41 S. Whiting, 'Mainstream Revolutionaries', 544.

Such intellectual dissenters, who resigned from the 'mainstream' movement but did not generally join one or other of the rival organizations of 'dissidents', objected vociferously to what they interpreted as the manipulation of the past, which was used to mask the effective renunciation of republicanism by Adams/McGuinness and their supporters. However, these dissenters faced the considerable difficulty of promoting counter-memories of the Provisionals' history, whilst seeking to maintain their own commitment to the republican community of remembrance. As with other revolutionary organizations, there was an inescapable dilemma for such dissenters: in order to criticize the contemporary trajectory of the movement (or, indeed, the use of the movement's collective memory to justify such a trajectory), activists had to put themselves outside the 'democratic centralist' organizational norms and cultural practice of the group.[42] From 'outside' these norms, however, the influence of such activists upon those who remained on the 'inside', umbilically attached to the organization, was negligible. The intense commitment of republican volunteers to the cause, their sense of solidarity and profound attachment to a 'resistance community', inhibited the willingness of many to openly question the leadership. Thinking and acting 'outside the movement's structures' entailed wrestling with hugely emotive issues; these individuals had lived together through some highly traumatic episodes, such as the blanket and dirty protests against criminalization of republican prisoners, and subsequently the deaths of ten republicans on hunger strike in 1981. Stepping outside these immensely strong bonds of memory entailed an irrevocable rupture, with a very high psychological cost.

Brendan Hughes, who died in 2008 and had been publicly critical of the direction travelled by the leadership since the GFA, nevertheless found it extremely difficult to put himself outside the movement to which he had devoted his adult life: 'Because I was a good Republican ... as the old cliché goes, "Stay within the Army lines, stay within the Army lines ... don't dissent" [...] – I was still of that calibre when I got released from prison.

42 For analysis of the dilemmas facing potential or actual dissenters in the French communist party, see F. Valentin-McLean, *Dissidents du Parti Communiste Français: Le Révolte des Intellectuels Communistes dans les Années 1970* (Paris: L'Harmattan, 2005).

So I didn't question'.[43] Anthony McIntyre took a more robust dissenting position, but the cost was enormous: on his release from jail after serving eighteen years, and having written a doctoral thesis about the strategy of the Provisionals, he delivered speeches and academic papers which dealt critically with the leadership's embrace of the 'peace process'. According to McIntyre, he was 'carpeted by the SF leadership', told in no uncertain terms that he could not deliver papers which 'contravened party policy', and warned that so long as he was a member of the 'Republican movement' he would be subject to 'party discipline'. As he refused to toe the party line, McIntyre found himself 'subjected to intimidation, being shunned, ostracised, marginalised, demonised, assaulted, my house picketed, my pregnant partner threatened [...]'.[44]

## Constructing and Challenging Republican 'Official Memory'

In general terms, SF's approach to 'dealing with the past' in Northern Ireland has involved a deep-seated contestation of the British and unionist interpretations of the *meta-conflict*, which concerns the primary causes of the conflict and which of the groups was primarily responsible for its outbreak and continuation. For many years, the republican collective memory of the communal sacrifices of the Troubles sustained a unifying and settled narrative of the recent past. However, since the GFA, the role of memory and commemoration in binding together the republican family has come under intense strain. There is now a good deal of analysis devoted to the role played by states and their agents, such as civil servants, teachers and historians, in establishing and defending an 'official memory' or master narrative of national events, especially in relation to the collective legacies

43 Brendan Hughes, cited in E. Moloney, ed., *Voices from the Grave*, 253.
44 A. McIntyre, 'Provisional Republicanism', 182.

of war.[45] There has been less attention to the ways in which political movements seek to engage in a similar process of the public management of memory. As Cossu has argued in relation to the Italian Communist Party's strategy of memorializing the wartime Resistance, the 'extensive use of biography as a peculiar genre of commemoration established continuities within the history of the party: it incorporated the young partisans in a more general and teleological framework of struggle and sacrifice [...]'.[46] For the Provisional movement, a similar effort to mobilize heroic biographies for the inter-generational transmission of political and cultural understanding has involved the 'canonisation' of certain life-stories as exemplary and emblematic.

The figure of Bobby Sands (the 'officer commanding' the IRA prisoners, and the first hunger striker to die in 1981) has been a compelling and iconic presence in SF's commemorative politics. His image has been reproduced on wall murals and all kinds of merchandise, suggesting a mnemonic exploitation of his legacy, and his writings have been regularly republished by the Bobby Sands Trust (established and effectively controlled by key SF personnel). As Jarman argued, Sands became the 'clearest symbol of republican determination to continue the struggle to the end'.[47] Once again, however, this determination of the leadership to maintain control of the historical narrative has been challenged by those who would question whether the 'struggle' has, in fact, been continued by SF into the post-GFA era. The Sands family, for instance, has criticized the commercial and *political* use made of Bobby Sands' legacy and image by the Trust: the family released a statement in February 2016 calling upon the Trust to 'disband and desist from using Bobby's memory as a commercial enterprise. Once again an opportunity to promote Bobby's ideals and sacrifice that he died

45 See T. Ashplant et al., *The Politics of War Memory and Commemoration* (London: Routledge, 2000); E. Jelin, *State Repression*, passim.

46 A. Cossu, 'Commemoration and Processes of Appropriation: the Italian Communist Party and the Resistance (1943–48), *Memory Studies*, 4(4) (2011), 386–400.

47 N. Jarman, *Material Conflicts: Parades and Visual Displays in Northern Ireland* (Oxford: Berg, 1997), 149.

for has been diminished by those who seek to promote different agendas – both personal and political'.[48]

Whilst SF has clearly sought to utilize the legacy of Sands to challenge the official British state narrative of the Troubles, which attempted to criminalize the republican struggle, the SF master narrative has, in its turn, been challenged by subaltern forces. These dissenting elements offer 'alternative narratives and meanings of the past', and what is required for such interpretations to gain a hearing, according to Jelin, are 'memory entrepreneurs'.[49] An absolutely critical example of this challenge to the dominant narrative within republicanism has occurred precisely in relation to the hunger strikes. Richard O'Rawe's allegation that the 'kitchen cabinet' outside the jail rejected an offer from the British which could have brought the protest to a conclusion after four deaths (but before six other republicans died), has prompted a huge rift in the 'republican family'. O'Rawe insisted in his first volume of memoirs that his objective was not to denigrate any individual within the leadership, that he was not a 'dissident', and that he agreed with the Provisional leadership's decision to end the 'unwinnable war'.[50] In other words, he remained committed to the culture of organizational loyalty and discipline that he had absorbed as a long-standing IRA volunteer. However, he quickly realized that this allegation marked such a fundamental rupture in the Provisional master narrative that, whether he liked it or not, it would not be allowed to go unchallenged. All of the leadership's resources were mobilized to denigrate O'Rawe's account, and he was personally vilified and accused of siding with the dissidents. Unsurprisingly perhaps, in his second volume of memoirs, devoted to defending his original position, and providing a detailed examination of the controversy which it had unleashed, O'Rawe appeared to recognize he was now definitively an 'outsider'. His critique of the leadership, both of

48 'Sands family responds to publication of book – "Bobby Sands Freedom Fighter"', *Pensive Quill*, <http://thepensivequill.am/2016/02/sands-family-responds-to-publication-of.html> accessed 25 February 2016.

49 E. Jelin, *State Repression*, 27.

50 O'Rawe cited in A. McIntyre, *Good Friday*, 88. For a fuller treatment of O'Rawe's allegations, see S. Hopkins, 'Chronicles of Long Kesh'.

its decisions in 1981 and its trajectory since 1998, went much further, and was inspired not just by disagreement over the historical record, but also by principled ideological objections to its current policy. In 2008, O'Rawe was scathing about SF's willingness to nominate DUP leader Ian Paisley as First Minister in the Northern Ireland executive: 'Jesus, what a debacle! Bobby Sands, socialist, secularist, Republican bears no resemblance to any of this. None of the boys did'.[51]

## Conclusion

The official Sinn Féin memory of the Troubles – and particularly its rendition of critical legacies such as the hunger strikes – is no longer pristine and uncomplicated. Whilst the Provisional movement continues to attempt to mobilize support through its commemorative strategies, not least surrounding the centenary of the Easter Rising and the annual celebration of the hunger strikers' sacrifice, this memory field has been the site of traumatic contestation, since the signing of the Good Friday Agreement in 1998, and all the more so since the definitive end of the IRA's campaign in 2005. There is a lively debate amongst scholars regarding the contemporary ideological character of SF: on one hand, there is a view that the SF leadership 'remain fundamentally unaltered at an ideological level [...] for SF, their self-proclaimed long march towards Irish unity does indeed go on'.[52] On the other hand, Bean has argued that 'the apparent continuity within the movement's discourse might have quite another purpose, namely to *conceal* the ideological retreat of the Adams-McGuinness leadership and the *defeat* of its political project'.[53] It is possible that both judgments have elements of truth: the leadership might be deceiving themselves, as well

51 Ibid., 104.

52 M. Frampton, *The Long March: The Political Strategy of Sinn Féin 1981–2007* (Basingstoke: Palgrave Macmillan, 2009), 77, 192.

53 K. Bean, 'Book Reviews', *Irish Political Studies*, 25 (1) (2010), 136.

as their members, with regard to their fidelity to republican goals, but in reality the movement has had to adjust its political outlook to a non-revolutionary era.

How does a self-proclaimed revolutionary party make such an adjustment? One answer is to deny that it is happening or has happened, and this can only be done through exercising firm control over the movement's collective memory or master narrative. In such circumstances, it is not surprising that there has been an effort to retain an organizational culture associated with 'military discipline'. As revolutionary goals recede, the maintenance of internal power becomes its own justification for the leading group. McIntyre argued soon after the signing of the GFA that 'there seemed no reason why all the clandestine activity, tight secrecy and whispering in dark corners should not be replaced with a culture of open discussion where the free flow of ideas rather than clipped militaristic commands would become the norm. In practice it was not to be. Rather the party [SF] was to be ruled with the ethos of the Army [IRA]'.[54] However, it should also come as no surprise that, increasingly, as the leadership ages, and the traditional goal of republicanism appears no closer to achievement, dissenters will seek to challenge and expose this grip on power.

## Bibliography

Adams, G., *Before the Dawn: An Autobiography* (Dingle: Brandon, 1996).

Ashplant, T., et al., *The Politics of War Memory and Commemoration* (London: Routledge, 2000).

Bean, K., *The New Politics of Sinn Féin* (Liverpool: Liverpool University Press, 2007).

'Bobby Sands Freedom Fighter', *Pensive Quill* <http://thepensivequill.am/2016/02/sands-family-responds-to-publication-of.html> accessed 25 February 2016.

Bradley, Gerry, and Feeney, B., *Insider: Gerry Bradley's Life in the IRA* (Dublin: O'Brien Press, 2009).

54 A. McIntyre, 'Provisional Republicanism', 183.

Brown, K, and Viggiani, E., 'Performing Provisionalism: Republican Commemorative Practice as Political Performance', in L. FitzPatrick, ed., *Performing Violence in Contemporary Ireland* (Dublin: Carysfort Press, 2009), 229.
Collins, Eamonn, and McGovern, M., *Killing Rage* (London: Granta, 1997).
Cossu, A., 'Commemoration and Processes of Appropriation: The Italian Communist Party and the Resistance (1943–48), *Memory Studies*, 4(4) (2011), 386–400.
English, R., *Armed Struggle* (London: Macmillan, 2003).
Frampton, M., *The Long March: The Political Strategy of Sinn Féin 1981–2007* (Basingstoke: Palgrave Macmillan, 2009).
Hopkins, Stephen, 'The Chronicles of Long Kesh: Provisional Irish Republican Memoirs and the Contested Memory of the Hunger Strikes', *Memory Studies*, 7(4) (2014), 425–39.
——, '"Our whole History has been Ruined!" The 1981 Hunger Strike and the Politics of Commemoration and Memory', *Irish Political Studies*, 31(1) (2016), 44–62.
——, *The Politics of Memoir and the Northern Ireland Conflict* (Liverpool: Liverpool University Press, 2013).
*Irish Times*, (29 March 2016).
Jarman, N., *Material Conflicts: Parades and Visual Displays in Northern Ireland* (Oxford: Berg, 1997).
Jelin, Elizabeth, *State Repression and the Struggles for Memory* (London: Latin American Bureau, 2003).
Kearney, R., 'Myth and Martyrdom: Foundational Symbols in Irish Republicanism', in R. Kearney, ed., *Navigations: Collected Irish Essays, 1976–2006* (Dublin: Lilliput Press, 2006).
Keena, C., *Gerry Adams: A Biography* (Cork: Mercier Press, 1990).
MacBride, S., 'Introduction', in B. Sands, *One Day in My Life* (Cork and Dublin: Mercier Press, 1983).
McDowell, S., 'Armalite, the Ballot Box and Memorialization: Sinn Féin and the State in Post-Conflict Northern Ireland', *The Round Table*, 96 (2007), 725–38.
McIntyre, Anthony, *Good Friday: The Death of Irish Republicanism* (New York: Ausubo Press, 2008).
——, 'Provisional Republicanism: Internal Politics, Inequities and Modes of Repression', in F. McGarry, ed., *Republicanism in Modern Ireland* (Dublin: University College Dublin Press, 2003).
McKearney, Tommy, *The Provisional IRA: From Insurrection to Parliament* (London: Pluto Press, 2011).
McKittrick, D., et al., *Lost Lives* (Edinburgh: Mainstream, 1999).
Moloney, E., *A Secret History of the IRA* (London: Allen Lane Penguin, 2002).

Moloney, E., *Voices from the Grave: Two Men's War in Ireland* (London: Faber and Faber, 2010).
O'Driscoll, D., *Stepping Stones: Interviews with Seamus Heaney* (New York: Farrar, Strauss and Giroux, 2008).
O'Hearn, D., *Bobby Sands: Nothing but an Unfinished Song* (London: Pluto Press, 2006).
O'Rawe, Richard, *Afterlives* (Dublin: Lilliput Press, 2010).
——, *Blanketmen: An Untold Story of the H-Block Hunger Strike* (Dublin: New Island, 2005).
Patterson, H., 'Beyond the "Micro Group": The Dissident Republican Challenge', in P. M. Currie and M. Taylor, eds, *Dissident Irish Republicanism* (London and New York: Continuum, 2011).
Rieff, D., *In Praise of Forgetting: Historical Memory and its Ironies* (New Haven, CT; London: Yale University Press, 2016).
Sharrock, D., and Devenport, M., *Man of War, Man of Peace? The Unauthorised Biography of Gerry Adams* (London: Macmillan, 1997).
Tiersky, Ronald, *Ordinary Stalinism: Democratic Centralism and the Question of Communist Political Development* (London: Allen & Unwin, 1985).
Valentin-McLean, F., *Dissidents du Parti Communiste Français: Le Révolte des Intellectuels Communistes dans les Années 1970* (Paris: L'Harmattan, 2005).
Waller, M., *Democratic Centralism: An Historical Commentary* (Manchester: Manchester University Press, 1981).
Whiting, S, 'Mainstream Revolutionaries: Sinn Féin as a "Normal" Political Party', *Terrorism and Political Violence*, 28(3) (2016), 541–60.

PART III

# Memory and Trauma

NIAMH NICGHABHANN

# 8 Memory, Public Space and the Body in Ireland: Locating and Negotiating the Asylum in Edna O'Brien's Short Fiction

Throughout the many discourses that engage with spatial experience, the concepts of place and memory are continually interwoven, forming the warp and weft of how the world is considered to be known and experienced.[1] Landscapes, whether urban, suburban or rural, are negotiated and inhabited according to inherited ideas of ownership, status and association. The names given to places or areas, both formal and informal, refer in a particular way to specific meanings associated with place, altered as they may be through the accretions of time and the processes of translation and interpretation.[2] Memorials and other commemorative artworks, as well as the names given to streets and notable public buildings, bridges, or amenities such as parks or sporting grounds, impact everyday spatial experience through their reference to particular historical narratives or individuals. Moreover, they form part of the visual and spatial language used by people to negotiate the landscape, becoming part of the weave of everyday life, and reinterpreted and used in multiple ways.[3] The relationship between

1 The interplay between these concepts informs many different critical discourses across a range of disciplines, but the works of Doreen Massey and Yi-Fu Tuan play a pivotal role in describing and theorizing these ideas. An overview of key developments in theories of space and place is given in David Crouch, 'Flirting with Space: Thinking Landscape Relationally', *Cultural Geographies*, 17(1) (2000), 5–18.

2 Questions of space, place and Irish studies are explored in Gerry Smyth, *Space and the Irish cultural Imagination* (Basingstoke: Palgrave, 2001).

3 These issues are explored in relation to Dublin city in Yvonne Whelan, *Reinventing Modern Dublin: streetscape, iconography and the politics of identity* (Dublin: University College Dublin Press, 2003).

memory and specific places has been the subject of sustained scholarly attention within Irish Studies scholarship, with many scholars focusing on the resonance and relevance of individual monuments, places or buildings that function as sites of memory due to their particular associations.[4] This essay shifts attention from individual monuments or sites to the negotiation of space throughout, around and between these sites, exploring the dynamics of memory and public space as it is experienced and negotiated within specific temporal-historical contexts. In my consideration of these ideas through Edna O'Brien's short fiction, I am particularly interested in exploring this experience as part of an exploration of the past, within the contexts of urban and architectural historiography. While these discourses often rely on archaeological evidence of the built environment in exploring the formation and histories of towns, cities and other settlement types, attention to the negotiated experience of streets, squares, alleys, forecourts and other public environments requires alternative and complementary interpretative methods and approaches to evidence.

In his exploratory essay on potential research directions emerging from the Irish Historic Towns Atlas project,[5] Tadhg O'Keeffe reflects on the methodological challenges involved in documenting urban environments and their histories, as well as 'connecting with historic experiences of living in urban places'.[6] In what he describes as a 'twin-track

4 Emily Mark-FitzGerald's *Commemorating the Irish Famine: Memory and the Monument* (Liverpool: Liverpool University Press, 2013) explores the role of Famine memorials as sites of memory nationally and internationally, and provides one example of recent work exploring the significance of specific locations, sites or monuments as bearers of meaning. Similarly, Emilie Pine's essay '"Coming Clean?" The Art of Representing the Magdalen Laundries', in Oona Frawley, ed., *Memory Ireland, Volume 1* (Syracuse, NY: Syracuse University Press, 2010) explores the social position of the Magdalen Laundries as sites with particular resonance within Irish society.

5 For information on this project, please visit <https://www.ria.ie/research-projects/irish-historic-towns-atlas> accessed 28 February 2016.

6 Tadhg O'Keeffe, 'Being-in-the-Town, Being-in-the-Atlas: Some Reflections on Capturing the Experiences of Town-Dwelling in the Historic Past', in Howard B. Clarke and Sarah Gearty, eds, *Maps and Texts: Exploring the Irish Historic Towns Atlas* (Dublin: Royal Irish Academy, 2013), 266–77.

approach – documenting the tangible, seeking the intangible', he argues that the former involves inventorying and collecting archaeological and architectural evidence, and the analysis of traces through various documentary and cartographic sources. The latter, described by O'Keeffe as 'more subjective and by its nature less easily represented', requires an attention to the relational quality of spatial experience. While O'Keeffe comments on the difficulty of locating appropriate evidence for the experience of a late medieval Irish town, sources are more abundant for the more recent past. These might include, for example, public order and disorder records, policy documents or plans for the development, management and regulation of public spaces, or written testimonies relating individual experiences.[7] These, like maps, census returns and architectural drawings, are representations, produced within and for a certain audience and directed towards one form of reading or interpretation, reflecting particular voices and structures of authority. In many cases, these forms of documentation explicitly state their aim for objectivity and accuracy in their recording processes. However, as J. B. Harley argued, all representations reflect their context, and 'far from holding up a simple mirror of nature that is true or false, maps redescribe the world – like any other document – in terms of relations of power and of cultural practices, preferences and priorities'.[8]

Literary texts, works of visual art or responses to places in sound or choreography can be considered as alternative maps, as documents that also redescribe the world in terms of power relations and cultural practices. Artistic representations of specific spaces and places, or of the experience of those spaces and places, occupy a particular position in relation to that which they record, refract or explore. The engagement with and interpretation of space, place or experience by different artists varies according to

7 Anne Wohlcke's exploration of street life during festival and fair time in eighteenth-century London provides a case study in the use of official source material in writing historical spatial experience. Anne Wohlcke, *The 'Perpetual Fair': Gender, Disorder, and Urban Amusement in Eighteenth-Century London* (Manchester: Manchester University Press, 2014).

8 J. B. Harley; in Paul Laxton, ed., *The New Nature of Maps: Essays in the History of Cartography* (Baltimore, MD: Johns Hopkins University Press, 2001), 35.

multiple factors, including genre, patron, intended audience, intended location, intention of the artist, as well as the reception and interpretation of the work by others. As interpretations located in relation to specific places and times, however, they provide a further register of evidence in the process of engaging with the less tangible, or immediately evident, histories of urban environments. These expanded methodologies for engaging with spatial experience and landscape history are particularly appropriate in considering the lives of those who occupy marginalized positions within society in order to reconsider how more familiar processes of recording have privileged certain ways of knowing, looking, and negotiating places, and have shaped the subsequent production of knowledge about them.[9] The relationship between literary text and historical source has been critically debated in several different contexts, and it maintains a somewhat fractious position within contemporary discourse, particularly within the contexts of debates on affect, historical narrative, authority and authenticity and similar subjects in current historiographical theory.[10]

9 These questions of evidence and power were explored in relation to female experience in the 2015 'Silence in the Archives' conference, held at Wolfson College, Oxford, on 7 November 2015. <https://silenceinthearchives.wordpress.com/conference-details/> accessed 1 March 2016.

10 In an Irish context, the issues around using literary texts as sources have been explored throughout a range of different contexts, and some texts from these debates include Tom Dunne, ed., *The Writer as Witness: Literature as Historical Evidence* (Cork: Irish Conference of Historians, 1985); Cormac Ó Gráda, 'Irish literary sources and Irish economic history', *Studies: An Irish Quarterly Review*, 80(319) (1991), 290–9; and Margaret Kelleher, '"Factual Fictions": Representations of the Land Agitation in Nineteenth-Century Women's Fiction', in Heidi Hansson, ed., *Nineteeth-Century Irish Women's Prose: New Readings and Contexts* (Cork: Cork University Press, 2008), 78–92. The relationship between narrative fiction and historiographical practice has also been explored by Margaret Kelleher in her essay 'The "Affective Gap" and Recent Histories of Ireland's Great Famine', in Marguérite Corporaal, Christopher Cusack, Lindsay Janssen and Ruud van den Beuken, eds, *Global Legacies of the Great Irish Famine: Transnational and Interdisciplinary Perspective* (Oxford: Peter Lang, 2014), 19–38. Questions of narrative, historical fiction, authority and authenticity were at the heart of a Radio 4 Start the Week debate in October 2015 between Jane Smiley and Niall Ferguson (for a report of this debate: <http://www.telegraph.co.uk/news/

## Negotiating the Asylum in Edna O'Brien's Short Fiction

In order to ground this exploration of the dynamics of memory and public space within a specific context, this essay explores the representation of public space within several short stories by Edna O'Brien. In particular, it focuses on the way in which the environment represented by O'Brien is shaped by the institutional architectural context, and how this is mediated, understood and negotiated by the characters in the stories. It is possible to recognize small, West of Ireland towns as the setting for most of the stories mentioned here. O'Brien rarely names the places within which her stories take place precisely, but there is sufficient contextual information to locate them both in terms of place and time, as most are set in the mid-twentieth century. The term 'institutional architectural context' is used here to signify the urban landscape of the early twentieth century that had been largely shaped by the sharp increase in institutional building projects from the mid-nineteenth century onwards. These building projects include state and civic buildings and services, as well as buildings associated with religious organizations and churches, or financial institutions such as banks.[11] Central to this exploration of the representation of the negotiation of public space is the action of memory, or inherited perceptions and ideas about particular spaces and places encountered and negotiated by the characters in her stories. The importance of individual memory in relation to place is evident across O'Brien's oeuvre, but here, I am concerned with collective meanings, recognized and shared by a community. These collective meanings are produced and transmitted verbally

---

bbc/11926281/Niall-Ferguson-historical-novelists-make-stuff-up.html> accessed 28 February 2016), while the Storying the Past website and #storypast project on Twitter explore the potential for creative registers in historiographical practice <https://storyingthepast.wordpress.com/> accessed 28 February 2016.

11 The growth of these institutions and their associated buildings are charted across a range of publications, but comprehensive overviews and relevant bibliographies per building type are provided in Andrew Carpenter et. al., eds, *Art and Architecture of Ireland Volume IV* (Dublin; New Haven, CT: Yale University Press in Association with the Royal Irish Academy, 2015).

as well as physically, often through gossip, rumour and nicknames, particularly when they relate to the presence of institutions of control and containment, specifically the lunatic asylum.[12] While the asylum features throughout several of O'Brien's short stories and novels, and also figures in her autobiography, *Country Girl* (2012), it is almost always referred to from the position of outsider, rather than from the perspective of one who has had direct experience of the institution as patient, service user or employee. Within the negotiation of everyday life, therefore, the asylum was present as a signifier of social control and order, but also as a physical reality in the landscape.[13] As will be explored in more detail below, the extent to which characters physically locate themselves in relation to the

12 O'Brien's use of rumour and gossip in her representation of the asylum in Irish social life reflects what Kathryn Crameri describes as the ways that 'close-knit communities construct stories – or even myths – of themselves as a collective, using tools such as rumour, gossip, unfounded assumptions, autobiographical memory, selective amnesia and the consensual creation of versions of events which may or may not be strictly accurate in their detail' in her essay 'Forging the Community: Explorations of Memory in Two Novels by Jesús Moncada', *The Modern Language Review*, 98(2) (2003), 353.

13 I use the term 'asylum' throughout this paper in order to reflect O'Brien's use of the term throughout her fiction. The institutions built to provide treatment for those with mental illness throughout the nineteenth century in Ireland were referred to as lunatic asylums. The names of the institutions were changed in the early decades of the twentieth century, and became known as mental hospitals. The names changed again in the mid-twentieth century and many hospitals were renamed, often associating them with local saints. For example, the Cavan and Monaghan District Lunatic Asylum opened in 1868, and was renamed as the Monaghan Mental Hospital in 1924, and renamed again as St. Davnet's Hospital in the late 1950s: Anne MacLellan, Niamh NicGhabhann and Fiona Byrne, *World Within Walls: From Asylum to Contemporary Mental Health Services* (Monaghan: Health Service Executive Cavan-Monaghan Division, 2015). For an overview of the asylum and mental hospital system in Ireland, see Mark Finnane, *Insanity and the Insane in Post-Famine Ireland* (London: Croom Helm; Totowa, NJ: Barnes & Noble Books, 1981); Catherine Cox, *Negotiating Insanity in the Southeast of Ireland, 1820–1900* (Manchester: Manchester University Press, 2012); and Pauline Prior, *Asylums, Mental Health Care and the Irish: Historical Studies 1800–2010* (Dublin; Portland, OR: Irish Academic Press, 2012).

asylum is telling. Rather than being referred to in a general way, details of distance and location are precisely noted, and in several cases, shared knowledge is presumed through the reference to the building by a nickname or town name.[14] An interesting relation, therefore, is established between the public spaces negotiated by the characters in O'Brien's stories and the asylum. The presence of the asylum structures their experience of public space, evident through their spatialized references to it, but their understanding of its meaning is almost entirely formed through inherited ideas and shared memories associated with the buildings and the experiences of individuals within the asylum system.[15] These memories are not explicitly stated in many cases, but emerge in the form of threats, gossip or rumour, creating a substratum of shared understanding based on stigma and fear.[16] O'Brien's deployment of the register of gossip to signify the production and presence of stigma reflects Bruce Link and Jo Phelan's observation on the dynamic of stigma in society, which is often understood as something 'in the person' rather than 'a designation or tag that others affix to the person'.[17]

14 The relationship between the asylum and small urban centres in the west of Ireland has been explored by A. J. Saris, 'Producing Persons and Developing Institutions in Rural Ireland', *American Ethnologist*, 26(3) (1999), 690–710, and A. J. Saris, 'Mad Kings, Proper Houses, and an Asylum in Rural Ireland', *American Anthropologist*, 98(3) (1996), 539–54.

15 This reading of the representation of gossip, rumour and stigma with reference to a specific institution reflects Mary Daly's exploration of the role of stigma in the negotiation of the workhouse in nineteenth-century Ireland <http://historyhub.ie/final-years-of-the-workhouse> accessed 28 February 2016, and Catherine Cox's examination of the Irish asylum system and mental illness through the nineteenth century in *Negotiating Insanity in the Southeast of Ireland, 1820–1900* (Manchester: Manchester University Press, 2012).

16 A useful overview of social psychological developments in the study of stigma and its construction can be found in Bruce G. Link and Jo C. Phelan, 'Conceptualizing Stigma', *Annual Review of Sociology*, 27 (2001), 363–85. Mary Douglas's theories of pollution and social order also provide important context for these ideas of stigma and space in Mary Douglas, *Purity and Danger: An Analysis of the Concepts of Pollution and Taboo* (London: Kegan Paul, 1966).

17 Link and Phelan, 'Conceptualizing Stigma', 366.

O'Brien's 1978 short story, titled 'A Rose at the Heart of New York', shifts between locations and times, tracing the troubled relationship between a mother and daughter from difficult birth to death. The story has a fragmented structure, following the thread of the mother-daughter relationship, but also mapping that relationship as the story moves through the different spaces inhabited by the family. These spaces include private and domestic spaces, as well as the broader social and physical landscapes of mid-twentieth-century Ireland. In an early part of the story, one of the monumental institutions within that landscape looms large, crossing the boundary between the private and the public spheres. In the violent scene that ensues when the mother figure in the story has hidden money from the father, he threatens to shoot her with a shotgun. In response, she threatens him with the spectre of the asylum.

> [T]he mother said that if he did so there was a place for him. That place was the lunatic asylum. It was twenty or thirty miles away, a big grey edifice, men and women lumped in together, some in strait-jackets, some in padded cells, some blindfolded because of having sacks thrown over their heads, some strapped across the chest to quell and impede them. Those who did not want to go there were dragged by relatives, or by means of rope, some being tied on to the end of a plow or a harrow and brought in on all fours, like beasts of the earth. Then when they were not so mad, not so rampaging, they were let home again, where they were very peculiar and given to smiling and chattering to themselves, and in no time they were ripe to go off again or be dragged off. March was the worst month, when everything went askew, even the wind, even the March hares. Her father did not go there. He went off on a belter and then went to a monastery.[18]

The asylum is characterized as a place of brutal incarceration, and committal was seen as marking patients for life. The description in 'A Rose at the Heart of New York' has the character of hearsay, reported sightings and hushed gossip, dramatizing the transition into the asylum itself, as well as the harsh treatment and restraining methods used within its walls. The

18 Edna O'Brien, 'A Rose at the Heart of New York', in Edna O'Brien, ed., *The Love Object: Selected Stories* (London: Faber and Faber, 2013), 141–2. I refer to the stories as published in this recently published selection. Full information on the original publication date and location of the stories is given in this volume.

knowledge enters the community as something like a collective memory, albeit one shaped by rumour and fear rather than direct experience. Despite the long and violent description, the father does not go there, and instead goes to a monastery to try to deal with his alcoholism and violent anger. It is, however, used as a threat, one that would undermine both his social status as well as his mental and physical autonomy and control. The position of the asylum in the story reflects its broader position within a landscape of public and private life in Ireland in the twentieth century.

The sudden and shocking appearance of the asylum in 'A Rose at the Heart of New York' is part of the strange violence of the story, but also points to the position of the asylum within the mental, domestic and physical landscape of Ireland in the period. The 'big grey edifice' mentioned by O'Brien in the story refers to the large institutional buildings constructed throughout the nineteenth century as part of a campaign across the British Empire to contain and treat mental illness more effectively.[19] Most of these buildings retained their original function following independence, and they currently exist in various levels of preservation, some restored for alternative use and some used by the Health Service Executive (HSE) for the provision of community health services. It is important to note that external perceptions were considered within the initial plans for these asylum buildings and their construction. As Claire Hickman has demonstrated, the public perception of the asylum building was a key consideration in their design.[20] They were often constructed on an elevated site, and were described initially as symbols of civic virtue and pride. In her exploration of the asylum at Ballinasloe, Co. Galway, Oonagh Walsh has also highlighted the importance of appearances and perceptions of the

19 Mark Finnane, *Insanity and the Insane in Post-Famine Ireland*. Architecture for the treatment of mental illness in an American context has been explored by Carla Yanni, *The Architecture of Madness: Insane Asylums in the United States* (Minneapolis, MN: University of Minnesota Press, 2007).

20 Claire Hickman, 'The Picturesque at Brislington House, Bristol: The Role Of Landscape in Relation to the Treatment of Mental Illness in the Early Nineteenth-Century Asylum', *Garden History*, 33(1) (2005), 47–60.

institution from outside.[21] The importance of the public statement made by the institution was clearly detailed in a pamphlet on *District Lunatic Asylums as approved by the Board of Control* (1863), which stated that 'the purpose of these buildings and the class of persons for whose reception they are intended, renders it necessary that economy of construction should be kept in view as a matter of primary importance; yet, looking to the prominent position they usually occupy, and their public nature, it is desirable that they should be of as pleasing and attractive a form and character as possible'.[22] The institutions were clearly conceived of as relational in terms of their presence in urban settings, acting as a powerful signifier for those within, and those without the walls of the institution.

With the exception of the brutal imagery in 'A Rose at the Heart of New York', quoted above, O'Brien usually associates the asylum with women and girls. The institution features several times in association with what is portrayed as an excess of female sexual desire, activity and emotional need within the social context of mid-twentieth-century Ireland. These stories reflect the position of the asylum as part of the apparatus of state and religion which policed the bodies and desires of women, to the extent that they were part of everyday gossip as well as being at the centre of intimate familial decisions. The stories also map the exchange of shared memories and understandings about the asylum and its role in managing and regulating behaviour. In O'Brien's short story 'The Connor Girls' (1982), 'it was suggested that Miss Amy had gone berserk and was shut up in an asylum', following the break-down of her engagement.[23] The asylum also features in 'A Scandalous Woman' (1974), in which the young protagonist, Eily, begins a relationship with a young man and is forced to marry him. As part of the proceedings during which her fate is decided by her father, the local

21 Oonagh Walsh, 'Landscape and the Irish Asylum', *Land and Landscape in Nineteenth-Century Ireland*, in Glenn Hooper and Una Ní Bhroiméil, eds, (Dublin: Four Courts Press, 2008), 156–70.

22 *District Lunatic Asylums as approved by the Board of Control* (Dublin: HMSO, 1863).

23 Edna O'Brien, 'The Connor Girls', in Edna O'Brien, ed., *The Love Object: Selected Stories* (London: Faber and Faber, 2013), 34.

priest and, to a lesser extent, her mother, the institution is mentioned as part of the threats made to her to ensure her compliance with the marriage.

> One minute they were asking her kindly, another minute they were heckling, another minute her father swore it was to the lunatic asylum that she would be sent, and then at once her mother was condemning her for not having milked for two weeks.[24]

The role of the families in committing young women to punitive institutions is clearly revealed within O'Brien's stories, and is reflected in her own memoir. In *Country Girl*, O'Brien recalls the moment when her family learned of her relationship with Ernest Gébler, and travelled to force her return. Reflecting the threats made to Eily in 'A Scandalous Woman', O'Brien recounts overhearing a conversation between her employer and his wife at the chemist shop that she worked in during her first years in Dublin. In her memoir, she writes that they said her family were coming to bring her home, and if necessary to put her away. According to O'Brien, 'the "putting away" meant nothing other than the lunatic asylum, and I had a momentary image of Mad Mabel'.[25] This section of the autobiography reflects the complex interweaving of sexual propriety, social control, physical incarceration and the public sphere. O'Brien's parents have heard of her behaviour – her private actions made public – via an anonymous letter on her mother's bicycle seat following Mass. Further to this, when her brother had questioned whether she was pregnant she recalled Mamie Cadden, an infamous abortionist at the time who was sentenced to death by hanging in 1957, but who was declared insane and who died in the Criminal Lunatic Asylum in 1959.[26] The figure of Mad Mabel reiterates the consequences with

24 Edna O'Brien, 'A Scandalous Woman', in Edna O'Brien, ed., *The Love Object: Selected Stories* (London: Faber and Faber, 2013), 76.

25 Edna O'Brien, *Country Girl* (London: Faber and Faber, 2012), 113–14. Mad Mabel was a frightening figure from O'Brien's childhood, associated with ill health. O'Brien and Gebler fled to the writer J. P. Donleavy to escape her family, but they were also followed by her brother, father, a Cistercian abbot and others.

26 In the confrontation that ensued, O'Brien recalled that 'my brother asked me if I was pregnant [...] the word "abortion" was not spoken but it was implied, and before my eyes ran images of Mamie Cadden, the dingy room, the bucket of Jeyes Fluid and

which those who deviated from the accepted norm were threatened. Mabel features earlier in *Country Girl* as a figure associated with mental illness, a women who exposes herself in public and is homeless, a very vulnerable figure abandoned and stigmatized by society.

The asylum features in several other short stories by O'Brien, and also in her novels. It features prominently in 'Oft in the Stilly Night' (1990), where the central character repeatedly visits the asylum due to her religious fervour.

> Very early, she was brought to the asylum, where she spent the best part of a year and took to sucking her cheeks, refusing to speak to anyone and having to be barred from the chapel because the sight of flowers drove her into a frenzy [...] In about a year, when she was calm with tablets and shock treatment, they brought her home, and from then on she avoided people, growling at anyone who spoke to her, even the priest or the doctor. Being alone now, she does the farm work and has taken to wearing her brother's old clothes and Wellington boots. She is always forking manure, or washing out the cow house, or carrying buckets of feed and water up the hills to the store cattle. 'There goes the one with the roastings', people say. She is like a landmark, one bucket in either hand, either going up the hill or returning to have them filled. Children say that she curses them, and those who knock on her door are likely to be met with a pitchfork or a saucepan of hot stirabout.[27]

As is evident from this excerpt, the figure is marked within society by her time in the asylum, and the descriptive tone employed by O'Brien conveys pity, but also distance and judgement. This tone reflects the view of the figure from the perspective of one within society who adheres to the accepted social norms. This descriptive technique reiterates the importance of visibility of the deviant within the public sphere, associated with the

---

fatality'. Ibid. Mamie Cadden was a midwife who carried out abortions illegally in Ireland, and who was convicted for murder following the death of a woman in 1957. She was sentenced to death by hanging, but this was commuted to life imprisonment. She began her term in Mountjoy Prison, but was declared insane and moved to the Criminal Lunatic Asylum, where she died of a heart attack in 1959. Clíona Rattigan, '"No worse and no better": Irish Women and Backstreet Abortions', *History Ireland*, 21(1) (2013), 42–3.

27 Edna O'Brien, 'Oft in the Stilly Night', in Edna O'Brien, ed., *The Love Object: Selected Stories* (London: Faber and Faber, 2013), 275.

asylum and reinforcing its threat through her isolation. While the physical heft of the institution itself exerted social control in its monumental presence in the landscape, this character represents the reality of institutional treatment and confinement in her very person. She takes on an almost fairy-tale, scarecrow or witch-like character, associated with curses, and the fact that she features in the stories of children reflects the way in which this spatial and social knowledge became part of a shared and collective memory through gossip, hearsay and rumour.

In several other stories, O'Brien locates the asylum geographically in relation to the main character. In 'Sister Imelda' (1982), a story about a convent girl's love and lust for a young nun, the girl notes that 'Baba said that saner people were locked in the lunatic asylum which was only a mile away. We saw them in the grounds, pacing back and forth, with their mouths agape and dribble coming out of them, like icicles. Among our many fears was that one of those lunatics would break out and head straight for the convent and assault some of the girls'.[28] In 'My Two Mothers' (2011), the main character recalled an exchange with her mother where she reads aloud a quotation from Voltaire, saying that 'Illusion is the queen of the human heart'. In response her mother 'looked at me as if I had escaped from the lunatic asylum twenty miles away'.[29] These references can also be found within O'Brien's novels – in *Night* (1972), the winner of a fancy-dress competition is described thus: 'the prizewinner turned out to be none other than a brown paper parcel. A parcel, within it another, and another and another, until at the very centre, a very underfed creature came peering through, screeching. Her mother before screeched, which is why they eventually took her to the Castle, the choice name for the loony bin'.[30] In her essay on Jamaica Kincaid and O'Brien, Eve Stoddard has explored the role of the asylum in O'Brien's novel *Wild Decembers* (1999), in which a young

28 Edna O'Brien, 'Sister Imelda', in Edna O'Brien, ed., *The Love Object: Selected Stories* (London: Faber and Faber, 2013), 116.

29 Edna O'Brien, 'My Two Mothers', in Edna O'Brien, ed., *The Love Object: Selected Stories* (London: Faber and Faber, 2013), 446.

30 Edna O'Brien, *Night* (London: Faber and Faber, 2014), 39.

woman is committed as a result of a love affair.[31] The asylum, therefore, occupies a powerful position within the regulation of social order across O'Brien's work, and maintains an important place in her fiction as part of the social structures navigated by her characters, and indeed, by herself, in her autobiography.

## Public Space, Morality and the Female Body

While some of the references to the threat of the asylum are located within the private domestic sphere, they more frequently relate to public behaviour or public perception. While individual life in the home was certainly circumscribed by the increasingly narrow strictures of church and state in mid-twentieth-century Ireland as well as economic conditions, the negotiation of the public sphere by men and women was also determined and policed according to social convention and custom. Many of the references to the asylum, therefore, relate to public shame or a public loss of bodily or mental control, or the threat of the asylum relates to unacceptable behaviour in public. As Diarmaid Ferriter has written, 'evidence of female "madness" was often scant; asylums sometimes served as another way of silencing and hiding women who did not conform, particularly when it came to sexuality'.[32] This is reflected in Una Crowley and Rob Kitchin's reading of the asylum structures as part of a legal landscape aimed at regulating sexual

31 Eve Stoddard, 'Sexuality, Nation, and Land in the Post-Colonial Novels of Edna O'Brien and Jamaica Kincaid', in Kathryn Laing, Maureen O'Connor and Sinead Mooney, eds, *Edna O'Brien: New Critical Perspectives* (Dublin: Carysfort Press, 2006), 104–21.

32 Diarmaid Ferriter, *Occasions of Sin: Sex and Society in Modern Ireland* (London: Profile Books, 2009), 202–3. The evidence for criminal committals of women between 1868 and 1948 is explored in Brendan Kelly, *Custody, Care and Criminality: Forensic Psychiatry and Law in Nineteenth-century Ireland* (Dublin: The History Press Ireland, 2014).

conduct and morality, particularly in relation to women and girls.[33] Crowley and Kitchin argue that the policies and laws introduced in the early years of the Free State in Ireland resulted in a 'highly gendered project aimed predominately at regulating the lives of women' that was that 'thoroughly spatial in its conceptualization', producing 'moral geographies' that led to a 'network of new interlinked institutional spaces' which were designed to reform and discipline those seen as deviant.

The institutional spaces listed by Crowley and Kitchin include Magdalene Laundries, the county homes, Mother and Baby homes, reformatories and industrial schools, but also include the lunatic asylum system, which had been established in the previous century by the British administration. They argue that the legal structures enacted between 1922 and 1937 aimed to produce 'decent' women by increasingly limiting their access to public spaces, and confining them to the home, in tandem with the 'ever-present threat of new sites of reformation'.[34] The behaviour and image of the ideal female body was promoted through these structures, but its 'other', and the consequences of deviation inform the negotiation of public spaces, and the performance of identity and individuality in the public sphere. Within O'Brien's stories, it is possible to observe the interaction between these institutions and shared social memories and associations with them, and how they operate in tandem in order to circumscribe experience throughout public and private space, regardless of physical proximity or visibility. While O'Brien does represent the threat of the asylum within the domestic context, there is an emphasis on the experience of women in public contexts and in the public sphere in relation to their awareness of the institutions, and their threat to their own social and physical security and autonomy.

In his study of the dynamics of nationalism in society, Michael Billig explored the concept of 'banal nationalism', and the role of everyday signs and symbols encountered as part of daily life in the constitution and

33 Una Crowley and Rob Kitchin, 'Producing "Decent Girls": Governmentality and the Moral Geographies of Sexual Conduct in Ireland (1922–1937)', *Gender, Place and Culture*, 15(4) (2008), 355–72.

34 Ibid.

enactment of the nation state.[35] Drawing on this concept, it is possible to view the asylum in twentieth-century Irish society as performing a similar function in the way in which moral ideology was produced and expressed spatially. Attention to the symbol and physical reality of the asylum, therefore, in representations of urban landscapes provides us with an insight into how its rhetorical power was negotiated and experienced on the level of the individual. It also provides a way into reading the public spaces and the public sphere as represented within O'Brien's work – moving beyond an emphasis on individual discrete locations towards an understanding of how the different elements work in concert in order to create a social sphere that individuals had to physically and verbally negotiate successfully in order to maintain social status and some degree of bodily autonomy.[36] These public spaces were far from neutral, and were created and constructed through bodily practices, rituals and daily interaction as well as official legislation and architectural interventions. In considering these spaces as enacted, and active, rather than passive, they can be explored as zones that were understood and negotiated differently according to class, age, gender, as well as the time of day or night. These zones are created emotionally and symbolically, as well as physically, and memory – shared, social memory, as well as individual memory – patterns public space, creating a strongly coded map that is intimately linked to identity, status, gender and power.

## Bibliography

Billig, Michael, *Banal Nationalism* (London: Sage, 1995).

Cox, Catherine, *Negotiating Insanity in the Southeast of Ireland, 1820–1900* (Manchester: Manchester University Press, 2012).

35 Michael Billig, *Banal Nationalism* (London: Sage, 1995).

36 Similar issues are considered in James M. Smith, *Ireland's Magdalene Laundries and the Nation's Architecture of Containment* (Notre Dame, IN: University of Notre Dame Press, 2007).

Crameri, Kathryn, 'Forging the Community: Explorations of Memory in Two Novels by Jesús Moncada', *The Modern Language Review*, 98(2) (2003), 353–66.

Crouch, David, 'Flirting with Space: Thinking Landscape Relationally', *Cultural Geographies*, 17(1) (2000), 5–18.

Crowley, Una, and Kitchin, Rob, 'Producing "Decent Girls": Governmentality and the Moral Geographies of Sexual Conduct in Ireland (1922–1937)', *Gender, Place and Culture*, 15(4) (2008), 355–72.

*District Lunatic Asylums as approved by the Board of Control* (Dublin: HMSO, 1863).

Douglas, Mary, *Purity and Danger: An Analysis Of The Concepts Of Pollution And Taboo* (London: Kegan Paul, 1966).

Dunne, Tom, *The Writer as Witness: Literature as Historical Evidence* (Cork: Irish Conference of Historians, 1985).

Ferriter, Diarmaid, *Occasions of Sin: Sex and Society in Modern Ireland* (London: ProfileBooks, 2009).

Finnane, Mark, *Insanity and the Insane in Post-Famine Ireland* (London: Croom Helm; Totowa, NJ: Barnes & Noble Books, 1981).

Harley, J. B., *The New Nature of Maps: Essays in the History of Cartography*, ed. P. Laxton (Baltimore, MD: John Hopkins University Press, 2001).

Hickman, Claire, 'The Picturesque at Brislington House, Bristol: The Role of Landscape in Relation to the Treatment of Mental Illness in the Early Nineteenth-Century Asylum', *Garden History*, 33(1) (2005), 47–60.

Kelleher, Margaret, 'The "Affective Gap" and Recent Histories of Ireland's Great Famine', in Marguérite Corporaal, Christopher Cusack, Lindsay Janssen and Ruud van den Beuken, eds, *Global legacies of the Great Irish Famine: Transnational and Interdisciplinary Perspectives* (Oxford: Peter Lang, 2014), 19–38.

——, '"Factual Fictions": Representations of the Land Agitation in Nineteenth-Century Women's Fiction', in H. Hansson, ed., *Nineteenth-Century Irish Women's Prose: New Readings And Contexts* (Cork: Cork University Press, 2008), 78–92.

Kelly, Brendan, *Custody, Care and Criminality: Forensic Psychiatry and Law in Nineteenth Century Ireland* (Dublin: The History Press Ireland, 2014).

Link, Bruce, and Phelan, Jo, 'Conceptualizing Stigma', *Annual Review of Sociology*, 27(1) (2001), 363–85.

MacLellan, Anne, NicGhabhann, Niamh, and Byrne, Fiona, *World within Walls: From Asylum to Contemporary Mental Health Services* (Monaghan: Health Service Executive Cavan Monaghan Division, 2015).

Mark-FitzGerald, Emily, *Commemorating the Irish Famine: Memory and the Monument* (Liverpool: Liverpool University Press, 2013).

Ó Gráda, Cormac, 'Irish Literary Sources and Irish Economic History', *Studies: An Irish Quarterly Review*, 80 (1991), 290–9, 319.

O'Brien, Edna, *Country Girl* (London: Faber and Faber, 2013).
——, *The Love Object: Selected Stories* (London: Faber and Faber, 2013).
——, *Night* (London: Faber and Faber, 2014).
O'Keeffe, Tadhg, 'Being-in-the-Town, Being-in-the-Atlas: Some Reflections on Capturing the Experiences of Town-Dwelling in the Historic Past', in H. Clarke and S. Gearty, eds, *Maps and Texts: Exploring the Irish Historic Towns Atlas* (Dublin: Royal Irish Academy, 2013), 266–77.
Pine, Emilie, '"Coming Clean?" The Art of Representing the Magdalen Laundries', in Oona Frawley, ed., *Memory Ireland, Volume 1* (Syracuse, NY: Syracuse University Press, 2010), 157–71.
Prior, Pauline, *Asylums, Mental Health Care and the Irish: Historical Studies 1800–2010* (Dublin; Portland, OR: Irish Academic Press, 2012).
Rattigan, Clíona, '"No worse and no better": Irish Women and Backstreet Abortions', *History Ireland*, 21(1) (2013), 42–3.
Saris, A. J., 'Mad Kings, Proper Houses, and an Asylum in Rural Ireland', *American Anthropologist*, 98(3) (1996), 539–54.
——, 'Producing Persons and Developing Institutions in Rural Ireland', *American Ethnologist*, 26(3) (1999), 690–710.
Smith, James M., *Ireland's Magdalene Laundries And The Nation's Architecture Of Containment* (Notre Dame, IN: University of Notre Dame Press, 2007).
Smyth, Gerry, *Space and the Irish Cultural Imagination* (Basingstoke: Palgrave, 2001).
Stoddard, Eve, 'Sexuality, Nation, and Land in the Post-Colonial Novels of Edna O'Brien and Jamaica Kincaid', in Laing, O'Connor and Mooney, eds, *Edna O'Brien: New Critical Perspectives* (Dublin: Carysfort Press, 2006), 104–21.
Walsh, Oonagh, 'Landscape and the Irish Asylum', in G. Hooper and U. Ní Bhroiméil, eds, *Land and Landscape in Nineteenth-Century Ireland* (Dublin: Four Courts Press, 2008), 156–70.
Whelan, Yvonne, *Reinventing Modern Dublin: Streetscape, Iconography and the Politics of Identity* (Dublin: University College Dublin Press, 2003).
Wohlcke, Anne, *The 'Perpetual Fair': Gender, Disorder, and Urban Amusement in Eighteenth Century London* (Manchester, Manchester University Press, 2014).
Yanni, Carla, *The Architecture of Madness: Insane Asylums in the United States* (Minneapolis, MN: University of Minnesota Press, 2007).

EMILIE PINE

# 9 The Witness and the Audience: Mary Raftery's *No Escape* (2010)

Documentary theatre 'is created from a specific body of archived material: interviews, documents, hearings, records, video, film, photographs, etc'.[1] Documentary is not a new dramatic form but it has experienced a recent resurgence perhaps due to the testimonial turn, or what Shoshana Felman and Dori Laub have called the 'era of testimony', which itself broadly constitutes an attempt, through giving testimony, to represent some of the ethical crises of recent decades.[2] In particular, documentary theatre has tended to represent the voices and perspectives of the oppressed other who has suffered violence, such as we see in the Tricyle theatre's cycle of 'tribunal' plays, for example *The Colour of Justice* (Richard Norton Taylor, 1999), which dramatized the Stephen Lawrence Inquiry on institutional racism. Through the relatively direct form of documentary, which remediates the past, combining factual material with directed performance, audiences are given the opportunity to encounter these perspectives and to engage with experiences beyond their own. Additionally, documentary theatre offers the opportunity to audiences already engaged with these experiences, to see and hear their own perspectives in a format and space that enables them to be shared respectfully and without interruption.[3] While documentary

1 Carol Martin, 'Bodies of Evidence', *TDR: The Drama Review*, 50(3) (2006), 8–15; 9. This essay introduces this special issue of *TDR* devoted to documentary theatre.

2 See Shoshana Felman and Dori Laub, *Testimony: Crises of Witnessing in Literature, Psychoanalysis and History* (London: Routledge, 1992), 5.

3 Therese, one of the witnesses in Theatre of Witness's production *I Once Knew A Girl* (Theatre of Witness, dir. Teya Sepinuck, Playhouse Derry, 2010), suggests that for her being listened to was an enormously important aspect of the documentary process.

theatre maintains an emphasis on creating and performing an absorbing piece of theatre-art, it is very often equally motivated by a desire for social good or social change. The advantage of using the theatrical form for this purpose is the communal nature of theatre – as with other commemorative practices, the audience constitutes a community listening to and watching narratives that are held to be important for that community, in order to either validate that community or effect a change within that community's own perspectives. The individual witness, in being chosen as the subject of a documentary play, is seen to relay something that is significant for, or revealing of, the wider group. The witness's testimony moreover functions, as Freddie Rokem argues, as part of 'the cathartic processes activated by the theatre performing history [...] where the victim is given the power to speak'.[4] In joining an audience, we gather together to share stories and memories and, in doing so, we experience collective witnessing and we become what Feldman terms 'communities of witness',[5] thereby extending what Sara Jones terms the 'remembering community'.[6] This act of secondary witnessing by the audience is necessary, as Paul Ricoeur puts it, there are 'witnesses who never encounter an audience capable of listening to them or hearing what they have to say'.[7]

In this essay I will discuss one example of documentary theatre produced in Ireland in 2010, Mary Raftery's *No Escape*. This play is significant for two main reasons – firstly, it was the first ever commissioned piece of documentary theatre staged by the Abbey, the National Theatre in Dublin and, secondly, it is based entirely on the documents of the Ryan Report, which was the 2009 official State report into seventy years of child abuse in Irish residential institutions run by the Catholic Church. To give a sense of the scale of this institutional history, in thirty-four years, between 1936

4 Freddie Rokem, *Performing History: Theatrical Representations of the Past in Contemporary Theatre* (Iowa City, IA: University of Iowa Press, 2000), 205.

5 Allen Feldman, 'Memory Theaters Virtual Witnessing and the Trauma-Aesthetic', *Biography*, 271 (winter 2004), 163–202; 164.

6 Sara Jones, *The Media of Testimony* (Basingstoke: Palgrave, 2014), 187.

7 Paul Ricoeur, *Memory, History, Forgetting*, trans. Kathleen Blamey and David Pellauer (Chicago, IL: University of Chicago Press, 2004), 166.

and 1970, when the system began to decline, approximately 170,000 children and young people were confined across fifty institutions. The Report found conclusively that these institutions had constituted an emotionally, physically and sexually abusive system in which thousands of children were seriously damaged. Claims by the Catholic orders, and indeed the State (via the Department of Education), that the children, now adults, were lying about the abuse, were completely swept away as the Report revealed the scale of the cruelty, violence and exploitation of children; the Report described in great detail the 'climate of fear' in the institutions and the 'systemic' nature of abuse.[8] The Ryan Report was thus a major undertaking and its findings directly affected a significant proportion of the Irish population and diaspora, and impacted Irish society in general.[9]

There are few moments in a social history when it can be said that social memory experiences a complete sea change, and the publication of this Report is one of those moments, when attitudes to the history of institutional care run by the Catholic Church were completely inverted.[10] This major change followed decades of individual cases of abuse or religious misconduct gaining media attention, and these cases, combined with the gradual secularization of Irish society, created a receptive audience for the publication of the Report in 2009. Nevertheless, the scale of the Report's findings – of systemic abuse – were deeply shocking to the majority of the population, while at the same time being a vindication of those isolated voices who had long claimed this to be the case. The 'Letters to the Editor' pages of the national newspapers reflected the shock of the population and

8 'Executive Summary', *The Report of the Commission to Inquire into Child Abuse* (Ryan Report: 2009), 20.

9 For example, the One in Four support group for survivors of abuse reported a 'huge surge in new client numbers in the three months since the Ryan Report', Press Release (15 September 2009).

10 In using the term 'social memory' here I follow the example of Guy Beiner to reflect the role of social processes in the formulation and circulation of this particular set of memories. See Beiner's discussion and definition of social memory (versus, for example, 'collective memory' and 'cultural memory') in the opening chapter of *Remembering the Year of the French: Irish Folk History and Social Memory* (Madison, WI: University of Wisconsin Press, 2007).

the impact on their perceptions of the history of the treatment of children in Ireland, with one writer declaring that she was 'ashamed to have been gullible enough to believe the Big Lie peddled [...] by members of religious orders [. ...] I cannot understand why these religious professionals hated these children so much'.[11]

At the publication of the Report, the Abbey directors met to discuss how best the National Theatre could respond to its significance; their decision was to commission a documentary play. The Abbey had previously hosted the Tricycle's touring production of Richard Norton-Taylor's *The Saville Inquiry* (2005), but had never commissioned documentary work before. Aideen Howard, then Literary Director of the Abbey, suggested that the documentary form would do justice to this history in a more direct way than a fictionalization could.[12] Though the media coverage was widespread and thorough, the Report itself was 2,600 pages long and so this theatre piece was a chance for audiences to access in more depth some of the detail and individual testimony of the Report. The play represents a meeting point for two different discourses of memory, the legal and the theatrical, a fusion that enables different forms of witnessing to be performed both onstage and off.[13]

The Abbey was astute in asking Mary Raftery to create the play, as Raftery, an investigative journalist whose work over the previous two decades had pioneered and championed the case of survivors of abuse, had an exhaustive knowledge of the institutional system and an authoritative public identity as a campaigner. Raftery's television documentary series in 1999,

11 Kennedy, Maeve, 'Letter to the Editor', *Irish Times* (27 May 2009), 16. The newspaper extended the 'Letters to the Editor' section over two pages on the day after the Ryan Report's publication to reflect the volume of correspondence they received (the section is usually less than one page).

12 Aideen Howard, in conversation as part of the *Ways of Representing the Past* event at UCD (November 2014) <http://irishmemorystudies.com/index.php/memory-cloud/>.

13 See Aleida Assmann for a discussion of different forms of testimony, including a discussion of the distinctions between legal and oral testimony. Assmann, 'History, Memory, and the Genre of Testimony', *Poetics Today*, 27(2) (Summer 2006), 261–73.

*States of Fear*, was a major factor in the government's official recognition of the abuse and the Taoiseach Bertie Ahern's historic apology for the state's 'failure', which led to the establishment of the Commission to Enquire into Child Abuse.[14] Though the resulting play is a tiny slice of the vast data in the Report – the play is just under ninety minutes long, but represents the Report's 2,600 pages, detailing evidence from 1,712 complainants and 1,090 witness statements – Raftery was a trusted pair of hands, working with the National Theatre team, including Howard and director Róisín McBrinn, to make this tiny slice feel representative of the whole history.

In order to achieve this, Raftery used a dramatic editing technique (familiar from television editing), combining information from different sections of the Report into unified scenes, and alternating testimony between survivors, the religious congregations, and civil servants. Raftery also thematized the sections of the play so that, roughly speaking, the first scenes are concerned with physical abuse of boys and girls, the next section is concerned with sexual abuse, and the final sections are concerned with the institutional system and its legacy on the lives of the survivors. Raftery also concentrates the play on a small number of institutions which are held to be representative of the general system.

Documentary theatre is always a mediated form. The material being presented comes from an archive, whether oral or written, but, in its presentation on stage, it is always a limited version of that archive. Though Raftery is a trusted pair of hands, it is still worth emphasizing that her role in compiling the play was a highly interventionist one, manipulating the particular archive of the Ryan Report, as discussed below. The purpose of this manipulation was to create an accessible format, or digest, of the Report itself, which Catriona Crowe, in one of the few reviews of the play,

14 Ahern's apology was read out outside government buildings just hours before the final episode of *States of Fear* was due to be broadcast. The apology begins: 'On behalf of the State and of all the citizens of the State, the Government wishes to make a sincere and long overdue apology to the victims of childhood abuse for our collective failure to intervene, to detect their pain, to come to their rescue' (11 May 1999).

argued was achieved as *No Escape* is 'a very successful way of dealing with a huge public issue that convulsed [...] and is still convulsing the country'.[15]

In thinking about documentary theatre, there are some key questions to consider: What is the role of the play in mediating the original archive? What is the relationship of an audience to an archive which does not belong to them? How does the action on stage address itself to the audience? Is the audience somehow implicated, or are they simply distanced listeners? How do the experiences onstage which are particular to one social group become memorable to a much larger community via performance? Behind this group of questions is an overarching interest in asking whether documentary theatre can transform the audience from bystanders, listening to testimony, into witnesses themselves?

## Mediating the Report

*No Escape* takes all of its material from the 2009 Ryan Report and adapts it to the stage by using an actor to play Judge Sean Ryan, who narrates and guides the audience through the entirety of the play. Upstage, behind Judge Ryan, the stage is stacked with boxes of files that give a clue to the numbers of individual histories that the Report represents. These histories are compiled from the original interviews and research conducted by the Commission, but they also indicate the bureaucratic historical records of the religious congregations and the State, so that the audience see onstage part of the material history of the abuse. The boxes also contribute to the status of *No Escape* as a piece of documentary theatre, underlining its research base and suggesting the amount of archival material that Raftery had to sort through and condense. Raftery's condensing techniques perform a route through

15 Catriona Crowe reviewing *No Escape*, alongside Padraig O Morain, on *Arena with Sean Rocks*, RTÉ Radio One (15 April 2010) <http://www.rte.ie/radio1/arena/programmes/2010/0415/352096-arena-thursday-15th-april-2010/> accessed 10 May 2016.

this archive that is otherwise relatively inaccessible, given the scale of the report. In the first production at the Abbey's Peacock theatre (directed by Roisin McBrinn), the stage area was also bisected by two glass screens, reminiscent of classroom blackboards, onto which the judge wrote place names, dates and figures. This school-room aesthetic affirms the didactic approach of the play (and documentary theatre generally), with Ryan playing the role of teacher as much as Judge. In this production, the onstage action varied depending on who is speaking: Ryan and survivor witnesses directly addressed the audience, while the representatives of religious congregations and the civil service were interviewed by the Commission's legal team (the cast for the play consisted of seven actors, playing multiple roles, with the exception that one actor consistently played Judge Ryan).[16] The inclusion of the interview format is important for the insight it grants into the process of constructing the Report itself. The process of an inquiry is private and closed and, in opening it up, the play informs the audience not only about the abuse documented in the report, but shows how that document was formed.

The alternation between interview and direct address, however, is most important for its role in creating an affective relationship between the audience and selected witnesses.

| | |
|---|---|
| MCMAHON | Perhaps you would like to begin your evidence by giving a broad outline of your own position in the Sisters of Mercy and how you come to be the person who is giving evidence in relation to the issues under review? |
| SR O'DONOGHUE | Thank you. I am Sr. Helena O'Donoghue, a member of the Sisters of Mercy. I currently hold the position of Provincial Leader of the South Central Province of the Sisters of Mercy. |

*[shift to report extract mode]*

| | |
|---|---|
| SEAN RYAN | A witness who was in Goldenbridge for nine years in the 1960s described her time there: |

16 In the original production, Ryan was played by Lorcan Cranitch. The ensemble was: Jane Brennan, Michelle Forbes, Eamonn Hunt, Eleanor Methven, Donal O'Kelly and Jonathan White.

WITNESS (FEMALE) *I mean the first sentence that always comes to me is that it was a reign of terror, it was a terrifying place for any child to be. Speaking for myself I found it utterly terrifying, it was vicious, it was so full of fear, it was so full of tension. It was indescribably terrifying.*[17]

[...]

SEAN RYAN One witness spoke of arriving at Goldenbridge as a six-year-old child [...] after her mother had died [...]. She said she used to lie in her bed at night and wished that she didn't wake up in the morning. She said that she would sob her heart out crying for her mother.

[...]

SR O'DONOGHUE Well [...] from all of the material that we have examined and all of the people that we have talked to over the past ten years we are of the conviction that Goldenbridge was [...] a reasonably efficient and caring school, that the managers and Sisters there were committed and worked long and hard in the interest of children, and that it was both committed and dedicated and progressive in very many ways. We believe that having examined some of the, certainly, serious allegations we have not been able to find grounds that would convince us that they were part of the reality.[18]

There is a clear dichotomy here between the points of view of Witness 1 and Sr O'Donoghue and the choice of who to believe is shaped by Raftery, who leaves little room for any audience to sympathize with or identify as truthful Sr O'Donoghue's statement that Goldenbridge was caring and progressive. Raftery's editing strategy works to validate the survivor testimony and to set up an emotional connection between the survivor-witnesses and the audience. There is always the option that an audience, individually or as a group, may identify with the religious congregations and choose not to believe in the survivors' testimony of abuse. In order to do so, however, an audience will have to resist the structure of the play which, as above, works

17 Quotations from the play are taken from an unpublished script, courtesy of the Abbey Theatre and the Raftery Estate, 22–3.

18 Ibid., 24.

to support the truth claims of the testimony of the survivors over those of the religious congregations and the State. The readiness of an audience to trust the survivor witnesses is hugely influenced by the publication of the Report a year previous to the play's first production and its authoritative endorsement of allegations of abuse. The testimony of the survivor witnesses may also be authenticated as, as Ashuri and Pinchevski argue, 'being a victim may count as [...] a form of capital in producing testimony'.[19] The empathetic engagement of the audience for the sufferings of the survivors may further override concerns about historical recall as, in terms of the relative accuracy of memory, the survivor witnesses seem far more truthful than the religious congregations, and therefore more trustworthy.

The structural bias towards establishing children as victims of the system is apparent from the first scenes of the play, when Raftery alternates between Department of Education regulations on punishment, which are humane and relatively lenient, and lists of modes of punishment which were actually inflicted on children.

| | |
|---|---|
| DEPT OF EDUCATION | 1933 Department of Education Rules and regulations for Certified Industrial Schools – Rule 13<br>[...]<br>(c) Chastisement with the cane, strap or birch.<br>Referring to (c) personal chastisement may be inflicted by the Manager, or, in his presence, by an Officer specially authorised by him [...] No punishment not mentioned above shall be inflicted. |
| SEAN RYAN | [...] hosed down with cold water before being beaten, beaten while hanging from hooks on the wall, being set upon by dogs, being restrained in order to be beaten, physical assaults by more than one person, and having objects thrown at them.<br>[...] |
| SEAN RYAN | There were accounts of boys being hit or beaten with a variety of sticks, including canes, ash plants, |

19 Tamar Ashuri and Amit Pinchevski, 'Witnessing as a Field', in Paul Frosh and Amit Pinchevski, eds, *Media Witnessing: Testimony in the Age of Mass Communication* (Basingstoke: Palgrave, 2014), 138.

blackthorn sticks, hurleys, broom handles, hand brushes, wooden spoons, points, batons, chair rungs, yard brushes, hoes, hay forks, picked and piece of wood with leather thongs attached [...] bunches of keys, belt buckles, drain rods, rubber pram tyres, golf clubs, tyre rims, electric flexes, fan belts, horse tackled, hammers, metal rulers, butts of rifles, t-squares, gun pellets and hay ropes.

DEPT ED circular No11/1946 – 'Discipline and Punishment in Certified Schools': Corporal punishment should only be used as a last resort, where other forms of punishment had been unsuccessful as a means of correction.'[20]

Several implications emerge here – the first is the very obvious contradiction between the regulations on the one hand and the experience of punishment on the other. The duration and extent of the lists being read onstage is an early shock to the audience and signifies the seriousness of the subject; in teaching this play, I find that students are always shocked and overwhelmed by these lists. In each class there is also at least one student who, as a result, cannot read any further or who finds the material so disturbing that they choose not to come to class. This is always the risk in teaching emotionally difficult material; the rewards of this are equally notable, with students actively participating in noticing the use of editing and positioning and critically reflecting on the processes of mediation at work. For the Abbey presenting this work as a live performance, the risk of alienation and secondary trauma was, of course, also a risk. My reading of this is necessarily subjective, but on the two nights that I saw the play, the auditorium seemed preternaturally quiet and concentrated, a quiet broken by the standing ovation of the audience at the end. As Crowe put it in her review, though the play is depressing, it is also 'quite an uplifting experience', one that Padraig O'Morain declared everyone should go and see.[21] That the play only ran for two weeks during its first production,

20 Raftery, *No Escape*, 14–16.

21 Catriona Crowe and Padraig O'Morain reviewing *No Escape*, on *Arena with Sean Rocks*, RTÉ Radio One (15 April 2010) <http://www.rte.ie/radio1/arena/

however, is also testament to the draining effect the performance had on its cast and crew.

The inclusion of the regulations is also significant for the years that they date from; these seem included as a response to the justification made by some[22] that accusations of abuse are anachronistic misjudgements of an earlier era's socially accepted punishment regime. The regulations, in their careful elucidation of what represents acceptable punishment, and what does not, demonstrate that punishment by beating with the rung of a chair, for example, would have been seen as equally unacceptable in 1933 as it is eighty years later.

## Status of the Witness

Survivor witnesses also testify to the hierarchy of power and truth, asserting that both as children and as adults their abuse testimony was disbelieved because of the status of the religious congregations. One witness's statement puts this in stark terms. When discussing how a priest sexually abused her, she says:

> I couldn't tell anyone. They were Gods, the priests were God, no one would believe you. I was about 11.[23]

As the play suggests, when family members outside the institution made official complaints on the part of institutionalized children, their allegations were also discredited.

---

programmes/2010/0415/352096-arena-thursday-15th-april-2010/> accessed 10 May 2016.

22 For example, John Waters, 'A stick to beat the past with', *Irish Times* (20 October 2003).

23 Raftery, *No Escape*, 37.

SEAN RYAN In February 1963, a Mrs McCarthy brought her grandson to Artane to discuss with the Superior his difficulty in keeping jobs and to see if he could help in finding employment. What happened in the course of this meeting is in dispute. The grandmother gave her version of what happened in a letter written later that month 26 February to the Minister for Education.

MRS MCCARTHY I could not believe my eyes, without word or warning the Superior, closed his fist + struck the boy a most brutal blow on the side of the jaw [...].[24]

When Mrs McCarthy went to her local priest for 'assistance with her complaint' the priest wrote to the Superior General of the Christian Brothers (who ran Artane), but the priest represented Mrs McCarthy as 'a mental case with a strong antipathy against Artane School [...]. It is easy to note that she is a very dangerous type of woman' (*NE*, 9). Mrs McCarthy followed up her complaint with a letter to the Department of Education, but that investigation concluded 'it is clear that the charges of brutality and sadism made by Mrs McCarthy are without foundation' (*NE*, 10).

Again, in teaching this play, this is one of the moments that we focus on in class. I ask students (who have only read the script of the play and not seen a production, so are not basing their reactions on perceptions of the authenticity or power of particular performances) who in this scene they believe. Many of these students will have been educated in religious schools (given that the Catholic Church still administer approximately 92 per cent of primary education in the state). Despite this background, the students unanimously declare for the grandmother. This is one indication of the complete about-face in Ireland in the past twenty years, as the status and credibility of the Catholic congregation – as a group, rather than as individuals – has radically decreased, to the point where it is now normative for accusations of wrongdoing against priests, nuns and brothers to be automatically believed. As Judge Ryan put it in 2004, 'The Taoiseach's

24 Ibid., 9.

apology [...] marked a transformation in attitudes'.[25] Prior to this major cultural shift, the reverse would have been true, with allegations of religious wrongdoing automatically discredited or rejected, as suggested by the example of Mrs McCarthy above. However, in the immediate wake of the publication of the Report, there was widespread public acceptance of the occurrence, and impact, of abuse, as evidenced by the respectful space given to survivors of abuse across the media, for example, Michael O'Brien's harrowing testimony of his experience of rape and abuse in an industrial school on the main public television station and in the print media the following day.[26]

The Ryan Report's single most important finding: that abuse was systemic rather than a rarity has thus had a major cultural and social impact; an impact underlined by additional reports such as the Ferns Report (2005) and Murphy Report (2009). This impact can be most easily identified in students' routine assumption that the allegation of abuse in this instance is, of course, true. This represents not only a shift in the perception of the Catholic congregations, but a complete shift in the relative status of witnesses, so that the formerly discredited witness now becomes the arbiter of truth, and the formerly authoritative witness now becomes the perpetrator and liar.

This shift also enables perpetrators to 'come clean' themselves – Raftery includes, for example, the admission by the Rosminian order, who 'accept their responsibility for what happened in their industrial schools' (*NE*, 47); and one Brother in Letterfrack, Brother Sorel,[27] who testifies that he forced one very young boy to eat his own excrement (*NE*, 32). The relative honesty of these responses further alerts the audience to the equivocation

25 Judge Sean Ryan, quoted in Bruce Arnold 'What was real reason for Bertie's apology to State's victims?', *Irish Independent* (18 February 2007).

26 Michael O'Brien described his experiences of abuse in St Joseph's Industrial School in Clonmel on Newstalk Radio, and RTÉ's *Questions and Answers* (25 May 2009). His comments on *Questions and Answers* were printed in full in the *Irish Times*, 'Ex-Mayor tells of abuse' (27 May 2009), 9.

27 He does admit wrongdoing, but his identity is protected by the Report's use of pseudonyms.

of the other religious witnesses, and this contradiction thus also functions to underline the truthfulness of the survivor witnesses. However, though the survivor witness is now accorded cultural credibility, they can still struggle to be heard. This struggle includes the difficulty of testifying about abuse and the painful memories this forces the witness to remember/re-experience, and the fraught dilemma of what to do with that testimony once it has been given. The Irish State's establishment of a Redress Board, to award compensation to those who had been victimized, represents one form of response; in putting elements of the Report onstage, Raftery suggests other responses and memories are possible.

Throughout *No Escape*, Raftery prioritizes the survivor witness above all so that the audience hear fully what they have to say. The final section of *No Escape* is a series of thirteen witness statements, each of which testifies to the personal – and national/international – legacy of institutional abuse. For witnesses, this legacy includes the inability to form close family relationships and the continuation of feelings of shame and stigma: 'It's a darkness they gave me. I live alone, my family don't come near me' (*NE*, 54). Other witnesses are no longer able to testify: 'I knew seven people who after leaving [the school] committed suicide. [...] I know an awful lot of people who just cannot come forward to this day, an awful lot are dead' (*NE*, 55, brackets and ellipses in original).

Within this final section, however, Raftery also strikingly includes some positive memories. Ryan introduces this mini-section:

| | |
|---|---|
| SEAN RYAN | Many witnesses who complained of abuse nevertheless expressed some positive memories: small gestures of kindness were vividly recalled. A word of consideration or encouragement, or an act of sympathy or understanding had a profound effect. Adults in their sixties and seventies recalled seemingly insignificant events that had remained with them all their lives. |
| WITNESS (MALE) 1 | That Br [...] he seemed to have an understanding of us, he was the best one I met in my life. I felt safe with him [...] He was able to help with my reading and he would put a mark saying 'well done!' |
| WITNESS (FEMALE) 2 | The kindest thing that ever happened to me was a nurse [...] we were all around saying the Rosary and she put |

> a sweet in my hand, one sweet. I didn't want to eat the sweet I wanted to hold on to it, somebody gave me something, somebody was kind.[28]

These statements are tiny moments of kindness and consideration in these survivors' memories, and in the Report and play. Their emotional impact is huge – students often comment that, in their seeming insignificance, these statements bring into focus the relentlessness of abuse in these children's lives. Raftery's choice to include these memories in the final moments of the play is also important because of how these few statements enable survivors to remember alongside their painful memories, a narrative of pleasure or humanity. Without these statements, we would not get a full and complex picture of what it was like in the institutions, and without these statements, the survivors would not be allowed to narrate or remember a full and complex memory of their lives in the institutions. The trauma label can be empowering in identifying an experience of suffering that has a lasting impact on the victim; yet it can also be an overpowering label, which engulfs the variations within an experience.

## Collective Witnessing

The final statement reads:

> WITNESS (MALE) 13: They all said 'that couldn't have happened' but they can't say that to 5000 of us when we all have a similar story to tell.[29]

The citing of the thousands of other victims at the end of *No Escape* is a powerful reminder of absence – of what cannot be represented on stage, both because of the scale of victimization versus the limited scale of the

28 Raftery, *No Escape*, 54.
29 Ibid., 55.

theatre's stage (and the equally limited scale of the audience's capacity) and because of the loss of those victims to death, to emigration and to silence. Within this brief statement, the witness acknowledges that previous attempts to tell his story met with disbelief, but that this response is no longer tenable in the face of *collective witnessing*. The historical attempt at bearing witness to the abuse, underlines the fact that the survivor witnesses do not constitute new subjects, but constitute newly valued subject positions – newly valued, that is, by the audience. They were always there, but it is only now that they are being listened to and, through their power as a collective, their individual testimony is being authenticated.

The role of the witness is obviously in giving testimony, telling the story. The role of the audience is to respond adequately to that performance. This is not, as we have seen, a straightforward invitation. As Paul Ricoeur's formulation reminds us: for the 'witnesses who never encounter an audience capable of listening to them or hearing what they have to say',[30] the problem of silencing is not due on any failure on the part of the witness to speak, but rather on the part of the audience to attend to, or believe, what is being spoken. The audience's failure to believe the stories of survivors was, itself, a further act of abuse, compounding the physical, sexual and emotional abuses. The purpose of the Ryan Report, and of *No Escape*, is to combat the dispossession of the abused, to return to them the right and the power to tell their own stories and, in doing so, to build a new collective memory, or a community of witnessing. And the purpose of the audience to *No Escape* as a theatrical event is to make up for the failures of previous audiences by according these acts of testimony with the respect denied to them by the dominant community for so many decades.

This brings me back to my initial overarching question: Can documentary theatre transform the audience from bystanders, listening to testimony, into witnesses themselves? Rather than giving an easy conclusion, I would follow this query up with a few more, angled at understanding how the audience functions in regard to being witnesses themselves, and what

30 Ricoeur, *Memory, History, Forgetting*, 166.

the audience does with these intolerable images.[31] Does the intolerable become integrated into a form of prosthetic memory, to paraphrase Alison Landsberg, a privately felt memory held publicly?[32] And if they do, that is, if the audience members construct a personal archive which remembers these stories, even those from anonymous witnesses, what impact does it have on them? Is an empathic catharsis the most that can be hoped for, or is there a transformative moment in listening to and accepting these stories?

I have discussed how Raftery carefully edits and constructs *No Escape*, so that the play prioritizes the perspectives of the survivor witnesses and that the purpose of bringing this archive to further public attention is so that this testimony becomes part of a new collective memory. I believe that in order for that to happen, the audience must be active and cannot be mere bystanders or passive listeners, but must in turn become witnesses themselves, first of all witnessing the testimony onstage, and second of all continuing to act as witnesses after they have left the theatre, by advocating for social and cultural change and acceptance.

The necessity for the audience to critically reflect on their role to this play is underlined if we consider the major absence onstage and in the list of the play's witnesses of any representative of the communities that housed these institutions, and that allowed children to be sent to them. This absence raises the question of whether the audience stands in for these communities. Audiences may not even notice this as an absence, with the structure of the play and the Report as a binary opposition of the Department of Education and the religious congregations on the one hand and the survivors on the other. In this binary model, in large part, the community escapes responsibility and censure.

The revolution in understanding the past, brought about by the Ryan Report, should invoke a similar shift in the future identity of Ireland and

31 See Jacques Rancière for a discussion of the 'intolerable image' in *The Emancipated Spectator*, trans. Greg Elliott (London: Verso, 2009), 83.

32 Alison Landsberg's term 'prosthetic memory' can be defined as: 'privately felt public memories [...] based on empathy for the experiences of others'. See Landsberg, *Prosthetic Memory: The Transformation of American Remembrance in an Age of Mass Culture* (New York: Columbia University Press, 2004), 3.

its treatment of the vulnerable. In many ways, however, the absence of the community from the Report – and from *No Escape* – and the primary assignation of guilt to the religious congregations fail to tackle the social problems and the structural callousness which created an environment that acquiesced to the continuation of an abusive childcare system.

If the Ryan Report, and *No Escape*, are to have a real and lasting impact on the Irish future, as much as they have had on Irish cultural memory, each audience member, and the audience as a collective, must take on the more active role of audience-witness. This is more than simply expressing an empathic response to the spectacle of the play; perhaps the most dynamic element of theatrical witnessing and memory formation actually occurs after the show itself, when the audience talk over the performance and *re-mediate* its meaning so that its message is both interpreted and disseminated. The validation that the audience offers in witnessing and accepting these memories cannot change the violence of the past, yet this moment of active witnessing nevertheless represents a form of social change in the sense of recognizing the status and validity of the survivor witnesses and their personal histories of suffering. This social change can then be built on, so that memories of abuse have a progressive social legacy, as Ricoeur argues 'the duty to remember consists not only in having a deep concern for the past, but in transmitting the meaning of past events to the next generation'.[33]

## Bibliography

Arnold, Bruce, 'What was real reason for Bertie's apology to State's victims?', *Irish Independent* (18 February 2007).

33 Paul Ricoeur, 'Memory and Forgetting', in Richard Kearney and Mark Dooley, eds, *Questioning Ethics: Contemporary Debates in Philosophy* (London; New York: Routledge, 1999), 5–11; 9.

Ashuri, Tamar, and Pinchevski, Amit, 'Witnessing as a Field', in Paul Frosh and Amit Pinchevski, eds, *Media Witnessing: Testimony in the Age of Mass Communication* (Basingstoke: Palgrave, 2014), 133–57.
Assmann, Aleida, 'History, Memory, and the Genre of Testimony', *Poetics Today*, 27(2) (Summer 2006), 261–73.
Beiner, Guy, *Remembering the Year of the French: Irish Folk History and Social Memory* (Madison, WI: University of Wisconsin Press, 2007).
Carroll, Stephen, 'Ex-Mayor tells of abuse', *Irish Times* (27 May 2009), 9.
Crowe, Catriona, and O Morain, Padraig, 'Review of *No Escape*', on *Arena with Sean Rocks*, RTÉ Radio One (15 April 2010) <http://www.rte.ie/radio1/arena/programmes/2010/0415/352096-arena-thursday-15th-april-2010/> accessed 10 May 2016.
Feldman, Allen, 'Memory Theaters Virtual Witnessing and the Trauma-Aesthetic', *Biography*, 27(1) (Winter 2004), 163–202.
Felman, Shoshana, and Laub, Dori, *Testimony: Crises of Witnessing in Literature, Psychoanalysis and History* (London: Routledge, 1992).
Howard, Aideen, in conversation (November 2014). <http://irishmemorystudies.com/index.php/memory-cloud/> accessed 10 May 2016.
Jones, Sara, *The Media of Testimony* (Basingstoke: Palgrave, 2014).
Kennedy, Maeve, 'Letter to the Editor', *Irish Times* (27 May 2009), 16.
Landsberg, Alison, *Prosthetic Memory: The Transformation of American Remembrance in an Age of Mass Culture* (New York: Columbia University Press, 2004).
Martin, Carol, 'Bodies of Evidence', *TDR: The Drama Review*, 50(3) (2006), 8–15.
Raftery, Mary, *No Escape*, unpublished script, courtesy of the Abbey Theatre and the Raftery Estate.
Ranciére, Jacques, *The Emancipated Spectator*, trans. Greg Elliott (London: Verso, 2009).
Ricoeur, Paul, 'Memory and Forgetting', in Richard Kearney and Mark Dooley, eds, *Questioning Ethics: Contemporary Debates in Philosophy* (London and New York: Routledge, 1999), 5–11.
——, *Memory, History, Forgetting*, trans. Kathleen Blamey and David Pellauer (Chicago, IL: University of Chicago Press, 2004).
Rokem, Freddie, *Performing History: Theatrical Representations of the Past in Contemporary Theatre* (Iowa City: University of Iowa Press, 2000).
Ryan, Judge Sean, 'Executive Summary', *The Report of the Commission to Inquire into Child Abuse* (Ryan Report: 2009).
Waters, John, 'A stick to beat the past with', *Irish Times* (20 October 2003).

NELSON BARRE

# 10 Perpetual Stagnation and Transformation: *Ballyturk* and *The Walworth Farce* as Memorial (Re)Inscription

Enda Walsh's plays often hinge upon the repetition of a single event which requires the characters to enact a sort of memorial process of re-membering the past, that is to say a physical reconstruction. Characters in his plays choose to embody versions of the past to confront notions of truth within a necessarily subjective process of memorialization. In her introduction to *Memory Ireland*, Oona Frawley claims: 'memory seems to be something that we need to make concrete, that we need to *realize* in the world [...] memory has been transformed over and over again from an ether, an energy, into a tangibility that we want to see'.[1] To frame memory in this way introduces two issues in performance: the claim that there is no single version of the past that is considered truth, and that memory is under constant transformation. Emilie Pine similarly argues that 'Cultural artefacts, such as photographs, are useful ways of illuminating history, yet they are only ever representations, versions of the actual event [...] cultural representations do not always tell the 'truth' they appear to tell'.[2] For Pine, historical memorialization is synonymous with appropriation of the past through physical performance. Even seemingly static objects can be framed so that they speak to a particular experience or view. That is to say, memory relies on archives that constantly revise narratives of the past based on present beliefs and needs. Therefore history may appear stagnant but the retelling of these events re-members them into a new meaning. These two arguments also imply considerations of subjective versions of history as created in the

1 Oona Frawley, *Memory Ireland* (Syracuse, NY: Syracuse University Press, 2011), xxiv.
2 Emilie Pine, *The Politics of Irish Memory* (Basingstoke: Palgrave Macmillan, 2011), 2.

present, which are necessary to an analysis of Walsh's plays. In this chapter, I will examine two of Walsh's plays using memory and performance studies methodologies.

In *Ballyturk* (2014), two characters select various people from their memories to re-enact as they try to correctly impersonate the past and its associated actions. The onstage figures present an extreme version of selective amnesia that is juxtaposed against their zealous desire for precision in performance. *The Walworth Farce* (2006) follows a father and his two sons re-performing the story of the father's departure from Cork to London. In the play, characters take on roles to retell a piece of history but each daily repetition moves them further from the original events. Both plays depict worlds where characters know no other way of living beyond the performed memory that dictates their daily routine. The terror for these onstage figures appears only when the stories break down and become subject to the reshaping of the past due to dropped lines or misremembered actions. Through a firm belief in the reality that has been created by ritualized repetition, the characters demonstrate the complex workings of memory and its constant evolution over time – even when strict adherence to details is required of those performing the past.

By analysing how performance and memory highlight the instability of re-membered pasts within these two plays, this article will undertake questions of reality and belief as a form of world-making. The idea that no two versions of a story can ever be the same will be juxtaposed against the process of developing a ritual so precisely enacted that even the smallest deviation can create a significant change. And yet, it is impossible to escape the citationality of memory in performance. Marvin Carlson explains that,

> like the memory of each individual, [theatre] is also subject to continual adjustment and modification as the memory is recalled in new circumstances and contexts. The present experience is always ghosted by previous experiences and associations with these ghosts are simultaneously shifted and modified by the process of recycling and recollection.[3]

3 Marvin Carlson, *The Haunted Stage: Theatre as Memory Machine* (Ann Arbor, MI: University of Michigan Press, 2001), 2.

Carlson's concept of 'ghosting' portrays performance as a memorial act – a palimpsestuous layering of performativity. In this way, any previous action, dialogue, or object becomes infused with the implications of historical precedent. Enda Walsh's works depict these experiences of the past and their reinscription in the present as a source of identity creation. The transformation of memorial space across time and bodies becomes the site of contested certainty, especially when authority figures have constructed a specific story as a function of their power. Characters then use these malleable boundaries of truth as a form of agency and finding security in personal values and social definitions. In this way, Walsh's onstage figures grapple with these issues showing the personal and communal inevitability of (mis)remembering one's place in history and society.

Walsh's works often portray mediated versions of reality that have been re-performed many times prior to the version an audience sees onstage. The play-within-a-play has been rehearsed countless times with the intention of crafting a finished product, and therefore mirrors the theatrical form. As performers who rehearse, revise, and repeat their performances, actors create a reality out of an idealized vision of enactment. Some aspect of the performance will necessarily change, given the fact that an act cannot be repeated precisely as it existed before, no matter the attention to detail or mechanical nature of the performance. Walsh's works rely on this aspect of re-membered, reality-creation as a present act. *The Walworth Farce* and *Ballyturk* demonstrate the multi-layered reality that marks a performer's attempts to create a character. The plays depict characters who simultaneously see the manufactured citationality of their reality while also blinding themselves to the falsity behind their intents and actions. Walsh's plays fracture the possibility of a singular history being told, even in a highly structured performance venue and style. The world that is created in performance becomes its own reality even if the space is only realized in the actors' embodiment of the play.

The first part of the essay will consider *The Walworth Farce* as theatrical invention and performative reality-making. This play depicts an extreme case of revising personal history. Theatre in general creates a space for invented fictions where the performances are accepted as simultaneously real and not real. Christopher Morash and Shaun Richards's *Mapping Irish*

*Theatre* provides a useful consideration of spaces where reality is at odds with the perceived world outside the walls of the play. They state: '[Irish stage realism] remains available for use in the theatre as a resource in a present in which the 'real' is an increasingly slippery concept'.[4] Granted, Walsh's plays deviate from the traditions of realism but they reflect a version of reality that critiques memorial representation. *The Walworth Farce* blends these two expectations by giving agency in performance to the characters. The three men who begin onstage are actors portraying characters whose only purpose is to follow the script that has been rehearsed for the past seventeen years. Dedication to precision and exact replication of the reality of the story is paramount to the construction of their world. The breakdowns in storytelling appear similar to when an actor drops a line in a stage production, but each lapse in performance highlights the constructed nature of their lives, the structures of power, and the subjective truth on which they depend for meaning. Schneider and Roach's theories of re-enactment and surrogation provide insight into the play's layering of storytelling. For Roach, the attempt to revisit the past requires knowing substitution. In his *Cities of the Dead*, he claims that 'By means of restored behaviors, which gave form to events at which a certain amount of improvisation was necessarily required, the independent dramas of surrogation and sacrificial expenditure could be staged. Their staging featured the performance of memory [...] condoled by the rejuvenating imperative of legitimate succession'.[5] The re-membering of the past by a new participant, therefore, resonates in the present and lends continuing importance to the story. In a similar way, Schneider argues: 'To witness a re-enactment is to be a bystander [...] in a syncopated temporal relationship to the event that (some) participants hope will *touch the actual past*, at least in a partial or incomplete or fragmented manner'.[6] The act of repeating the past would exceed reality to literally transform the present. Reality becomes a representation of times past and also a constant referent to the present

4 Christopher Morash and Shaun Richards, *Mapping Irish Theatre*, (Cambridge: Cambridge University Press, 2013), 123.

5 Roach, *Cities of the Dead*, (New York: Columbia University Press, 1996), 168.

6 Rebecca Schneider, *Performing Remains: Art and War in Times of Theatrical Re-enactment* (London: Routledge, 2011), 9.

which is meant to maintain a particular frame of control for the author of the piece, Dinny. Instead of hindering the plot, the breaks in character reveal what reality resides beneath the façade of historical remembrance and appropriation of cultural power.

The second part of the essay examines an even more explicit invention of reality, exemplified in Walsh's *Ballyturk*. The later play creates a more complexly layered attempt of symbolic action as it comes up against socially constructed boundaries. For the two men stuck in the room, they are both the creators of their own confinement and also prisoners against their own will. Theories of performance, agency, and identity illuminate issues of reality-creation as a struggle for control over one's self and the worldview a person inhabits. In this play, the boon of existence relies on words and images that are not one's own. Particularly, Judith Butler's idea of performativity as repetition of cultural and phenomenological precedents frames the performed act of remembering certain aspects of one's past and actively forgetting others.[7] This act leads the characters to re-member and perform the identities of Ballyturk's many inhabitants at the expense of their own names. The power structure of repetition and routine crafts a version of the outside world that is both entirely imagined and embodied through physical enactment. Walsh's plays present characters who meta-theatrically recognize this process but also find it impossible to break from the author's scripted performance, which depicts the literal and metaphysical power structures that govern all acts of performed re-membering.

## 'Remember nothing! Say the line!'

Enda Walsh's 2006 play *The Walworth Farce* presents a repeated, reconstructed history taken to the extreme – a fusion of form and content. The play follows a single day in which Dinny and his two sons, Blake and Sean, perform the story of Dinny's final day in Cork, Ireland before he fled to

7 Judith Butler, 'Critically Queer', *GLQ*, 1 (1993), 17–32.

London (where the play occurs). Dinny created the farce as a story to tell his sons, but over time Dinny scripted a performance for the trio to re-enact the events exactly as he remembered them. The play-within-the-play frames what exists as an everyday occurrence for the family, as they have performed this version of the story every day for seventeen years. The routine has become so ingrained into their being that it occasionally becomes automatic, causing minor lapses due to boredom or lack of focus.

On this particular day, Sean mistakenly brings home the wrong shopping leading to several problems with the 'facts' of the piece (crackers instead of bread and sausage instead of chicken, for example). Each inconsistency further confirms the theatrically constructed nature of Dinny's memories and the story's truth, which is meant to vindicate the father for his double murder of his brother and sister-in-law for the family inheritance. These mistakes culminate in the ultimate disruption of their performance, the introduction of another person, Hayley, the Tesco checkout girl who brings the correct bag of shopping to their flat. In an attempt to regain control, Dinny forces Hayley to participate in the farce, but her very presence breaks the carefully constructed barriers of who plays each role as well as the precision he requires. The play's climax depicts Dinny's belief that a series of farcical 'twists and turns and ducks and dives and terrible shocks' led to his departure from Cork.[8] Blake breaks the final tableau when Dinny shares a loving goodbye with his wife Maureen (played by Blake in a dress and a wig), and the elder son stabs his father. Then Blake tricks Sean into killing him so the younger brother has a chance to escape the farce. Instead Hayley flees and Sean remains in the apartment and recreates his own version of that day's memory, complete with Hayley's unexpected entrance.[9]

In performance, *The Walworth Farce* moves at a frantic pace as the characters attempt to keep up with the reality-making process of enacting

8 Walsh, *The Walworth Farce*, 82.

9 This summary of the plot is necessarily brief and omits many of the finer points and nuances that are integral to the farce's complex structure. For an extensive exploration of the play's plot and information about the original production, see Eamonn Jordan, '"Stuff From Back Home": Enda Walsh's *The Walworth Farce*', *Ilha do Desterro*, 58 (January/June 2010): 333–56.

a precisely timed performance of the past. But the play-within-the-play constantly evolves due to additional lines supplied by Dinny, the author of the story. The playing space may appear as a dingy London flat (which in fact it is), but the characters create a theatrical production complete with a soundscape, stage props, costumes, and lighting made from various household items and clothes left behind by Dinny's murdered brother and sister-in-law. While the characters attempt to bring the past into reality via performance, they also necessarily recognize and accept the theatricality of their invented world. Although the script only exists in Dinny's head, the words are still prescribed and memorized for their daily enactment along with the quick-changes and slamming doors for which farce, as a genre, is noted. The story's importance is upheld because of the repetition and adulation of specificity in performance. As such, the characters represent exaggerated versions of human ritual for the perpetuation of dominant narratives of power. Even with this dedication to precision, the expectation for exact replication day after day proves unsustainable as the characters all experience small breaks in their character and storytelling. The lapses in performance eventually lead to questions about the truth of Dinny's story.

These issues implicate reality as a construction of mind, body, and social interactions similar to Judith Butler's assertions about socially inscribed repetitions. Butler states: 'social action requires a performance which is repeated. This repetition is at once re-enactment and re-experiencing of a set of meanings already socially established'.[10] A character's identity, therefore, is based on the structures of social performativity. When there are deviations, an authority figure notes the inconsistency between expected performance and actual embodiment of a certain persona. In this case, every part of the performance is under Dinny's scrutiny, and neither Sean nor Blake are safe from his expectations of precision. This includes the preparatory moments as Dinny turns on a recording of 'An Irish Lullaby' which is timed perfectly to placement of props, donning of costumes, and warm-up exercises by the performers.[11] These characters are actors creating

10 Judith Butler, 'Performative Acts and Gender Constitution', *Theatre Journal* 40(4) (1998), 526.

11 Walsh, *The Walworth Farce*, 5–7.

a decidedly theatrical rendition of history. Throughout the performance, both Blake and Sean make quick-changes in wardrobes at the back wall, flanking the main entrance to the space. Between them, the sons play eight roles in the farce, each with a change of costume and characterization. Dinny only plays one character throughout – a younger, idealized version of himself. This singular focus allows Dinny to arguably play the most nuanced character and win the acting trophy (the prize won for the best actor of each day – a prize Dinny has never lost[12]).

In Dinny's strict daily re-creation of his memories, he embodies what Rebecca Schneider says of historical re-enactors: 'if they repeat an event *just so*, getting the details as close as possible to fidelity, they will have touched time and time will have recurred'.[13] A similar belief drives Dinny throughout the performance; and he forces this adulation of belief onto his sons. Similar to those re-enactors, the men in the Walworth Road flat perform in an effort to display the events of the past accurately. Schneider analyses war re-enactment, but her study provides a useful lens for considering Walsh's play when considering the layered theatrical apparatus in the piece. In Walsh's work, the characters may not have access to the real, historical items that were present during the Dinny's final day in Cork, but they use their creativity to construct a reality of what they remember. That is, theatrical illusion takes precedent in making the reality. Of course, memory is an imperfect way of recreating an imagined, accurate re-enactment of a past event. However, the story has been repeated so many times that it has slipped into the realm of myth; that is, Dinny's version of the story relies on the performed reality as a representation of something that has a historical precedent. But these re-enactments always imply an imperfect system of performance and re-membering because of the distance from the original event.

Dinny insists on the tight structure (as a farce requires) in order to limit time to do anything except react with what each person has learned as the correct performance of their role. The men simply respond with what Dinny demands to attain an accepted re-membering of history. For example, when Dinny asks if Sean enjoys his daily outing to Tesco, Sean

12 Ibid., 23–4.
13 Schneider, *Performing Remains*, 10.

responds: 'Only if you want me to'.[14] Rather than argue, Sean goes through the story each day without questioning. The characters do all they can to relive the performances exactly as they appear in Dinny's idealized version of reality. They must not only remember the lines, but they must perform to a certain standard of 'realistic' truth. Joseph Roach's claims for the reliving of past in the present are instructive here. He says: 'To act well is to impart the gestures of the dead to the living, to incorporate, through kinesthetic imagination, the deportment of once and future kings'.[15] The surrogation that takes place when these characters re-enact the story each day illuminates Roach's assertion: the one who acts best is the one who knows the story best. In short, Dinny will always win the acting trophy because it is his memory, but the expectation for creating the most detailed reality is the same. In fact, the need to correctly interpret the surrogated role limits the ability to rebel against Dinny's story and provides little opportunity for Turner's suggestion that performance questions power structures.

However, the importance placed on correct repetition can never be fully accepted in the daily performance. Dinny, as author of the piece, always reserves the right to change parts of the script (which is engrained in his embodied memory, not written down).[16] There is a clear process and narrative structure to the farce in Dinny's mind. Furthermore, if anyone questions his authority, he uses it as a moment to express his dedication to the story. After Sean's improper delivery of a line, Dinny corrects him at length to which Sean responds:

SEAN That's a new line, Dad.
DINNY So?
SEAN You want me to use it?
DINNY Getting lazy on me?[17]

14 Walsh, *The Walworth Farce*, 23.
15 Roach, *Cities of the Dead*, 80.
16 This concept of an unwritten narrative reflects what Oona Frawley discusses in her article 'Toward a Theory of Cultural Memory in an Irish Postcolonial Context' as 'scripts'. She describes it as follows: 'Repeated exposure […] creates a different type of memory, one in which we are able to predict patterns and narrative direction' (26).
17 Walsh, *The Walworth Farce*, 13–14.

Dinny sets himself apart as a performer who never gets lazy and always places the reality he creates at the pinnacle of achievement. Dinny also uses the line change as an opportunity to demean his son's attempt to question the reality of the performance. Instead of valuing others' memory, Dinny attributes the ability to perform on cue as the performer's first responsibility.

| | |
|---|---|
| DINNY | Details, details! |
| SEAN | I remember, Dad. |
| DINNY | Remember nothing! Say the line![18] |

This moment appears in the middle of the second act as the plot continues to unravel and expose the fabricated nature of the performance. Dinny highlights the reactive nature of performance; it is not something which is consciously considered and then enacted but rather a constructed set of words. The dedication to the performance comes before what is actually remembered. That is, the re-membered history supersedes memory because it supposedly does not allow for criticism of the past. But instead, the body constantly misremembers and revises the story, much to Dinny's dismay, and he does everything he can to stop such deviations.

Later when Sean's shopping mistake provides a salami sausage in place of the expected whole roast chicken, Dinny flings the sausage against the wall and says:

| | |
|---|---|
| DINNY (*quietly*) | It's not working with the sausage. It's not right. |
| SEAN (*instinctively*) | Is any of it? ... Is any of this story real?[19] |

Dinny continues to dismiss these challenges to his authority because he rejects the possibility of their version of reality over his own. The process of world-making for these characters has occurred over seventeen years, but the characters actively view the revisionist aspect of storytelling in the breakdowns of the story. To this end, Sean and Blake question their father's motives, but Dinny repeats the daily performance and continues moving forward in the farce rather than accepting the possibility that he may have

18 Ibid., 65.

19 Ibid., 29.

lied. Instead, Dinny maintains the need to get the details correct, according to his story, and this influences the way in which all the characters act. As Dinny believes that if he performs the constructed version of his memory enough times, the embellishments he has inserted will eventually become true because of their precise repetition and the ritualized (almost religious) fervor with which the characters pursue this version of truth.

Bodily re-enactment as a process of re-membering leads to a more effective way for the performers to understand their relationship to the past and also to the specificity that Dinny requires. When Dinny questions Sean and Hayley about the conversation that led her to the flat, he demands a performance rather than a verbal retelling: 'Don't be lying to me and tell me what was said. Show me exactly how it was. The same words. Play it'.[20] Hayley mimes scanning the items while they re-enact their meeting from that day, and Dinny directs the performance when it does not conform to the idea he had of their encounter. Again, Dinny aligns himself with the need to correctly repeat and perform the past as a way of constructing reality in the present. Hayley's difficulty in taking on the performance does not prove she is a bad actor but rather demonstrates her outsider status and her limited understanding of the story's circumstances. Dinny corrects the mistake of her appearance when she takes on the role of Maureen (Dinny's wife, and the one who makes the meals); he uses some moisturizer cream to whiten her face.[21] This reverse blackface revises any shameful connection to outsiders such as Hayley; in short, he attempts to expunge any reference to his story's inconsistencies. Dinny's need for control continues to show his need for stability in memory. He defends against the outside forces that threaten the reality of the farce by physically and mnemonically implementing his own fixes, even in the present.

In the end, Dinny seeks truth in the creation of his re-enacted world. Dinny describes the performance as 'A story to be retold, no doubt, and

20 Ibid., 60.

21 Ibid., 78. According to Thomas Conway, this is the moment when 'the laughs leave the room' for an American audience. A shameful reference to the history and racial tensions of blackface (or in this case, its reverse) is enough to silence the comedy.

cast in lore. For what are we, Maureen, if not our stories?'[22] To some extent, his assessment is correct. If the ultimate rendition of reality is that which can be physically manifested in performance and storytelling, then the fact that his farce has been played every day for the last seventeen years lionizes the expectation that performance of self ritualizes the past and imbues it with importance for the present and future. When Blake stabs Dinny in an attempt to end the farce, the father redoubles his efforts to recognize his performance as real. As he dies, Dinny says 'Fuck it, that's some acting. Real blood'.[23] Dinny rightly acknowledges the enactment of his life and his death as a performance. Even without the meta-theatrical critique that this moment will be performed again the next night, Dinny's final words are indicative of the importance that he places on legacy. He clings to the acting trophy as he bleeds out, and embodies his purest performance by dying not just in the farce but in real life.

This moment seems to provide Blake and Sean the ability to leave the flat and the farce, but Blake fears a life without the safety and meaning of the performance. Dinny's question – 'what are we if not our stories?' – terrifies Blake and forces him to realize that he does not want to live in a world where he does not have the repeated ritual of the story. Instead, Blake tricks Sean into stabbing him in the hopes that Sean can escape with Hayley and leave the memory of the re-enacted world behind. But instead, the final moments of the play depict Sean replaying 'An Irish Lullaby' and restarting the performance of that day. The play ends '*as we watch him calmly lose himself in a new story*'.[24] For Sean, a precisely codified world cannot be escaped, even when the other members have departed or died. Although Sean seemed to be the most liberated and memorially free, the ending depicts a terrifying reality that the unknown can cripple a person's understanding of personal identity. Instead, the safety of the routine protects Sean (and the family's) sense of selfhood and reason for being. Sean no longer needs to remember the past for the sake of perpetuating his

22 *Walworth Farce*, 82.

23 Ibid., 83.

24 Ibid., 85.

father's story, but instead he must repeat the story because it is the only thing that he lives for any more.

These re-performances reflect what Frawley defines as a 'shifting subject that depends on present positioning and, to a large degree, on the revelation of and subsequent lightening of trauma'.[25] In this sense, *The Walworth Farce* represents a specifically Irish context not only due to the content of the story but the ways in which the Irish characters revise their version of the past. Walsh's play disrupts the expectation that memory works effortlessly, instead portraying the labour of performance as a world-making process. Frawley's claim highlights these aspects of subjective reality along with the imprecise nature of remembering. The Irish characters have left their home country, but they keep a version of Cork, Ireland alive by replaying the story. By precisely re-enacting the farce, they hope to maintain a static script – one that allows the memory of Ireland to stay longer. However, in their attempts to correctly portray the past according to Dinny's version of the story, the men instead move further away from the original with each passing day. This realization of memorial dependency and symbiosis manifests itself not only in this play but in many of Walsh's other works.

## 'You'll learn to forget – we did before'

The connections to re-membered world-making continue in Walsh's 2014 play *Ballyturk*. In this play, the playwright extends the layered performance apparatus to include more than a creative process. The characters onstage also recognize the possibility for world-making from nothing but words. The play starts with 1 (a man in his thirties) delivering a monologue under a spotlight, directly out to the audience. But when he comes to the climax of his speech and fails to find a purpose for the heightened emotion, the lights return and we see 2 (a man in his forties) watching the performance

25 Oona Frawley, *Memory Ireland*, xv.

by 1. These two men are stuck in this room creating the world of Ballyturk (an imagined outside world) with words and impersonations of its denizens. The name does not refer to an actual place in Ireland – unlike his reference to Inishfree in *misterman* – but instead to a metonymy of Irish stereotypes that can be easily replayed through performative re-enactment. We learn later that these characters have forced themselves to forget any reality beyond the walls of this room, and in their imaginations bunny rabbits have five legs and flies are buzzing specks of dust. This invented world is upset when the back wall comes down and exposes the outside world and 3 (a man in his sixties), who destroys the false reality 1 and 2 had previously committed their lives to creating. In a god-like turn, 3 tells 1 and 2 that he has been there before but they have forgotten, and now one of them must go to the outside world to die while the other continues to seek meaning by re-membering Ballyturk.

The confrontation between invention and reality leads to a frantic attempt to forget what 3 has revealed. In the final imagined exploration of Ballyturk, 1 questions the reality of their daily performances while 2 attempts to reassert its primacy as their purpose for being – the need to create. 2 re-members the first time he and 1 created Ballyturk with just twenty-seven words, and from there it expanded into an all-consuming performance. However, 1 seeks an end to the repetition and departs at the end, his absence leaving a vacuum where the space was once filled with energy and words. 2 is stunned into silence because he knows what happens next: a hidden door in one of the walls opens, and through it comes a young girl who will take up 1's place in remaking the performed reality of Ballyturk.

Everything within the room has been carefully calibrated and designed to enhance their performances, from drawings of the faces and places they remember outside, to carefully timed preparation routines set to music and sound effects. Even ambient sounds introduce a sense of realism in the performance. In this way, the space itself acts as an emblem of memory and reality. Illustrating Joseph Roach's observations on how '[performances] make publicly visible through symbolic action both the tangible existence of social boundaries and, at the same time, the contingency of those boundaries on fictions of identity',[26] the world these men create is framed as both necessarily real and imagined. That is to say, the re-membered version of Ballyturk is made manifest through their performance of it, but it is also necessarily fictional because it has no reference beyond their imaginations. But the characters have lived in the space (and the imagined world) so long that they cannot remember whether the performances exhibit reality at all. When small glitches in the story-telling appear (the discovery of a fly, or the dropping of a line), the narrative of their imagined past, and the space itself, begins to break down. The appearance of a third character forces the men to confront issues of constructed power over the past and the spaces in which it can be revived. Neither of the two original men can fully accept the truth of their invented existence, and yet one must continue the constant invention process. In this way, Walsh's play challenges notions of stable worlds and offers a new dialogue not simply about place within the play but within theatre and performance generally. *Ballyturk* crystallizes the meeting between theatrical and real-world instances of performance which confront the fractured realm of human experience and memory.

*Ballyturk* uses a similar set of rules as *The Walworth Farce* regarding the performances the characters undertake to find a purpose in their re-membering of the outside world. The men restart each performance at the sound of cuckoo clock, and they exercise and warm up for each performance to pop hits of the 1980s. 1 and 2 have memorized every character in the story they have created so they choose by chance who they will include by throwing darts at a wall of hand-drawn pictures depicting the inhabitants of Ballyturk. The two men perform a story of small-town insults and intrigues relating to Bingo, small shops and yellow jumpers. But as they go through the story, the pair find a fly, which is an entity from the outside world and something that could destroy the illusion of Ballyturk and what they think they remember. The memory for words' definitions fails, highlighting Frawley's claim about Irish memory's changeability. The appearance of the fly leads 1 to say 'How many other things have wings that I didn't know

26 Roach, *Cities of the Dead*, 39.

had wings? (*2 doesn't answer*) I thought we knew everything there was to know'.[27] Like the *The Walworth Farce*, the men in *Ballyturk* create a world of safety from what they choose to remember. The process of repeating, rehearsing, and, in a way, reviving these other characters in the bodies of these two men is an act of re-membering in the sense of assembling body parts. The memory they cling to for safety represents stability even as the pair must attempt an impossible feat of exact duplication of actions and physical remembrance. In fact, their performance is more about forgetting what was originally performed than correctly performing the present.

The men's re-enactment of the story supposedly provides a reason for their lives; they seek a similar connection to the reality as the historical re-enactors Rebecca Schneider discusses. But these men are a fictional creation themselves, highlighting the same issues of re-performance and the slippage of time and meaning. The importance of the re-membering process holds the same weight as those who seek historical accuracy. For the men, the brief moments where they finish the story and feel as if they have fully created a reality provide a sense of completion. After the first performance of the story, the stage directions read:

> *Everything suddenly stops and a single light snaps down on 2 as he looks upwards. A sudden blast of air almost sucks him off his feet and for a brief three seconds it is the most glorious release. We crash back into the room's normal state – and how harsh and bare everything looks now*'[28]

In these moments, 2 achieves a sense of fulfilment and finds a purpose in this daily ritual. The feeling of a 'glorious release' accompanies his best performances; that is, he comes close to Schneider's re-enactors who attempt to 'touch time' through precise repetition. However, 2 and 1 can only convince themselves of this illusion as long as the story is in motion. The performed version of their lives produces a more satisfying sensation than the one in which they only play themselves. Like Dinny, Blake, and Sean, 1 and 2 seek truth in performance rather than their lived reality outside the story.

27 Enda Walsh, *Ballyturk* (London: Nick Hern, 2014), 30.
28 Ibid., 25.

In other words, the imagined world they create, complete with costume changes and dramatic lighting, inhabits a presence in their daily reality which infuses their lives with meaning. The enacted memory produces a stronger feeling of connection to one's past than the process of remembering the self.

In this way, the memorized performance supposedly provides an opportunity for stasis. They have chosen to codify certain aspects of their memory in lieu of remembering as much as possible. This could be construed as an admirable mnemonic device for collectively maintaining a historical sense of one's world, to which one no longer has access. However, there is no proof in the play that Ballyturk is real or that it has ever existed beyond their imaginations. Over the course of the play, it becomes clear that these men enact these stories based almost entirely on the wisps of memory that 1 draws to cover the walls. The memory is made manifest through physical action, and for them that is enough to give it substance and meaning. In a similar argument, Samuel Beckett says in his essay 'Proust', 'In extreme cases memory is so closely related to habit that its word takes flesh, and is not merely available in cases of urgency, but habitually enforced'.[29] The words for Beckett as well as the characters in these plays are imperative to the memory, and yet it is the action and habit which asserts dominance over the past. Upon the appearance of the outside character, 3, 1 and 2 finally acknowledge their memorial inundation through action over words.

| | |
|---|---|
| 3 | Did ya give each other names by the way? ... |
| 1 | No. |
| 3 | And why not? |
| 1 | I don't know why. |
| 3 (*to 2*) | Can't ya remember deciding that? |
| 2 | No. |
| 3 | Did ya know I would come back? |
| 2 (*Slight pause*) | Maybe once I did.[30] |

29 Samuel Beckett, 'Proust', in Paul Auster, ed., *Samuel Beckett, The Grove Centenary Edition: Poems, Short Fiction, Criticism* (New York: Grove Press, 2006), 521.

30 Walsh, *Ballyturk*, 43.

2 recognizes his active forgetting even as he re-membered other aspects of the outside world. This moment in the play presents something that seems impossible. How could something as essential as a name be forgotten, especially in a world where the entire performance is an attempt to recall what they have not encountered in many years? The constructed meaning of words then disintegrates as the men realize that more and more of their memories are based on actions. As 2 tries to explain his own forgetting, he says 'There was nothing to start with – and out of that me and you pushed words'.[31] He revels in the creative process rather than in the alleged finality of meaning and purpose. That is to say, the re-membering act – the performance – takes precedence over stability in story.

In performance, when 3 leaves after introducing these references to reality beyond their room, the two men desperately seek a way to escape the reality of death that accompanies the outside world. The men put on a playlist and perform an exercise montage to Yazoo's song 'Situation'. Walsh's stage directions for this section provide a strict set of physical activities, but the times are the most instructive. The author has created such a specific vision of the events for the characters to perform that he includes '*On 0:10… On 0:25… On 0:34… On 1:02… On 1:25… On 1:47… at 2:23 they and Yazoo stop*'[32] which creates an exacting and rigorous regimen of expected actions associated with time. Beyond that, Walsh provides the men with an explanation for what the exercise provides for their actions. The stage directions read, '*All this time and their minds are racing with what 3 has left in the room – perhaps the exercise will expel these thoughts of life and death*'. In short, Walsh has provided the subtext of the action for anyone who reads or views his play. While this may seem overly prescriptive, this specificity further highlights the level of detail and precision that is required to depict a world such as Ballyturk.

As the play reaches its conclusion, definitions of what an invented reality entails come to light. 1 and 2 actively discuss their process of forgetting certain details of their lives only to replace them with the tale of Ballyturk. At one point, 1 and 2 have an argument about their names:

31 Ibid., 50.
32 Ibid., 47–8.

1 Do you know my name?
2 I made myself forget it.
1 Why would you do that!?
2 I don't know why …
1 Do you know your name?
2 None of that's important …![33]

The importance of self is lost amongst the need for a communal remembering of constructed social significance. By replaying the same story over and over from different perspectives, the characters enact a sort of homogenization of belief. But by the ending, 1 realizes that what they re-member only serves to highlight what they have chosen to forget. The strict adherence to certain performance expectations takes precedence over personal experience in this world. Again, 2 prefers the world of the imagined memory to what might actually provide a fixed identity. By performing the different cast of characters each time they repeat the story of Ballyturk, the world changes enough that it gives meaning to 2. But 1 instead chooses to leave and join 3 outside, where he will die shortly after experiencing the world for the first time since his childhood. On the other hand, 2 stays and meets 1's replacement as a girl of seven enters the space through a secret door in the wall to begin the process of teaching her to forget. The cycle repeats and the memory will live on in a new generation.

The implication made by the play's conclusion appears to present a rather dark interpretation of daily performance and people who find meaning in these repeated routines. However, one of 3's final lines before he leaves depicts a positive outlook for the men stuck in this cycle of remembering Ballyturk. 3 states, 'Everything you've imagined – it is. All life. It's out there. Everything'. This statement connects to the Performance Studies assertion, often attributed to Richard Schechner, that everything is performance. Except in this case, that which is performed is entirely imaginary and only becomes real through performance. 1 and 2 create a world that from an outside perspective is based entirely in fantasy, but because the characters embody their dreamt realities they move the idealized into a

33 Ibid., 50.

realm of realism. In this way, the words that 2 clings to as the way in which the pair first created Ballyturk actually become performative (in Judith Butler's sense) through repetition. The words transform into performance and thereby are made real, and the world 1 and 2 create is real because it exists in their bodies and lived memorial acts.

## Performance and the Creation of a World

Both *Ballyturk* and *The Walworth Farce* depict versions of performative world-building. The men trapped in the room who have ritualized their version of storytelling to forget their own names and create impersonations of an imagined group of townsfolk demonstrate the capacity for creating a world through performance. They literally and figuratively make the world each day, and the experience of living in that world is enough to satisfy them until the introduction of the reality beyond their four walls. For the characters in the flat on Walworth Road, they have made and re-made their memories so that the only correct way to live requires performance of the past. This belief changes their entire outlook so much that Dinny, Sean and Blake all refuse to leave the apartment – all three choose the safety of their performed world over the expectation of the unknown. The main object in these two plays relies more on the performed sense of world-making as a necessarily theatrical enterprise and practice. The everyday repetition for these characters mirrors that of the communal/cultural myth-making of the world at large. While the ritual storytelling may present an extreme or largely manipulated version of the outside world, the characters' daily re-membering of a world that never existed depicts a version of life that enables a connection to reality as a subjective experience. In this way, an imagined story can depict a stable version of personal truth to a community or individual.

Instead of worlds that are stable and remain unchanged, Enda Walsh depicts realms of constant motion and endless revision. In this way, Walsh's works depicts what Diana Taylor describes in her discussion of repertoire as a type of performed archiving. She argues that 'traditions are stored in the body, through various mnemonic methods, and transmitted 'live' in the here and now to a live audience. Forms handed down from the past are experienced as present'.[34] The micro-communities in *The Walworth Farce* and *Ballyturk* represent a heightened theatrical version of institutional control over cultural narratives. These types of power structures exhibit the possibility of world-making on an individual level as something contingent upon delusion. Those who (re)live tradition store the past for future re-memberings, but they also enact a commentary on the present. The performance of the past, in a theatrical sense, constantly refers to the present expectations, and this understanding manipulates the ways in which re-membering occurs. Therefore, performance in general relies on communal re-tellings and the revision that follows. Walsh's theatrical depiction of these theoretical underpinnings demonstrates the ways in which performed world-making can construct a manipulation of reality that can become real after enough repetitions. And without this understanding, theatre would be left to languish in the same interpretations and versions of plays that have ghosted the stage (to borrow Marvin Carlson's term) for centuries.

## Bibliography

Beckett, Samuel, 'Proust', in Paul Auster, ed., *Samuel Beckett, The Grove Centenary Edition: Poems, Short Fiction, Criticism*, 511–54. New York: Grove Press, 2006.

Butler, Judith, 'Critically Queer', *GLQ*, 1 (1993), 17–32.

——, 'Performative Acts and Gender Constitution: An Essay in Phenomenology and Feminist Theory', *Theatre Journal*, 40(4) (December 1998), 519–31.

Carlson, Marvin A., *The Haunted Stage: The Theatre as Memory Machine* (Ann Arbor, MI: University of Michigan Press, 2001).

Conway, Thomas, interview by Nelson Barre, (4 December 2013), transcript.

Frawley, Oona, *Memory Ireland: History and Modernity* (Syracuse, NY: Syracuse University Press, 2011).

34 Taylor, *Archive and the Repertoire* (Durham: Duke University Press, 2003), 24.

Jordan, Eamonn, '"Stuff From Back Home": Enda Walsh's *The Walworth Farce*', *Ilha do Desterro*, 58 (January/June 2010), 333–56.
Morash, Christopher, and Richards, Shaun, *Mapping Irish Theatre: Theories of Space and Place* (Cambridge: Cambridge University Press, 2013).
Pine, Emilie, *The Politics of Irish Memory: Performing Remembrance in Contemporary Irish Culture* (Basingstoke: Palgrave Macmillan, 2011).
Roach, Joseph, *Cities of the Dead: Circum-Atlantic Performance* (New York: Columbia University Press, 1996).
Schneider, Rebecca, *Performing Remains: Art and War in Times of Theatrical Reenactment* (London: Routledge, 2011).
Taylor, Diana, *The Archive and the Repertoire: Performing Cultural Memory in the Americas*, (Durham, NC: Duke University Press, 2003).
Walsh, Enda, *Ballyturk* (London: Nick Hern Books, 2014).
——, *The New Electric Ballroom/The Walworth Farce* (New York: Theatre Communications Group, 2009).

PART IV

# Theoretical Developments

MARGUÉRITE CORPORAAL, CHRISTOPHER CUSACK AND LINDSAY JANSSEN

# 11 From Restoration to Reinscription: The Great Famine in Irish North-American Fiction, 1847–1921

'The Emigrant Ship', an anonymous sketch that was republished in *The Literary Garland* in November 1850 and that had originally been written for a Protestant orphan bazaar raising funds to aid Irish children who had lost their parents on the transatlantic passage, emphasizes the severe despondency of the emigrants on deck who utter 'a wailing cry [...] so sad and plaintive' when they 'take a last look of their native land'.[1] Describing the fate of those Irish children who sought to escape imminent starvation and destitution during Ireland's Great Famine, the story, as a vehicle of transcultural memory, also bears witness to the transportability of recollection from one community to another, and raises questions like: how do recollections change when they are transported to other geographical and cultural spaces – for example through migration? Can we speak of a specific diasporic development of the cultural memories that are connected to the former homeland? Andreas Huyssen has argued in favour of a more thorough examination of 'the relationship between diasporic memory and the memory formations of the national culture within which a given diaspora may be embedded'.[2] More recently, Aleida Assmann and Sebastian Conrad have similary proposed a transnational approach towards investigating

1 'The Emigrant Ship', *The Literary Garland*, VIII/11 (1850), 508.

2 Andreas Huyssen, 'Diaspora and Nation: Migration into Other Pasts', *New German Critique*, 88 (2003), 151.

the dynamics of memory, as 'migrants carry their heritage, memories and traumas with them' which 'are transferred and brought into new social constellations and political contexts'.[3]

Our project *Relocated Remembrance: The Great Famine in Irish (Diaspora) Fiction, 1847–1921* (2010–2015) has engaged with the reconfigurations of the Famine across generations and national borders, thereby paying specific attention to the ways in which this fundamental episode from Ireland's past was remembered in the transatlantic host societies to which so many Irish emigrated. As our close examination of approximately 132 texts has revealed, during the earliest years, these diasporic recollections of the Great Hunger indeed developed in distinct ways, in response to the problematic integration of Famine emigrants in their American and Canadian host communities. Furthermore, our exploration of the recontextualizations of this diasporic famine memory across several decades serves to contest the long accepted notion that, in transgenerational transmission, homeland memories may gradually slip into the background, as ethnic attachments fade across generations and these memories are influenced by processes of intermarriage and assimilation.[4] Rather, the works of fiction that are central to our corpus show that one should rather speak of a growing 'multidirectionality' of Famine memory in North America:[5] the Famine past is more often placed in interaction with North-American cultural legacies and issues of American citizenship.

3 Aleida Assmann and Sebastian Conrad, 'Introduction', in Aleida Assmann and Sebastian Conrad, eds, *Memory in a Global Age: Discourses, Practices and Trajectories* (Basingstoke: Palgrave Macmillan, 2010), 2.

4 See, for example, Richard Alba, *Ethnic Identity: The Transformation of White America* (New Haven, CT: Yale University Press, 1990); and Joel Perlmann and Mary C. Waters, 'Intermarriage and Multiple Identities', in Mary C. Waters and Reed Ueda, eds, *The New Americans: A Handbook to Immigration Since 1965* (Cambridge, MA: Harvard University Press, 2007), 110–23.

5 Michael Rothberg, *Multidirectional Memory: Remembering the Holocaust in the Age of Decolonization* (Stanford, CA: Stanford University Press, 2009), 9.

## Displacing the Famine

While there are many similarities between the earliest Irish and Irish-North American works of fiction in the ways in which they represent the atrocities of the Great Hunger, there are also notable differences. Fiction written in the homeland often fully exposes the afflictions of Famine-related mortality, and ends with the main characters' death and the disintegration of family life. *A Tale of the Famine in 1846 and 1847* by 'Ireland' (1847) and 'The Black Potatoes' by Mrs Hoare (1851) follow similar narrative templates: the two mothers central to the story – Jude Mahoney and Honour McCarthy – are forced to live by the roadside, subsequently facing the starvation of their children before they succumb to hunger and disease themselves at the narrative conclusion. Moreover, in both cases the sense of tragedy is enhanced by the fact that help from the New World is untimely: in 'The Black Potatoes', Jude's brother-in-law James, who had made his fortune in America, returns too late to prevent Jude's death. Upon her death, James tries to console her remaining two infants Mary and Johnny, 'reconciling them to go with him' and engaging 'a passage for himself and them in a packet bound to America'.[6] In *A Tale of the Irish Famine in 1846 and 1847*, McCarthy speeds back from America to Ireland upon hearing about the Famine miseries, hoping to take his family 'to his adopted country [...] where he had made every provision for Honour's comfort'. However, to his 'bitter anguish' and 'deep disappointment', he arrives too late to save his wife and most of his infants. As in 'The Black Potatoes', Ireland fails to offer a future home: McCarthy and his son Mike, who has become 'sightless and speechless' due to malnutrition, leave for America together.[7] This recurring 'schematic narrative template' of personal loss and an irredeemable Ireland does not feature in early examples of North-American Famine

6 Mrs Hoare, 'The Black Potatoes', in *Shamrock Leaves: or Tales and Sketches from Ireland* (Dublin: J. M'Glashan, 1851), 385.

7 'Ireland', *A Tale of the Irish Famine in 1846 and 1847; Founded On Fact* (Reigate: William Allingham, 1847), 34–5.

fiction, often written by and for Irish immigrants of the Famine generation.[8] By contrast, these North-American Famine narratives frequently displace the harrowing recollections of hunger and bereavement from the narrative by inserting a sudden turn to a happy narrative conclusion that suggests the recuperation of an idyllic, pre-Famine Ireland. For example, Bessy Conway, the eponymous heroine of a novel by the well-known Irish North-American immigrant author Mary Anne Sadlier, returns from New York in time to prevent the brutal eviction of her family and to supply lavish food for her starving siblings. As a result, the family no longer has to sup on 'water and nettles, with a handful or so of oatmeal', and her dying sister Ellen, on whom 'the terrible fangs of hunger had fastened' is revivified as if by magic: 'and the ghastly paleness of the sweet features was tinted with a more lifelike hue'.[9]

The death of the main protagonist's only daughter, Kate, is likewise averted in Robert Curtis's *McCormack's Grudge*, which, though written by a county inspector in Ireland, was first serialized in *The Irish-American* in September 1862. Ironically, the fact that the girl was sent to the poorhouse after the family's eviction saves her life, for once there, she is sent off to hospital: 'Had the M'Cormacks not been turned out, Kate, without medical aid or warmth, would have pined away and died'. It is moreover Myles' heroic rescue of the land agent's son Harry from a raging bull, which almost costs him his own life, that forms the onset to further happiness: Myles receives back his land, and at the narrative conclusion he is a prosperous farmer. Furthermore, the 'sudden and destructive blight which the will of an inscrutable Providence cast upon the potato crops of Ireland in 1846' and which has resulted in an 'untenanted and comparatively waste' field, is transformed into a regenerated Ireland from which every scar of the Famine years appears to have disappeared: 'the potatoes were "getting the better" of the disease; and from this very period every year was better

8 James V. Wertsch, *Voices of Collective Remembering* (New York: Cambridge University Press, 2002), 57.

9 Mrs J. Sadlier, *Bessy Conway; or, the Irish Girl in America* (New York: D. & J. Sadlier and Co., 1862), 278.

for the farmer than the previous one. Corn, hay, potatoes, every article of produce began to look up, and the farmers began to look up after them'.[10]

In North-American Famine fiction written before the late 1860s, the Famine clearly represents an uncomfortable memory that is addressed but that also has to be 'resolved' by the return of an imagined, pre-Famine (if not almost prelapsarian) Ireland that was deeply rooted in Ireland's cultural memory, as becoms clear from James Arthur O'Connor's landscape paintings from the 1820s as well as by the poem *Mount Leinster* (1819), which casts the typical Irish landscape in pastoral terms of 'undulating fields' where the 'young potato-stems' display their green heads and the pleasant sight of 'white-washed cottages'.[11] This specific reconfiguration of the Famine past, which was at odds with post-Famine conditions in Ireland, can be clarified by the widely known problematic position of the Catholic Famine emigrants who encountered ferocious ethnic discrimination.[12] The broadside *What Irish Boys Can Do; an Answer to 'No Irish Need Apply'* (1864) takes issue with the prejudice towards Irish immigrants by American employers, who refuse to employ those who come from the Emerald Isle and continuously insult them: 'They'll call him green, an Irish bull: it happens every day'.[13]

As Andreas Huyssen claims, diasporic communities which constitute a minority culture with a 'tenuous and often threatened status within the majority culture' and are subject to 'stereotyping of otherness combined with [...] exclusionary mechanisms' therefore tend to 'create a unified or even mythic memory of the lost homeland' as a strategy to cope with present identity crises.[14] The outsider status of Irish emigrants of the Famine

10 Robert Curtis, *McCormack's Grudge*, in *The Irish American* (27 September 1862).

11 *Mount Leinster; a Poem, Descriptive of Irish Scenery* (London: Longman, Hurst, Ree, Orme and Browne, 1819), 8, 22.

12 See, amongst others, Kevin Kenny, *The American Irish: A History* (London: Longman, 2000), 115–16; David Roediger, *The Wages of Whiteness: Race and the Making of the American Working Class*, rev. edn (London; New York: Verso Books, 1999); Noel Ignatiev, *How the Irish Became White* (London: Routledge, 1995).

13 *What Irish Boys Can Do; an Answer to 'No Irish Need Apply'* (New York: H. De Marsan, 1864).

14 Huyssen, 'Diaspora and Nation', 149–50.

generation in British Canada and America may therefore have fuelled the cultural construction of an idealized version of Ireland from which all recent trauma of pestilence had to be dissociated. The emphasis on reappropriating the ancestral land among Irish-American communities – well illustrated by a nationalist article from 12 November 1864 in *The Irish People* whose objective is to 'sweep aristocracy away and give the land to the people' – is manifested in these Famine novels as an attempt to reclaim the Irish land from the painful past.[15] In fact, a close examination of cultural expressions produced in and for Irish transatlantic migrants of the Famine reveals that a nostalgic yearning for a distant Edenic Irish landscape was central to the early stages of this generation's emerging diasporic consciousness.[16] For example, the poem 'The Irish Emigrant's Lament', which was published in *The Pilot* of 23 January 1858 and which describes the heart-breaking departure of an Irishman from his native shore and his love Mary, celebrates the Irish scenery in May when the 'corn is green' and when 'the lark's loud song' can be heard.[17] Similarly, David O'Callohan's 'Will He Return' (1869), published in *The Emerald*, express a heart-felt longing to travel back to 'old Ireland' with its 'simple village church with its open campanile' and the 'quaint old cottage'.[18]

15 'The Irish Americans', *The Irish People* (12 November 1864).

16 It must be noted that a similar contrast between a pastoral Ireland and the urban climate of the host country can be found in Miss Mason's Irish-British novel *Kate Gearey; or Irish Life in London; A Tale of 1849* (London: Charles Dolman, 1853). Here Kate's childhood in a mud-cabin 'on the fair banks of the Awbeg, amidst the fertile valleys of Castletown Roche', while 'the blue sky of Ireland was above her, its soft green turf beneath her feet, its pure air around her' forms a stark contrast with Kate's habitation in a London tenement house: 'a square hole looking into a damp filthy passage'. See 9, 3. This not only shows that the living conditions of Irish Famine immigrants in Britain and America were similarly bleak, but also illustrates Andreas Huyssen's earlier quoted point about the idealization of the homeland in diaspora cultures.

17 'The Irish Emigrant's Lament', *The Pilot* (23 January 1858).

18 David O'Callohan, 'Will He Return', *The Emerald* (2 October 1869).

## Relocating Ireland

North-American Famine narratives which envisage a return of a pristine, Edenic Ireland become much rarer towards 1870. Instead, many works of Famine fiction place the Famine past at a comfortable distance from the New World, while reimagining new, pastoral Irelands at the North-American frontier. Dillon O'Brien's *The Dalys of Dalystown* (1866) can be regarded as an important transition text. On the one hand, the narrative suggests that the Irish rural community can be restored by the return of the wealthy Irish emigrant, just as Sadlier's *Bessy Conway* does: Henry Daly, the descendant of the former owners, an ancient Catholic family 'more Irish than the Irish themselves', manages to chase away the painful memories of famine and poverty and to 'wall round my property with the smiling cottages of as tenantry whose interests and mine shall be as one' when he reacquires his ancestral home through money made at the American frontier.[19] As the present owner, the Protestant O'Roarke, has squandered away the estate that once belonged to Henry's family during the Famine, and the property is up for sale in the Encumbered Estates Court, Henry can make sure that the house and grounds will not pass into English hands by the wealth he has gained through hard agricultural labour. Conforming to the North-American ideal of the self-made man who 'had determined to go to work, just as if he had no capital to rely upon' and bringing with him American experiences of equality and land management, Henry is even better suited to the task of reforming and recuperating the estate.[20] Similar to *Bessy Conway*, *The Dalys of Dalystown* imagines a democratization of the landowning class that appears to be typical for early North-American Famine fiction.

In many works of Famine fiction published in Ireland or England during the 1860s and 1870s, the narrative conclusion often underscores feudalism, as the negligent first generation of a landowning family is replaced

19 Dillon O'Brien, *The Dalys of Dalystown* (St Paul, MN: Pioneer Printing, 1866), 26, 526.

20 Ibid., 512, 524.

by the second generation, which is set upon introducing some agricultural reform without drastically changing the class system. For instance, in Reginald Tierney's *The Struggles of Dick Massey* (1860), the estate passes into the hands of the eldest son Dick, who drastically cuts back on the family's budget, and improves and reforms the farms. With the assistance of his betrothed Clara, Dick eventually succeeds in bettering the conditions of the tenants in Kilfarney '[i]n spite of sickness and famine'.[21] Likewise, in Annie Keary's *Castle Daly* (1875), the Irish Catholic landlord's daughter Ellen Daly becomes the new landlady of the estate after the Famine, and manages Good People's Hollow together with her kind-hearted English husband, sticking to 'the spirit of the old traditions' in the hope that changed circumstances and the lapse of time 'may bring alternations'.[22] By contrast, in *Bessy Conway*, the heroine eventually marries the converted English landlord's son, Henry Herbert, and thus becomes part of the landowning class herself. While O'Brien's Henry Daly descends from a genteel family, he first has to undergo the fate of the labouring farmer in the American West before he is reinstated into the inherited position: he has to become 'New Money' and move upwards on the social ladder before he can reclaim possession of the ancestral home and lands.

At the same time, *The Dalys of Dalystown* offers an alternative to this stock template of return, suggesting new Irelands on North-American soil that are not afflicted by blight or marred by class tensions. The novel thus geographically displaces the Famine from the narrative centre. After Henry, the son of a ruined landlord, has been in America for twenty years and has successfully settled in the Michigan territory with his relatives, he receives a letter from Ireland informing him of the dire food crisis and the Young Ireland rebellion: 'I have just had news from Ireland. It is as I expected; those few brave spirits, scorning life, whilst gaunt famine slays their countrymen by tens of thousands, have taken the field'.[23] While the narrative

21 Reginald Tierney, *The Struggles of Dick Massey; or, the Battles of a Boy* (Dublin: James Duffy, 1860), 400.

22 Annie Keary, *Caste Daly; The Story of an Irish Home Thirty Years Ago* (London: Macmillan, 1886), 574.

23 O'Brien, *The Dalys of Dalystown*, 498.

openly addresses the Great Hunger, it also throws up a barrier between the traumatic events and the scope of narrative consciousness, for the events take place in a world that is distinct and distant from the American West which is 'dotted with thriving settlements, growing cities, and cultivated fields' and provides opportunities for 'freedom' and social advancement.[24]

O'Brien's novel moreover implies that Irish immigrants who move into frontier areas can find and model landscapes that remind them of their motherland. When Henry Daly, his sister Emily and his spouse Rose set up a new home in Michigan, they plan and lay out a flower garden to make it 'resemble, as nearly as possible, the one it was intended to represent', namely the gardens they used to have at the Dalystown estate, even though 'to have it of the same extent was out of the question, for many a day to come'.[25] While the Daly family consciously seeks to recreate their former Irish garden on American soil, the narrator creates the impression that the landscapes of the American West are naturally imbued with traits which are inherent to the Irish scene: in the narrator's opinion, the fields of the West have always been 'the green-carpeted ballrooms of the fairies', for '[s]tretched, with closed eyes, in one of those grassy dells, on a dreamy summers evening, the little people have danced around me'.[26] The narrator thus enscripts the frontier landscape with Irish mythology, thereby representing it as a natural habitat for the immigrants from the emerald isle.[27]

This sense of a distanced Famine past and relocated, pastoral homelands is strongly present in North-American fiction of the following decades. Father Hugh Quigley's *Profit and Loss* (1873) focuses on the Mulrooneys who leave Ireland for America during the later years of the Famine. The horrors of the Famine years are not incorporated into the plotline, but

24 Ibid., 502.

25 Ibid., 484.

26 Ibid., 468.

27 See also Marguérite Corporaal, '"The recollection of sorrow and misery past": Memories of the Great Famine in Irish North-American Fiction, 1855–1870', in Christine Kinealy, Patrick Fitzgerald and Gerard Moran, eds, *Irish Hunger & Emigration: Myth, Memory and Memorialization* (Hamden, CT: Quinnipiac University Press, 2015). 83–95.

resurface in the song 'An Exile's Farewell to his Native Land', supposedly written by a compatriot in diaspora, Mr Haley. The verses engage with Ireland's oppression by England, and explicitly allude to the Famine as an era which exemplifies this 'Saxon' tyranny:

> Whole villages in ruins left,
> The only record keep,
> Of myriads of human being
> Displaced by flocks of sheep.
> Two million mouths, by famine shut,
> Have ceased to cry for bread,
> While landlords, like hyenas,
> Dig up the murdered dead,
> With blasphemous assurance.

This clearly Mitchelite interpretation of British governmental policies during the Famine[28] emphasizes the radical transformations in Irish agriculture at the cost of human labourers. The transition from tillage to cattle grazing – a measure adopted by many debt-ridden landlords both prior to and during the Famine in order to make the soil profitable when they would receive no income from agricultural cultivation –[29] leads to the eviction of the tenantry who can no longer dwell on or live from the land. While Haley's lines bring the painful recollections of the hunger years to the New World, testifying to the transportablity of memory, the narrative also clearly demarcates these atrocities as belonging to a distant land. The onset of the poem states that the speaker would 'sooner as an exile seek' homes elsewhere than 'come back again to view the woes/Of this most oppress'd

28 John Mitchel famously wrote in his *Jail Journal: Or Five Years in British Prisons* (1854) that 'four years of British policy, with the Famine to aid' resulted in 'killing fully two millions', as Parliamentary politics encouraged 'the extermination, that is, the slaughter' of tenantry on the part of the landlords. See *Jail Journal; Or Five Years in British Prisons* (New York: 'The Citizen', 1854), 17–18.

29 David S. Jones, 'The Transfer of Land and the Emergence of the Graziers during the Famine Period', in Arthur Gribben, ed., *The Great Famine and the Irish Diaspora in America* (Boston, MA: University of Massachusetts Press, 1999), 93.

of lands'.[30] Even if emigration entails exile, it is preferable to resettlement in the motherland; even more so, as Quigley's novel implies, because the frontier offers the opportunity to create microcosmic bucolic Irelands. The land farmed by the emigrant family is cast in characteristically Irish tones, as Mulrooney seeks to create

> a place that would bear some likeness, in imagination at least, to the old estate on the banks of the Blackwater in Ireland called 'Cappa,' out of which his forefathers were expelled by the ruthless hands of confiscation and war. The hills, the groves, the lake, as the St Croix is called, and the rivulet or creek, all carried his mind back to the memory of the past on the banks of the 'Avon Duv'.[31]

Indeed, Mulrooney is successful in bringing back his ideal Ireland, as 'the beauties of the surrounding sceneries' and 'the richness of the soil, which was equal to the "Golden Vale" in Ireland' remind him of home.[32] Likewise, '[Y]oung Mulcahy [...] who was out from the old sod but two years' remarks that 'these hillocks remind one of the fairy hills or mounds which we have in Ireland'.[33] It must be noted, however, that despite these opportunities for reconstructing an Irish home at the frontier, both father and son Mulrooney do not find happiness, basically because they are lured by the temptations of the New World: the elder Mulrooney dies as a result of excessive alcohol consumption, and his son loses his fortune by mingling with Methodists. These North-American Famine novels from a later period do not propose a return to a recuperated Ireland, then, but instead endow the new country of settlement with what Svetlana Boym calls a 'restorative nostalgia' which 'puts emphasis on *nostos*', or the possibility to 'rebuild the lost home'.[34] Furthermore, as these examples illustrate, North-American Irish Famine fiction bears witness to an emerging 'multidirectional memory': legacies of the homeland are transported to a North-American setting and intersect

30 Father Hugh Quigley, *Profit and Loss* (New York: T. Kane, 1873), 284–5.
31 Ibid., 29–30.
32 Ibid., 31.
33 Ibid., 279.
34 Svetlana Boym, *The Future of Nostalgia* (New York: Basic Books, 2001), 41.

with (and in turn even may have impacted) North American myths of the self-made pioneer, which also imagine a soil of plenty.[35]

This shift can be explained by the fact that the ongoing problems between landowners and tenants in the late 1860s and early 1870s and the threats of future famines thwarted any hopes of a peaceful, idyllic homeland.[36] As, for example, a series of sketches republished from *The Illustrated News* of 1870 suggests, in the 'spring of 1870' the farming population in the Kildare countryside was still weighed down by 'the wholesale eviction of forty-two families, numbering 152 individuals in all' and extreme poverty, as '[a] more starving, ragged, ill-housed community than the occupants of the wretched mud-cabins that lined one side of one of the principal streets in Kildare, was hardly possible to conceive'. Around the same time, 'the Relief Committee' in Co. Sligo supplied starving women with 'meal'.[37]

At the same time, the position of the Catholic Irish in North America changed around 1870: groups of Irish immigrants of the Famine generation

35 Michael Rothberg, *Multidirectional Memory: Remembering the Holocaust in the Age of Decolonization* (Stanford, CA: Stanford University Press, 2009), 9. For these representations of a fruitful soil which enables the establishment of prosperity, see, for example, William B. Dewees, *Letters from an Early Settler of Texas* (1854), which recurrently emphasizes that the land in Texas is 'very rich and fertile'; and Henry T. Newton, *Canada in 1864: A Hand-Book for Settlers* (1864) contends that in the West 'the variety of its products and abundance of its harvests attest, where proper care is bestowed on its cultivation, the excellence of its soil'. See William B. Dewees, *Letters from an Early Settler of Texas* (Louisville, KY: Hull and Brother, 1854), 218; and Henry T. Newton, *Canada in 1864: A Hand-Book for Settlers* (London: Sampson Low, 1864), 11.

36 In 1896, William Patrick O'Brien wrote that the potato failure of the late 1870s 'made people realize that there were still to be found, in certain districts of the country, a congested population always hovering, more or less, on the brink of destitution'; a population which was 'almost as liable as of old to be reduced to a state of absolute famine', although there had been 'many previous warnings' in the form of earlier famines. See W. P. O'Brien, *The Great Famine in Ireland and a Retrospect of the Fifty Years 1845–95, with Sketches of the Present Condition and Future Prospects of the Congested Districts* (London: Downey & Co., 1896), 276.

37 *Irish Pictures: Eighty Sketches taken on the Spot by F. Dadd, M. Fitzgerald, Harry Furiss, Wallis Mackay and Procter and R. C. Woodville. Republished from the Illustrated London News* (London: Vizetelly and Co., 1870), 1, 21.

who had initially setttled in urban areas such as Boston and New York gradually began to move to the Western interiors of America and Canada.[38] As such, the configuration of new Irelands at the frontier rather than in the old country in these works of fiction appears to mirror the changing living circumstances of many Irish-Americans and Irish-Canadians who moved westwards. The fact that some of these novels begin to conceptualize idyllic Irish communities in the New World can thus be attributed to the progressive integration of Irish immigrant communities. Not only did influential spokespersons for the Irish transatlantic diaspora, such as Thomas D'Arcy McGee, urge 'the Irish man in America to modify his manners [...] and to fit his habits as far as propriety demands to his new relations';[39] the end of the American Civil War also saw the gradual emergence of an Irish-American middle class.[40] These developments in the social conditions of Irish North-Americans are therefore mirrored by the changing approaches towards both homeland as well as host country in Irish North-American Famine fiction of the late 1860s, the 1870s, and early 1880s.

That it is indeed legitimate to analyse this development in Famine memory in North-American fiction as a response to social mobility and assimilation becomes clear when one examines the constellations of Famine memory in fiction written during subsequent decades. In such texts, we see a move away from the idea that, in response to the Famine past, Irish emigrants may successfully create pre-Famine little 'Irelands' elsewhere: in the Irish-Canadian Famine novel *The Days of a Life* (1883) by Margaret

38 David Emmons, *Beyond the American Pale: The Irish in the West, 1845–1910* (Norman, OK: University of Oklahoma Press, 2010), 212, 214.

39 Thomas D'Arcy McGee, 'The Social Duties of Irishmen in America', *The American Celt* (23 July 1853).

40 See Patricia Kelleher, 'Young Irish Workers: Class Implications of Men's and Women's Experiences in Gilded Age Chicago', in Kevin Kenny, ed., *New Directions in Irish American History* (Madison, WI: University of Wisconsin Press, 2003), 191. See also Timothy J. Meagher, *Inventing Irish America: Generation, Class, and Ethnic Identity in a New England City, 1880–1920* (Notre Dame, IN: University of Notre Dame Press, 2001), 134. Kerby Miller has argued that this process of an emerging Irish-American middle class was completed by 1895. See *Ireland and Irish America: Culture, Class, and Transatlantic Migration* (Dublin: Field Day, 2008), 299.

McDougall Dixon, for example, Mrs Mulrooney and her sons soon 'sowed and reaped on a snug little farm of their own'. The land that they cultivate fails to materialize as a New Ireland, however. The 'little whin bush' that Mrs Mulrooney had taken with her 'in a flower pot [...] as a reminder of the Donegal hills' is ill-fated in the country of adoption, for 'that prickly emigrant refused to flourish under the cloudless skies of Canada, and died of stove-heat the first winter'.[41] Eventually, her son Pat even gives up a farmer's life altogether, establishing himself as a shopkeeper instead and thereby moving away from the traditional Irish rural ideal.[42]

## The Famine Past as a Tool of Social Integration

In Irish-American Famine fiction of later decades, the Famine past is moreover increasingly represented as a shared memory that overcomes social division and promotes acceptance and assimilation. John Talbot Smith's story 'The Deacon of Lynn' (1891) shows how Deacon Lounsbury unsuccessfully tries to prevent the settlement of 'these God-forsaken Irish' workers and the establishment of a Catholic church in his Connecticut small town.[43] Ironically, the Deacon recalls that he had donated 'corn meal in '48' to the Irish, wondering 'who could refuse' to help the starving. This memory seems to facilitate Lounsbury's eventual reluctant acceptance of the Irish Catholic labourers. A Catholic church is built, and the deacon is even persuaded to attend Mass. While '[h]e had fought against the [cross]

41 'Norah' (Margaret Dixon McDougall), *The Days of a Life* (Almonte: W. Templeman, 1883), 208–9.

42 See also Lindsay Janssen, 'The Impossibility of Transporting Identity: The Representation of Diasporic Irishness in Transatlantic Irish Fiction, 1860–1900', in Eleftheria Arapoglou, Mónika Fodor and Jopi Nyman, eds, *Mobile Narratives: Travel, Migration, and Transculturation* (New York: Routledge, 2014), 162–73.

43 John Talbot Smith, 'The Deacon of Lynn', in *His Honor The Major. And Other Tales* (New York: Vatican Library, 1891), 4.

for what it represented to his mind – idolatry', in the end Lounsbury gazes upon the newly built church and realizes that he 'had failed in the fight'. He goes 'home to struggle no more', suggesting that even the staunchest opponent can be converted to Catholicism.[44] In another story by Talbot Smith, 'How the McGuinness Saved His Pride' (1891), long-time resident and French-Canadian labourer Jean-Baptise Nolin is ousted from the Irish-American town of Sundsbury when he marries Irish girl Anne, and thereby blemishes her family's respectable identity. However, Nolin is reaccepted into the community of emigrated Galway Irish when the townspeople find out that his mother in fact was Mary Cassidy, an Irish emigrant girl 'whose parents died of the fever' in the Grosse Île fever sheds, and who was subsequently adopted by a French Canadian family.[45] After hearing this news, Mrs McGuinness, Anne's mother, immediately exclaims: 'may the sowl o' Mary Cassidy forgive us that we iver trated her son so'.[46] The Famine legacy of Jean-Baptiste's mother thus serves as a bridge that transcends ethnic difference and cultural conflict.

Furthermore, in fiction from the late nineteenth and early twentieth century, the Famine becomes a rhetorical tool to emphasize the desirability of American citizenship and assimilation, rather than a unique past that is a marker of ethnic separatism. In Mary Synon's story 'My Grandmother and Myself' (1916), published in *Scribner's Magazine*, an elderly woman who escaped from Famine-stricken Ireland chastises her son John Stutton for his willingness to side with Germany in the First World War, as their victory would spell the end of the British Empire and would lead to Germany's liberation of Ireland from the English. The grandmother points out that his plots are a betrayal to the country of his birth, which has enabled him

44 Ibid., 31.

45 Many Irish orphans were adopted by French-Canadian families. See for example Jason King, 'Remembering the Famine Orphans: The Transmission of Famine Memory between Ireland and Quebec', in Christian Noack, Lindsay Janssen and Vincent Comerford, eds, *Holodomor and Gorta Mór: Histories, Memories and Representations of Famine in Ukraine and Ireland* (London: Anthem Press, 2012), 115–44.

46 John Talbot Smith, 'How the McGuinness Saved his Pride', in *His Honor The Major. And Other Tales* (New York: Vatican Library, 1891), 255.

to become an independent shop owner rather than a landless labourer; and to the country that offered the Irish a chance of survival at all: ''Tis not England you fight with your plots [...] 'Tis America you strike at when you strike here. And, as long as you stay here, be Americans and not traitors'.[47] As she states in reply to his yearning that he had been born in Ireland: 'And if I'd stayed in Ireland I'd have starved [...] and little chance you'd have had of being born anywhere'.[48] In this tale, the Famine past is evoked to address the importance of being a loyal American citizen, who is grateful for the chances that the country has given to the Famine-stricken Irish that fled to its shores. As the grandmother relates to her grandson Shauneen, America not only provided the family with relief after her father and sister Brigid had died of starvation, bringing food to the Connaught shores; but also gave her the opportunity to 'come here with the promise of work to do'. Learning there to save money, the grandmother could bring her siblings to the New World as well, where they were forever free from 'famine and persecution' by landlords.[49] As such, the narrative conjures up the Famine past to endorse the image of America as a land of freedom and second chances.

John Brennan's novel *Erin Mor* (1892) situates the Famine past in the present American political agenda.[50] As its title suggests, *Erin Mor* expounds an intricate – if not exactly sophisticated – transatlantic synthesis of Irish nationalism and American politics, with continuous reference to British trade policies and imperialism. This syncretic political ideology is developed against the background of the establishment of 'Erin Mor', or Greater Ireland, which in the novel both refers to the Irish diaspora in general and to the neo-Irish community which the protagonist Andy Dillon establishes in post-Gold Rush California. *Erin Mor* concentrates on the Irish-American community and ends with the pastoral idyll of a new 'Greater Ireland' in California, and the message it promulgates is that the

47 Mary Synon, 'My Grandmother and Myself', *Scribner's Magazine* (August 1916), 228.

48 Ibid. 224.

49 Ibid. 231.

50 A version of this argument has been published in Christopher Cusack and Lindsay Janssen, 'Famine, Home, and Transatlantic Politics in Two Late Nineteenth-Century Irish-American Novels', *Atlantic Studies*, 11(3) (2014), 403–18.

only way to be a good Irishman in the United States is by assimilating and moving beyond the faction interests of the Irish-American community.

The novel's moral centre, Andy Dillon, leaves Ireland with his parents during the Famine and grows up in an East coast tenement slum. After having led an itinerant life for several years, he ultimately discovers gold in California in a symbolic plot twist: his garden dam breaks, washing away his potato patch – which in Ireland would spell disaster, but in California lays bare a large amount of gold. As such, the novel also moves away from the idea that microcosmic Irelands can be reproduced in the New World, just like McDougall's aforementioned narrative. His newfound fortune enables Dillon to establish a utopian community in California for 'his exiled race in their trans-Atlantic home',[51] which he dubs 'Erin Mor'. While this plot development creates the impression that the novel underscores ethnic separatism, this is far from the case. Brennan states in his preface that he wrote the novel to stimulate political assimilation and influence the Irish-American vote in the 1892 presidential election: he has 'the earnest desire of creating in the minds of his Celtic brethren a deep, intense and fervid American National spirit'.[52] In fact, the preface suggests that ethnic legacies such as the Famine are instrumental to the development of an engaged American citizenship. The American Irish should be conscious of their Famine past, for this will help them making the right choice in elections: their inherited awareness of the damaging effects of political liberalism during the Famine should inevitably enlist them for the Republican party, as Brennan states in the preface:

> The writer of the following pages candidly believes that while foreign rule has been the crowning and all-comprehending curse of Ireland, and while landlordism has been among its most potent ills, yet *British free-trade*, identical with the *tariff reform* which the Democratic party now seeks to inflict upon the United States, was the underlying, if not the immediate, cause of the terrible famine of 1846–47.[53]

51 John Brennan, *Erin Mor; The Story of Irish Republicanism* (San Francisco: P. M. Diers and Co., 1892), 126.

52 Ibid., n.p.

53 Ibid., v. Interestingly, this is an almost verbatim reproduction from the 1888 pamphlet on the undesirability of Irish-American support for the free trade movement which

Interestingly, the novel appears to implicitly critique the earliest North-American tradition of Famine fiction, which displaced Famine suffering by representations of an idyllic, recuperated Ireland. Barney Devoy and his family were on the same coffin ship as the Dillon family, escaping the Famine. However, shortly upon his arrival, Devoy starts denying that the Famine happened, so as to gain support for his political ambitions by downplaying the impact of free trade on Ireland: 'Famine? Famine? Famine? Sure there never was a famine in Ireland. It's full and plenty they had at home, lashins and lavins galore, wid the finest English broadcloth and Scottish tweeds to wear'.[54] According to Devoy's description, Ireland is like a Land of Cockayne, and has never been smitten by famine. In fact, he and his supporters 'stood ready to prove that the country roads were paved with four-pound loaves, and that three millions of the race had emigrated just for a change of air and relief from a diet of roast beef, turkeys and terrapin'.[55]

This repression of recollections of the Famine impedes responsible American citizenship, for if the Democratic Party have their way, 'the same calamity' might strike in the United States as well, Brennan implies.[56] Furthermore, the novel hints at a vision of ethnic assimilation in a broader American context: while Dillon seeks to create a miniature Ireland in anticipation of 'the brightening vista [of] a dismembered British Empire and an independent and united Ireland',[57] Greater Ireland is very much an inclusive Californian Dream that he shares with a black ex-slave, Tony Sexton, who becomes one of his closest friends, and with his Native American neighbours. Overall, then, Famine legacies are elicited in a broader context of

---

was co-written by Brennan. See *Why Irishmen Should Be Protectionists* (New York: American Protective Tariff League, 1888), 4.

54 Ibid., 32. Devoy's remark is interesting, because its emphasis on British produce – English broadcloth and Scottish tweeds – suggests that free trade did not actually affect Ireland. Thus, Brennan suggests that Devoy is lying in two respects, as the focus on non-Irish produce suggests that native industries cannot compete.

55 Ibid., 32–3.

56 Ibid., vi.

57 Ibid., 263.

American politics, thereby suggesting a strong integration of the Irish and their cultural memories in the United States.[58]

The fact that, by this time, the Famine past came to function as a more broadly shared American legacy than as a past that was revisited by those identifying with an Irish ethnicity, is moreover underscored by the frequency with which the Famine started to become an often discussed subject in periodicals that catered for a general American audience, such as *Harper's Weekly*, *Scribner's Magazine*, *Century*, and *McClure's Magazine*. This included Famine fiction, such as Octave Thanet's 'Stories of a Western Town: III – Tommy and Thomas', published in *Scribner's Magazine* in October 1892; as well as stories such as Kate McPhelim Cleary's 'The Mission of Kitty Malone' (*McClure's*, 1901) that strongly feature Famine tropes.[59] By contrast, in earlier decades, the Famine years had mainly been remembered in ethnically specific newspapers and magazines such as *The Pilot*, *The Irish American* and *The American Celt*, which had serialized Famine novels such as Alice Nolan's *The Byrnes of Glengoulah* (*The Irish Citizen*, 1869), *Rebel Scenes* (*The Irish American*, 1851) and short stories such 'Thade M'Sweeney; Or, A Tenant Farmer's Trials. A Story of the Great Famine' (*Irish American*, 1880).

## Conclusion

Recently, Lucy Bond and Jessica Rapson have sought to define the notion of transcultural memory as 'the travelling of memory *within* and *between* national, ethnic and religious collectives'.[60] Our analyses of fiction by and

58 See Christopher Cusack, 'Famine Memory and Diasporic Identity in US Periodical Fiction, 1891–1918', *Symbiosis: A Journal of Transatlantic Literary and Cultural Relations*, 19(2) (2015), 153–69.

59 Ibid., 168–9.

60 Lucy Bond and Jessica Rapson, 'Introduction', in Lucy Bond and Jessica Rapson, eds, *The Transcultural Turn: Interrogating Memory Between and Beyond Borders* (Berlin: De Gruyter, 2014), 19.

for North-American Irish has shown that recollections of the Great Famine indeed travelled within transcultural ethnoscapes of Irish emigrants and their descendants in the motherland and their host communities, but, in due course, also became increasingly embedded in North-American cultural contexts and legacies. The specific development of Famine legacies in North-American fiction demands a reassessment of Jason Byrne's recent allegation that 'there is 'virtually no significant engagement in nineteenth-century Irish-American literature with the personal and communal sufferings caused by the Famine',[61] as well as Mary C. Kelly's claim that Famine memory became increasingly repressed in the course of the century, as a shameful episode that could 'undermine Irish-American progress'.[62] In fact, as our research demonstrates, the Famine remained an important element both of Irish-American self-imaginations and of American explorations of Irish-American diaspora culture in periodical fiction. An important reason for the conviction that the Famine's memory became entirely sublimated is the sheer amount of under-researched material.

Furthermore, the novels that we examined in the context of our project indicate that as cultural groups are involved in processes of assimilation, their memories become more multidirectional as well as engrained within mainstream rhetoric and cultural heritage. Therefore, instead of speaking of an ethnic fade,[63] one should read any reconfigurations of a homeland memory in diasporic cultural traditions rather as specific responses to intersections with other cultural communities and their legacies. The trajectories of Famine memory in Irish North-American fiction have often been interpreted in connection to the negotiation of trauma; as this paper has shown, this is but one side of the coin, for one could also convincingly argue that in these texts, Famine horrors are strategically employed in processes of displaced identity construction.

61 Jason P. Byrne, 'Cultural Memory, Identity, and Irish-American Nostalgia', in Oona Frawley, ed., *Memory Ireland, Volume 2: Diaspora and Memory Practices* (Syracuse, NY: Syracuse University Press, 2012), 51.

62 Mary C. Kelly, *Ireland's Great Famine in Irish-American History: Enshrining a Fateful Memory* (Lanham, MD: Rowman and Littlefield, 2013), xviii.

63 Byrne, 'Cultural Memory, Identity, and Irish-American Nostalgia', 51.

# Bibliography


Alba, Richard, *Ethnic Identity: The Transformation of White America* (New Haven, CT: Yale University Press, 1990).

Assmann, Aleida and Sebastian Conrad, 'Introduction', in Aleida Assmann and Sebastian Conrad, eds, *Memory in a Global Age: Discourses, Practices and Trajectories* (Basingstoke: Palgrave Macmillan, 2010), 2.

Bond, Lucy and Jessica Rapson, 'Introduction', in Lucy Bond and Jessica Rapson, eds, *The Transcultural Turn: Interrogating Memory Between and Beyond Borders* (Berlin: De Gruyter, 2014).

Boym, Svetlana, *The Future of Nostalgia* (New York: Basic Books, 2001).

Brennan, John, *Erin Mor; The Story of Irish Republicanism* (San Francisco, CA: P. M. Diers and Co., 1892).

Byrne, Jason P., 'Cultural Memory, Identity, and Irish-American Nostalgia', in Oona Frawley, ed., *Memory Ireland, Volume 2: Diaspora and Memory Practices* (Syracuse, NY: Syracuse University Press, 2012), 51.

Corporaal, Marguérite, '"The recollection of sorrow and misery past": Memories of the Great Famine in Irish North-American Fiction, 1855–1870', in Christine Kinealy, Patrick Fitzgerald and Gerard Moran, eds, *Irish Hunger & Emigration: Myth, Memory and Memorialization* (Hamden, CT: Quinnipiac University Press, 2015), 83–95.

Curtis, Robert, 'McCormack's Grudge', *The Irish American* (27 September 1862).

Cusack, Christopher, 'Famine Memory and Diasporic Identity in US Periodical Fiction, 1891–1918', *Symbiosis: A Journal of Transatlantic Literary and Cultural Relations*, 19(2) (2015), 153–69.

——, and Lindsay Janssen, 'Famine, Home, and Transatlantic Politics in Two Late Nineteenth-Century Irish-American Novels', *Atlantic Studies*, 11(3) (2014), 403–18.

D'Arcy McGee, Thomas, 'The Social Duties of Irishmen in America', *The American Celt* (23 July 1853).

Dewees, William B., *Letters from an Early Settler of Texas* (Louisville, KT: Hull and Brother, 1854).

Emmons, David, *Beyond the American Pale: The Irish in the West, 1845–1910* (Norman, OK: University of Oklahoma Press, 2010).

Father Hugh Quigley, *Profit and Loss* (New York: T. Kane, 1873).

Hoare, Mrs, 'The Black Potatoes', in *Shamrock Leaves: or Tales and Sketches from Ireland* (Dublin: J. M' Glashan, 1851), 385.

Huyssen, Andreas, 'Diaspora and Nation: Migration into Other Pasts', *New German Critique*, 88 (2003).

Ignatiev, Noel, *How the Irish Became White* (London: Routledge, 1995).
'Ireland', *A Tale of the Irish Famine in 1846 and 1847; Founded On Fact* (Reigate: William Allingham, 1847).
*Irish Pictures: Eighty Sketches taken on the Spot by F. Dadd, M. Fitzgerald, Harry Furiss, Wallis Mackay and Procter and R. C. Woodville. Republished from the Illustrated London News* (London: Vizetelly and Co., 1870).
Janssen, Lindsay, 'The Impossibility of Transporting Identity: The Representation of Diasporic Irishness in Transatlantic Irish Fiction, 1860–1900', in Eleftheria Arapoglou, Mónika Fodor and Jopi Nyman, eds, *Mobile Narratives: Travel, Migration, and Transculturation* (New York: Routledge, 2014), 162–73.
Jones, David S., 'The Transfer of Land and the Emergence of the Graziers during the Famine Period', in Arthur Gribben, ed., *The Great Famine and the Irish Diaspora in America* (Boston, MA: University of Massachusetts Press, 1999), 93.
Keary, Annie, *Caste Daly; The Story of an Irish Home Thirty Years Ago* (London: Macmillan, 1886).
Kelleher, Patricia, 'Young Irish Workers: Class Implications of Men's and Women's Experiences in Gilded Age Chicago', in Kevin Kenny, ed., *New Directions in Irish American History* (Madison, WI: University of Wisconsin Press, 2003), 191.
Kelly, Mary C., *Ireland's Great Famine in Irish-American History: Enshrining a Fateful Memory* (Lanham, MD: Rowman and Littlefield, 2013), xviii.
Kenny, Kevin, *The American Irish: A History* (London: Longman, 2000).
King, Jason, 'Remembering the Famine Orphans: The Transmission of Famine Memory between Ireland and Quebec', in Christian Noack, Lindsay Janssen and Vincent Comerford, eds, *Holodomor and Gorta Mór: Histories, Memories and Representations of Famine in Ukraine and Ireland* (London: Anthem Press, 2012), 115–44.
Mason, Miss, *Kate Gearey; or Irish Life in London; A Tale of 1849* (London: Charles Dolman, 1853).
Meagher, Timothy J., *Inventing Irish America: Generation, Class, and Ethnic Identity in a New England City, 1880–1920* (Notre Dame, IN: University of Notre Dame Press, 2001).
Miller, Kerby, *Ireland and Irish America: Culture, Class, and Transatlantic Migration* (Dublin: Field Day, 2008).
Mitchel, John, *Jail Journal: Or Five Years in British Prisons* (New York: 'The Citizen', 1854).
*Mount Leinster; a Poem, Descriptive of Irish Scenery* (London: Longman, Hurst, Ree, Orme and Browne, 1819).
Newton, Henry T., *Canada in 1864: A Hand-Book for Settlers* (London: Sampson Low, 1864).

'Norah' (Margaret Dixon McDougall), *The Days of a Life* (Almonte: W. Templeman, 1883).
O'Brien, Dillon, *The Dalys of Dalystown* (St Paul, MN: Pioneer Printing, 1866).
O'Brien, W. P., *The Great Famine in Ireland and a Retrospect of the Fifty Years 1845–95, with Sketches of the Present Condition and Future Prospects of the Congested Districts* (London: Downey & Co., 1896).
O'Callohan, David, 'Will He Return', *The Emerald* (2 October 1869).
Perlmann, Joel, and Waters, Mary C., 'Intermarriage and Multiple Identities', in Mary C. Waters and Reed Ueda, eds, *The New Americans: A Handbook to Immigration Since 1965* (Cambridge, MA: Harvard University Press, 2007), 110–23.
Roediger, David, *The Wages of Whiteness: Race and the Making of the American Working Class*, rev. edn (London and New York: Verso Books, 1999).
Rothberg, Michael, *Multidirectional Memory: Remembering the Holocaust in the Age of Decolonization* (Stanford, CA: Stanford University Press, 2009).
Sadlier, Mary Anne, *Bessy Conway; or, the Irish Girl in America* (New York: D. & J. Sadlier and Co., 1862).
Smith, John Talbot, 'The Deacon of Lynn', in *His Honor The Major. And Other Tales* (New York: Vatican Library, 1891), 4.
——, 'How the McGuinness Saved his Pride', in *His Honor The Major. And Other Tales* (New York: Vatican Library, 1891), 255.
Synon, Mary, 'My Grandmother and Myself', *Scribner's Magazine* (August 1916), 224–33.
Tierney, Reginald, *The Struggles of Dick Massey; or, the Battles of a Boy* (Dublin: James Duffy, 1860).
——, 'The Emigrant Ship', *The Literary Garland*, VIII/11 (1850).
——, 'The Irish Americans', *The Irish People* (12 November 1864).
——, 'The Irish Emigrant's Lament', *The Pilot* (23 January 1858).
Wertsch, James V., *Voices of Collective Remembering* (New York: Cambridge University Press, 2002).
*What Irish Boys Can Do; an Answer to 'No Irish Need Apply'* (New York: H. De Marsan, 1864).

GRAHAM DAWSON

# 12 Memory, 'Post-Conflict' Temporalities and the Afterlife of Emotion in Conflict Transformation after the Irish Troubles

> It is simply too soon to ask Northern Ireland to set about an official and systematic exploration of the history of the Troubles. Even now [...] the wounds are still too sore, the divisions too deep and the past too hotly contested. Just talking about how Northern Ireland might deal with the conflict's legacy generated scenes of anger and bitterness of a type many dared to hope were themselves in the past.[1]

> [T]here comes a time when we should accept that no matter how many more investigations we hold, or how many witnesses we call, or how much money we spend, they are unlikely to achieve anything more of use. That time has come. For Northern Ireland, the path to lasting peace lies in looking to the future, not raking up the past.[2]

These two statements, both contributions to debates in the Northern Ireland peace process about how to 'deal with the legacies' of past conflict during the Troubles, were made within five years of each other, in 2009 and 2013 respectively. Read together, they give rise to two questions that will be addressed in this chapter. Firstly, they invite consideration of how conflicting understandings, evaluations and activations of temporality play out in 'post-conflict' culture and complicate efforts towards 'dealing with the past' in Northern Ireland. Secondly, they prompt reflection on the

1 Noel Whelan, 'Wounds Still Too Raw for an Examination of the Troubles', *Irish Times* (31 January 2009) <http://www.irishtimes.com/opinion/wounds-still-too-raw-for-an-examination-of-the-troubles-1.1238308> accessed 20 June 2016.

2 David Davis, 'John Larkin is Right – It's Time to Stop Probing the Killings of the Troubles', *The Guardian* (20 November 2013) <https://www.theguardian.com/commentisfree/2013/nov/20/john-larkin-troubles-killings-northern-ireland-david-davis> accessed 20 June 2016.

relation between these complex temporalities of conflict and its aftermaths, on one hand, and the domain of emotion, feeling and affect, on the other.[3]

The desire to put an end to violence and to create a new and 'better' future, characterized by social arrangements that improve on those of the past – recreating social relations and reworking historic conflict into constructive and democratic forms – motivates peace-making and conflict transformation initiatives like the Irish peace process. However, it is now widely recognized that the building of any imaginable alternative future involves a necessary engagement with the conflict that is – or is wished to be – 'in the past'. The temporalities of transition envisaged in this way are complex, contradictory, paradoxical, and riven with ambivalence. In Hamber and Kelly's influential practical definition of 'reconciliation', for example, this process of remaking damaged social relationships is understood to be 'both a backward- and forward-looking' one,[4] requiring both the creation of a shared vision of the future and a necessary reckoning with the emotional as well as political legacies of violent conflict. Emotion, feeling, and affect are central to the ways in which 'the past' is thought to live on after political violence; permeating 'post-conflict' memory, reproducing fear, antagonism and hostility within subjectivities and inter-subjective relations, and liable to erupt into the present in repetitions of intractable discord that subvert or complicate conflict management and efforts towards any conciliatory or reparative remaking of social relationships.

Something of this 'explosive power [of the past] to tear apart the present'[5] became manifest in the events that contextualize the two statements quoted above. My first quotation is from an article in the *Irish Times* reflecting on the Belfast launch of the Report by the Consultative Group

3 In what follows I have avoided rigorous distinction between these three terms in favour of the referential instability which keeps in play their different resonances within what I take to be a common domain. I am grateful to the Critical Studies Research Group at the University of Brighton for discussion of this issue.

4 Brandon Hamber and Gráinne Kelly, *A Place for Reconciliation? Conflict and Locality in Northern Ireland* (Belfast: Democratic Dialogue, 2005), 38.

5 Jonathan Gil Harris, quoted in Rebecca Bryant, 'History's Remainders: On Time and Objects after Conflict in Cyprus', *American Ethnologist*, 41(4) (2014), 683.

on the Past (CGP) on 28 January 2009.[6] The CGP, set up in 2007 by the British Government, made the most comprehensive proposals to date about how to address the 'legacy' questions of memory, truth and justice within a framework designed to promote reconciliation. Throughout its existence, from the announcement of its establishment and the conduct of the consultation process to the Report and subsequent discussion about what the British Government intended to do with it, the CGP was a focus of intense anxiety, hurt and anger. Amongst a number of issues sparking especially emotive controversy were the discussions about whether or not to offer amnesty to perpetrators; whether the conflict was, and should be described officially as, a 'war'; and the recommendation for a 'recognition payment' of £12,000 payable to the closest surviving relative of every individual killed in a 'conflict-related' incident, which was widely regarded as offensive 'blood money'.[7] Anger erupted at the public launch of the Report. Unionist victims of terrorism, protesting against the proposals, launched verbal attacks on Republicans attending the event, including the President of Sinn Féin, Gerry Adams. They focused on atrocities committed by the Provisional Irish Republican Army (IRA), while Republicans reciprocated with taunts about atrocities by loyalist paramilitaries. According to one press report, 'At one point the gathering threatened to descend into violence amid the welter of jabbing fingers and virulent insults'.[8]

6 See 'Full Text of Report Launch Address' (28 January 2009) <http://cain.ulst.ac.uk/victims/docs/consultative_group/cgp_280109_address.pdf>; 'Dealing with the Legacy', Report of the Consultative Group on the Past (23 January 2009) <http://cain.ulst.ac.uk/victims/docs/consultative_group/cgp_230109_report.pdf >; both accessed 25 February 2016.

7 See, for example, 'Victims' Anger at "Amnesty for Terrorists"', *News Letter* (9 January 2008)<www.newsletter.co.uk/news/.../victims-anger-at-amnesty-for-terrorists-1-1862850>; 'Inishowen Families Slam Report on Troubles', *Derry Journal* (2 February 2009) <http://www.derryjournal.com/news/inishowen-families-slam-report-on-troubles-1-2134212> both accessed 30 June 2016.

8 'Payment Plan for Northern Ireland Reconciliation Provokes Outrage', *New York Times* (29 January 2009) <http://www.nytimes.com/2009/01/29/world/europe/29ireland.html?_r=0>; see also 'Bitter Memories of the Troubles Threaten

The event provoked wide public concern and reflection in the media about its significance, which meshed with debates about the merits of the Report's various proposals. In this context, some commentators, including Noel Whelan in the *Irish Times*, drew the conclusion that it is 'too soon' to initiate a systematic exploration of the history of the Troubles, because 'even now' – in the 2009 present, fifteen years after the Provisional IRA's ceasefire of 1994 that kick-started the peace process – 'the wounds are still too sore, the divisions too deep and the past too hotly contested'; as was clearly apparent when discussion of how to 'deal with the conflict's legacy' generated 'scenes of anger and bitterness' that seemed to return from the past where, it had been hoped, the peace process had put them. Evidently the British Government agreed with this assessment. The Report itself was consigned to the past – 'binned', as Suzanne Breen put it[9] – and serious official efforts towards 'dealing with the past' did not begin again until the political consultations and negotiations of 2013 led by Richard Haass and Meghan O'Sullivan.

My second quotation is from the Conservative MP David Davis, writing in support of a controversial proposal to 'bring an end' to all judicial investigations of 'Troubles-related deaths' prior to the 1998 Agreement, made to the Haass/O'Sullivan consultation in November 2013 by Northern Ireland's foremost lawyer, the Attorney General John Larkin QC. Larkin explained his proposal to the *Belfast Telegraph:*

> More than fifteen years have passed since the Belfast Agreement, there have been very few prosecutions, and every competent criminal lawyer will tell you the prospects of conviction diminish, perhaps exponentially, with each passing year, so we are in a position now where I think we have to take stock [...] It strikes me that the time has come to think about putting a line, set at Good Friday 1998, with respect to prosecutions, inquests and other inquiries.[10]

---

to Wreck £300m Ulster Healing Plan', *Times* (29 January 2009). <http://www.thetimes.co.uk/tto/news/politics/article2026830.ece> both accessed 30 June 2016.

9 Suzanne Breen, 'Time to Take Care of Unfinished Business', *Belfast Telegraph* (3 March 2011) <http://www.belfasttelegraph.co.uk/opinion/news-analysis/time-to-take-care-of-unfinished-business-28593789.html> accessed 30 June 2016.

10 'Troubles Prosecution Call Slammed', *Belfast Telegraph* (20 November 2013) <http://www.belfasttelegraph.co.uk/news/northern-ireland/

Larkin's proposal envisaged Northern Ireland's violent past as too unruly and unmanageable for resolution by proper legal procedure, having due regard to statutes of limitation.[11] However, it was met with a largely critical reception and rejection by the Prime Minister, David Cameron, and leading politicians across the spectrum in Northern Ireland. The *Belfast Telegraph* also reported the 'outrage' expressed by relatives of those killed by Republican and loyalist paramilitaries and British state forces, 'that those perpetrators yet to be caught should not face justice'.[12] The paper cited two such reactions. Stephen Gault, whose father was killed in the IRA bombing in Enniskillen, Co. Fermanagh in 1987 ('for which no-one has ever been convicted'), was quoted as saying: 'How dare he [the Attorney General] airbrush the innocent people who were murdered at the hands of terrorists to move things forward. I just think it's totally disgusting'. Kate Nash, whose brother William was shot dead by British soldiers on Bloody Sunday in Derry/Londonderry in 1972, was reported to have 'challenged' Dr Haass with 'her family's opposition to the [Attorney General's] proposal' in an 'unscheduled encounter' in the foyer of a hotel in Derry, and '[a]fterwards [to have] explained her anger: 'What are they trying to

troubles-prosecution-call-slammed-29770237.html>. See also 'The Nuts and Bolts of a Plan Aimed at Drawing a Line under the Troubles', *Belfast Telegraph* (20 November 2013) <http://www.belfasttelegraph.co.uk/news/northern-ireland/the-nuts-and-bolts-of-a-plan-aimed-at-drawing-a-line-under-the-troubles-the-views-of-attorney-general-john-larkin-on-29768511.html> both accessed 26 June 2016.

11 Chris Lorenz's important distinction between judicial or jurisdictive time and historical time has a bearing on Larkin's proposal ('Blurred Lines: History, Memory and the Experience of Time', *International Journal for History, Culture and Modernity*, 2(1) [2014], 43–62). While 'the time of jurisdiction is always limited' by legal statutes of limitation (Lorenz, 'Blurred Lines', 47), this does not preclude, as Larkin recognized, 'historical research' by 'historians' or 'in terms of recent history, […] journalists' ('Attorney General John Larkin: It's Time to Call Halt to All Troubles Cases', *Belfast Telegraph* [20 November 2013] <http://www.belfasttelegraph.co.uk/news/northern-ireland/attorney-general-john-larkin-its-time-to-call-halt-to-all-troubles-cases-29770396.html> accessed 28 March 2015).

12 'Troubles Prosecution Call Slammed'.

do, draw a line under victims, draw a line under my brother? We are not going to let that happen.'[13]

In his support for the proposal, David Davis endorsed the Attorney General's observations about the deterioration of evidence (both forensic and eyewitness) over time, and the value of 'putting a line' under the violence of the past, 'set at Good Friday 1998', as – in Davis's words – 'a price worth paying for the long-term stability of a whole region'. Given the supposed unlikelihood of convictions, he concluded, 'Any investigation, inquest or inquiry would be a fruitless exercise in opening up old wounds.' The 'time has come' to draw that line, argued Davis, because: 'For Northern Ireland, the path to lasting peace lies in looking to the future, not raking up the past'.[14] Many people in Northern Ireland will have found David Davis's defence of the Attorney General's proposal deeply offensive, not only because he concurs with Larkin's suggestion of 'drawing a line' and demarcating the past as indeed over and to be set aside in the interests of 'long term' future goals, but also because of the imputation about themselves and their affective experience contained in this expression, 'raking up the past'. For those bereaved, injured or otherwise harmed by political violence in the Troubles, the conflicted past does not need to be 'raked up', as they live with its painful aftermath every day, and its emotional consequences not only permeate subjective and intersubjective experience but also motivate continuing campaigns to achieve understanding, acknowledgement and redress utilizing discourses of truth and justice.

The intimate relation between temporality and emotion emerges clearly in these two instances. In each case, we can witness the production of anger, bitterness, outrage or disgust stemming from a sense of violation experienced and conveyed by relatives of the victims of conflict-era violence, signalling the existence of 'unfinished business' from events which often took place forty or more years ago. We also see how emotional and affective currents with sources deep in the past percolate into and flow through present-day public discourse and politics. These particular responses evoke temporality in their unstable and contradictory assessments of when the

13 Ibid.
14 Davis, 'John Larkin Is Right'.

time might be ripe for a comprehensive inquiry into acts of political violence that currently lack explanation or redress. For the *Irish Times* in 2009, the time is 'too soon', whereas for Attorney General Larkin and David Davis MP in 2013, that time has already passed and it is now considered 'too late' for such inquiry.

In what follows, I explore further this relation between temporality and emotion, reflecting on how what I will call the temporal afterlife of emotion, and its relation to memory in a society marked by violent political conflict, might be understood. First, I draw on recent thinking on regimes of temporality and the politics of time to illuminate the complex relations between past, present and future, and concomitant struggles over their demarcation, in 'post-conflict' societies such as Northern Ireland. I go on to consider the temporalities of post-conflict emotion, feeling and affect, and explore the implications of these approaches and arguments for understanding struggles over memory, truth and justice in Northern Ireland during and after the Troubles.

## Regimes of Temporality and the Politics of Time in 'Post-Conflict' Northern Ireland

'Temporality' is defined by David Scott as 'the lived experience of time passing', involving the 'social relation [...] between past (the time of memory), present (the time of conscious awareness), and future (the time of anticipation)'.[15] His own theoretical interest in 'time-consciousness'[16] is driven by a concern to understand 'the temporality of the aftermaths of political catastrophe, the temporal disjunctures involved in living *on* in the wake of past political time, amid the ruins [...] of [...] futures past'.[17]

15 David Scott, *Omens of Adversity: Tragedy, Time, Memory, Justice* (Durham, NC; London: Duke University Press, 2014), 1.

16 Ibid., 1.

17 Ibid., 2.

In reflecting on the intensified experience of time 'out of joint' in societies undergoing 'catastrophic aftermaths',[18] his particular focus is the collapse and defeat of the Grenadan Revolution (1979–1983), and the consequent curtailment and loss, for the generations who lived its temporal possibilities for radical transformation of past sufferings, of any vision of a progressive future to give hope, meaning and purpose to politics in the present. Scott reads Grenada as a precursor of a more widespread phenomenon 'post-1989', involving the collapse not only of the Soviet bloc as a Communist alternative to Western liberal capitalism, but also of the progressive temporal structure and horizons of future betterment that have inspired progressive and revolutionary politics in the modern post-enlightenment era. Now stuck in the stasis of a neo-liberal present, those generations confronted by their 'lost horizon' of future possibility become enmeshed in forms of hope-less emotion and affect such as loss, despair, disillusionment, bitterness and anger. For Scott, the so-called 'memory-boom', and widespread post-1980s fascination with memory in relation to historical trauma and practices of transitional justice, are 'symptoms' of this loss of a viable alternative future.

Scott's argument suggests new ways of thinking about temporality in 'post-conflict' societies and cultures. This common but problematic expression – 'post-conflict'– is itself a site of ambivalence and contradiction; a term carrying two different meanings that exist in uneasy relation to one another. Used in a normative sense, in UN-speak – the dominant international discourse of peace-building and transitional justice deployed by UN agencies and NGOs – it refers to the time after conflict; meaning the time when conflict is ended and is now 'in the past', having given way to the new temporal era of peace-time. This new era is also described as 'transitional', moving towards a specific, anticipated future of securely achieved peace, democracy, and implementation of human rights for all. In this sense, 'post-conflict' is a concept that instrumentally 'draws a line' between the past and the future-oriented present. The drawing of this line necessitates and justifies measures to 'deal with the past', a temporal expression increasingly used to name a confident, prescriptive strategy of

18 Ibid., 2.

mastery deployed by international peace-building institutions and agencies (and now normalized by an acronym, DwP),[19] that signifies 'the past' as a problem to be managed and controlled by the powerful 'from above'.

Used in another, more ambivalent sense (sometimes marked by scare quotes), 'post-conflict' refers to the time after in the sense that we speak of 'post-structuralism' or 'post-industrial', to signify something that follows but whose characteristics are not only different from, but also deeply marked by – indeed inconceivable without – what came before; and requires understanding in its relationship to that specific 'past'. In this sense the past of conflict is *not* entirely over and finished, but has a complex temporal existence, sometimes described as 'the present past',[20] in which transition to a new future of peace, or democracy, is by no means guaranteed or secured, or even necessarily desired in the specific form on offer. In attempting to signify the complexities of post-conflict temporality in this second sense, a range of concepts and metaphors are commonly put to use. War and conflict are said to 'cast shadows' forwards in time – shadows that persist, into the future, long after the violence has ended. Violence 'in the past' may be described as 'haunting' the present and future, or be understood to have deposited 'remains' of affect, sometimes felt to possess a persistent 'afterlife'. We speak of the emotional or affective 'legacies' of the violent past – referring to anger and bitterness, fear, suspicion and distrust, sadness and loss, hatred, guilt, humiliation and so on – often regarded as something malign or problematic, that is handed down or passed on from a previous time to posterity, and may cross from older to younger, or future, generations. When we talk of drawing a line under the past; of stepping out of its shadow, laying it to rest, letting it go; of moving on, or forward; of establishing 'closure', these metaphors of mastery tend to have as their object states of emotion, feeling and affect, and to be driven, as Paul Gready puts

19 See, for example, the DwP programme run by Swiss Peace: <http://www.swisspeace.ch/topics/dealing-with-the-past.html> accessed 28 June 2016.

20 Graham Dawson, *Making Peace with the Past? Memory, Trauma and the Irish Troubles* (Manchester: Manchester University Press, 2007), 10; Andreas Huyssen, *Present Pasts: Urban Palimpsests and the Politics of Memory* (Stanford, CA: Stanford University Press, 2003).

it, by 'a need to liberate the present and future from the burden of the past that threatens to overwhelm them'.[21]

The term 'post-conflict', then, carries with it two fundamental questions of temporality: when, and by whom or what, is a conflict deemed to be 'over' and consigned to the past? and how should this ambiguous time 'afterwards' be described, characterized, and engaged with? Such questions have been illuminated in recent thinking on complex temporalities and the 'politics of time' by Berber Bevernage and Chris Lorenz.[22] Their conception of regulatory 'regimes' of temporality[23] underpins a constructionist account of the historical emergence into dominance during the late eighteenth and nineteenth centuries – and its calling into question in the later twentieth and early twenty-first centuries – of modern time; understood as objectively real, as measurable by clock and calendar, and as proceeding in an arrow-like linear trajectory into the future,[24] causing the past to become increasingly 'distant and absent from the present'.[25] Challenging the objectivity of what they insist is a social construct, Bevernage and Lorenz point to the 'performativity' of temporal distinctions produced by discursive practices (including history), as these actively create and institute the categories of 'past', 'present' and 'future', and the temporal boundaries that demarcate one from the other.[26] The purpose of their critique is to make visible the human agency that has produced and continues to sustain 'modern time'

21 Paul Gready, *Political Transition: Politics and Cultures* (London: Pluto, 2003), 2.

22 Lorenz, 'Blurred Lines'; Berber Bevernage, *History, Memory and State-sponsored Violence: Time and Justice* (Abingdon: Routledge, 2012); Berber Bevernage and Chris Lorenz, 'Breaking up Time: Negotiating the Borders between Present, Past and Future', *Storia della Storiografia*, 63(1) (2013), 31–51.

23 Bevernage and Lorenz, 'Breaking Up Time', 34; Bevernage, *History, Memory, and State-sponsored Violence*, 12–13.

24 Bevernage, *History, Memory, and State-sponsored Violence*, 11–16; Bevernage and Lorenz, 'Breaking Up Time', 37–41; Lorenz, 'Blurred Lines', 48–9.

25 Lorenz, 'Blurred Lines', 45.

26 Bevernage, *History, Memory, and State-sponsored Violence*, 15; Bevernage and Lorenz, 'Breaking Up Time', 34.

as a dominant regime of temporality. A 'politics of time' ensues,[27] centred on the reproduction of this temporal regime and the practices and subjects authorized by it, such as those seeking a break with the past in order to manage a political transition, and a fresh start in the new time of the post-conflict nation; or its contestation by, for example, those whose lives have been damaged through conflict and for whom the past is not over but compels attention, engagement and redress in the present.

Much discourse concerned with 'post-conflict' Northern Ireland can be read as an instrument of boundary-setting performative politics, a strategy for managing the unruly and unresolved past by casting conflict as precisely 'past', in the name of a new normative 'present' that is clearly separated from it and directed towards bringing into being a vision of a particular desired future. Consider this example of what Bevernage calls '"allochronistic" practice [...] that (symbolically) allocates into another time'.[28] In February 1994, at a crucial stage in establishing a credible basis for the peace process between the Downing Street Declaration of December 1993 and the Provisional IRA's ceasefire declaration in August 1994, the Prime Minister of the United Kingdom, John Major, wrote in the northern nationalist *Irish News*: 'We cannot live in the past. Dreadful deeds have been done by all sides in past centuries. We should all regret that, but those of us alive today are not responsible for them. Our generation must look to the future'.[29] In this performative speech act, Major sought to ground his future-oriented proposals towards a negotiated peace settlement on a double distancing, both temporal and moral. Temporally, it consigns to the comfortable distance of 'past centuries' a current and continuing armed conflict involving 'dreadful deeds' committed over the previous twenty-five years, leading to some 3,500 mostly unresolved deaths, tens of thousands of injuries, and hundreds of people in prison. Its consequent moral distancing enables 'those of us alive today' to 'regret' all this without shouldering any responsibility for it; while 'our generation' is required to 'look to the

27 Bevernage, *History, Memory, and State-sponsored Violence*, 11; Bevernage and Lorenz, 'Breaking Up Time', 36.

28 Bevernage, *History, Memory, and State-sponsored Violence*, 16.

29 Quoted in Dawson, *Making Peace*, 60.

future' in its peace-making, those who continue to focus on current 'dreadful deeds' in an unresolved conflict will be labelled as 'liv[ing] in the past'.

Attorney General Larkin's proposal to 'draw a line' under unresolved conflict-related deaths is a further example of performative reallocating of these issues 'to the past'. However, boundary-setting as a politics of time may also include endless deferral into the future of demands for reparation and redress for past wrongs, on grounds (like those utilized by Noel Whelan) that the time is not yet ripe. In both cases, the time for ameliorative action is not now. It is either deferred to a time of anticipation: that moment still to come when actors in the peace process will perhaps have matured sufficiently to carry out unemotional exchanges, when the wounds will not be as sore, the emotions will not be as hot, angry and bitter, when hope may more safely be dared. Alternatively, the possibility of ameliorative action is cancelled entirely by the 'irreversible time of history',[30] and consigned to the time of opportunity lost forever. 'Post-conflict' temporality, however, remains a site of ideological struggle between these practices dedicated to the policing of time and countervailing practices that seek to maintain a more fluid, open, dialogic relationship between the present, the past and the future. Temporal boundary demarcations, setting the border between what is over and done, what is now and what is to come, are vulnerable to being unravelled, perforated, or challenged on the basis of alternative temporalities in which the past appears as unfinished and the future as questionable.

Unresolved 'legacies' with deep emotional roots are widely perceived, and feared, as having the potential to unravel the peace settlement and even to ignite a return to violent conflict in the future. The arrest of the Sinn Féin President and Republic of Ireland TD, Gerry Adams, in April 2014 by police investigating the 'historic' case of Jean McConville's abduction, murder and disappearance by the Provisional IRA in 1972 provides a potent example. Among several Sinn Féin leaders fiercely critical of the arrest, the Deputy First Minister, Martin McGuinness, laid the blame on 'a dark side' within the Police Service of Northern Ireland (PSNI) engaged in 'political

30 Bevernage, *History, Memory, and State-sponsored Violence*, 5.

policing' in pursuit of a 'negative and anti-peace process agenda'. Pointing in this way to what he considered to be an unreconstructed minority remnant of the old Royal Ulster Constabulary (RUC) – hated and mistrusted by northern nationalists for its sectarianism and violence, whose abolition and replacement by the reformed PSNI was a key element in the implementation of the Good Friday Agreement leading to devolution in 2007 – and suggesting its continuing capacity to influence the political process and 'settle old scores', McGuinness threatened to 'review' Sinn Féin's support for the PSNI if Adams was charged.[31] This threat carried the potential to collapse the devolved, power-sharing Assembly and Executive, a stance that seriously exacerbated already deteriorating relations with Sinn Féin's partners in government, the Democratic Unionist Party (DUP). The DUP's leader, First Minister Peter Robinson, responded with accusations that Sinn Féin were reverting to anti-democratic 'republican bullyboy tactics' in 'a despicable, thuggish attempt to blackmail the PSNI': 'The threat now means that ordinary, decent citizens will conclude that the PSNI and the PPS [Public Prosecution Service] have succumbed to a crude and overt political threat if Adams is not charged.'[32] The crisis provoked public debate throughout Ireland and Britain about the 'fragility' of the peace settlement, accompanied by widespread expressions of anxiety about its future sustainability in a context where 'the past' had not been dealt with.[33] The furore

31 'McGuinness: SF Will Not Allow These Dark Elements to Succeed', *Irish News* (3 May 2014) <http://www.irishnews.com/paywall/tsa/irishnews/irishnews/irishnews//news/2014/05/03/news/mcguinness-sf-will-not-allow-these-dark-elements-to-succeed-90652/content.html> accessed 28 March 2016.

32 'Sinn Féin Trying to Blackmail Police over Gerry Adams, Says First Minister', *The Guardian* (4 May 2014) <http://www.theguardian.com/politics/2014/may/04/sinn-fein-police-gerry-adams-peter-robinson> accessed 28 March 2016.

33 'Adams' arrest, and the international furore that surrounded it, was one more reminder [...] of unfinished business. Everyone living on this island – whether we know it or not – depends on the peace process. We got a frightening reminder this past weekend of how fragile it is, depending as it does on the dark secrets of the past remaining uncovered'. Fergus Finlay, 'Adams' Arrest a Grim Reminder of Just How Fragile the Peace Process Is', *Irish Examiner* (6 May 2014) <http://www.irishexaminer.com/viewpoints/columnists/fergus-finlay/

surrounding Adams' arrest clearly demonstrates how assertions about the unfinished past and uncertain future may well take different forms and have different meanings for competing political communities according to their own well-established narratives of memory.

On his release Adams strove to quell talk of the peace process unravelling through a performative restoration of the temporal borders breached by his arrest: 'Everyone knows these are changed times, that things have changed, they will be treated fairly and we can have hope and confidence. To send any other signal is to encourage the bigots. I support the Police Service of Northern Ireland [...] The past is past. The IRA has gone. Its finished [...] This is not 1981, it is not 1972 [...] There is no possibility of going back'.[34] However, the disruptive potency of the Jean McConville case stems from its capacity to dissolve those very temporal borders so that the injustice of this act of political violence, the narratives that justify, lay blame for and absolve it, and the truth about responsibility for it, remain a present-day concern at the very centre of the peace-building project in Ireland. Among those unresolved 'historic cases' from the Troubles with the highest public profile, the Jean McConville case has generated persistent allegations about Adams' involvement in her disappearance from both the McConville family

fergus-finlay-adamsrsquo-arrest-a-grim-reminder-of-how-fragile-the-peace-process-truly-is-267577.html>. See also Nicholas Watt, 'Arrest of Gerry Adams Raises Fears for Northern Ireland Peace Process', *The Guardian* (1 May 2014) <http://www.theguardian.com/politics/2014/may/01/arrest-gerry-adams-northern-ireland-sinn-fein-stormont>; Alex Kane, 'Present and Future on Hold until We Deal with the Past', *News Letter* (5 May 2014) <http://www.newsletter.co.uk/news/present-and-future-on-hold-until-we-deal-with-the-past-1-6039969#ixzz44DCuW8eY>; David McKittrick, 'Adams Events Show the Fragility of Peace and How Easily It Can Fracture', *Irish Independent* (5 May 2014) <http://www.independent.ie/opinion/analysis/adams-events-show-the-fragility-of-peace-and-how-easily-it-can-fracture-30243396>; Arthur Beesley, 'Belfast Agreement is Fragile and Can Be Very Quickly Disrupted by Events', *Irish Times* (6 May 2014) <http://www.irishtimes.com/news/politics/analysis-belfast-agreement-is-fragile-and-can-be-very-quickly-disrupted-by-events-1.1784509> all accessed 28 March 2016.

34 'Freed Gerry Adams: the IRA Has Gone, the Past Is Past', *Telegraph* (4 May 2014) <http://www.telegraph.co.uk/news/uknews/northernireland/10808288/Freed-Gerry-Adams-the-IRA-has-gone-the-past-is-past.html> accessed 28 March 2016.

and their supporters, and also from political enemies of Adams seeking to expose a paramilitary past that he denies, in order to destroy his current public reputation and political authority. These allegations motivated the infamous injunction secured by the PSNI to subpoena recordings of oral interviews with former IRA Volunteers by researchers at Boston College, USA, that led to his arrest.[35] Tropes of haunting have been widely deployed to characterize Adams' inability to shake off his association with the case despite repeated protestations of his innocence: Jean McConville figures as the 'ghost' of a mother once considered insignificant and dispensable, that refuses to 'disappear' but persists to disturb the Sinn Féin narrative of peace-making by making visible the consequences of disreputable actions by Republicans that the party would perhaps prefer to let lie in the past.[36]

35 Gerry Moriarty, 'Boston Tapes Are Reason Why Adams Is Being Questioned over McConville Murder', *Irish Times* (2 May 2014) <http://www.irishtimes.com/news/ireland/irish-news/boston-tapes-are-reason-why-adams-is-being-questioned-over-mcconville-murder-1.1780908> accessed 28 March 2016. See also Boston College Subpoena News website <https://bostoncollegesubpoena.wordpress.com> accessed 28 March 2016.

36 Steven Alexander, 'Gerry Adams: Haunted by the Shadow of a Shocking Troubles Crime', *Belfast Telegraph* (1 May 2014) <http://www.belfasttelegraph.co.uk/news/northern-ireland/gerry-adams-haunted-by-the-shadow-of-a-shocking-troubles-crime-30234297.html>; Henry McDonald, 'Jean McConville Case Continues to Haunt Gerry Adams', *The Guardian* (1 May 2014) <http://www.theguardian.com/uk-news/2014/may/01/jean-mcconville-gerry-adams-sinn-fein>; Suzanne Breen, 'Gerry Adams' Arrest for Murder Is a First – but Its Still Hard To Believe He Won't Survive This', *Daily Mail* (1 May 2014) <http://www.nuzhound.com/articles/breen/arts2014/may1_Gerry_Adams_arrest_for_murder_is_a_first_SBreen_Daily-Mail.php>; Jenny McCartney, 'Jean McConville: The Stubborn Ghost Who Won't Go Away', *Telegraph* (2 May 2014) <http://www.telegraph.co.uk/news/uknews/northernireland/10802227/Jean-McConville-The-stubborn-ghost-who-wont-go-away.html>; Gail Walker, 'Why the Ghost of Murdered Jean McConville Won't Lie at Rest', *Belfast Telegraph* (6 May 2014) <http://www.nuzhound.com/archive/goto.php?id=244366> all accessed 28 March 2016. For an analysis of Republican politics of the past in the peace process, see Stephen Hopkins, 'Sinn Féin, the Past and Political Strategy: The Provisional Irish Republican Movement and the Politics of "Reconciliation"', *Irish Political Studies*, 30(1) (2015), 79–97. For a scholarly use of the concept of haunting in engagement with the unfinished business

The memory of this ghostly figure has been kept alive by her sons and daughters. Forty-two years after witnessing, as children, the abduction of their widowed mother at gunpoint by an IRA unit never to see her again, and enduring separation and dispersal around a number of different social care homes following the disappearance of their family's lynchpin, the McConville siblings were given space in the news media to voice support for the police investigation into Adams. Cowed at the time of the abduction into silence about the names of the Volunteers they recognized and know to be responsible, by the threat of IRA reprisal, the siblings continue to be fearful of intimidation by Republican splinter groups. Michael McConville, an eleven-year-old in 1972, challenged the public perception that in 2014 the time of violence is over when he told the BBC: '[E]verybody thinks this has all gone away [...] My blood boils in my body [...] I just can't stand those people for what they have done to us', yet still 'I wouldn't tell the police. If I told the police now a thing, me or one of my family members or one of my children would get shot'.[37] However, his sister Helen McKendry, who was ostracized and attacked by Republicans in her local community after co-founding the justice campaign Families of the Disappeared in the period following the IRA ceasefire in 1994,[38] indicated her willingness to overcome her fear and assist the police inquiry: 'I have been living in fear for far too long. I thought to hell with them, they did this, they killed my mother, now they have to pay for it [...] It would bring me closure and maybe then we will learn the full truth of what happened and why it happened'.[39] Now, if prosecutions were to follow Adams' arrest, 'I would be prepared to name names. To me that is not informing but doing my duty

of the past, see Avery F. Gordon, *Ghostly Matters: Haunting and the Sociological Imagination* (Minneapolis, MN: Minnesota University Press, 2008).

37 'Jean McConville Murder: Son Says He Knows Killers', *BBC News* (1 May 2014) <http://www.bbc.co.uk/news/uk-27236414> accessed 28 March 2016.

38 Dawson, *Making Peace*, 75–6.

39 'Jean McConville's Daughter Tells of Fears over 40 Years after IRA Killing', *Belfast Telegraph* (6 May 2014) <http://www.belfasttelegraph.co.uk/news/northern-ireland/jean-mcconvilles-daughter-tells-of-fears-over-40-years-after-ira-killing-im-relieved-my-mother-hasnt-been-forgotten-30246116.html> accessed 28 March 2016.

to my mother'.[40] The McConville siblings have lived with the emotional consequences of this experience, and this silence, for the entirety of their adult lives. Perhaps they are the ones who are best described as haunted;[41] but in any case their story testifies to the centrality of emotions – the fear of being treated as an informer in a repetition of the original violence; the blood 'boiling' in anger at being subjected to this power – within the 'post-conflict' politics of time in Northern Ireland.

## 'Post-Conflict' Temporalities of Emotion in the Northern Ireland Troubles

Where to locate the temporality of emotion in a 'post-conflict' context is a complex matter that prompts questions both of theory and of history.[42] In a fundamental sense, it is understood and experienced as the product of an originating event that occurred during violent conflict 'in the past'; as something caused by this event and that remains attached to it afterwards. According to one kind of account, this emotional and affective response then persists continuously after the event into any given present moment, where it is always ready to be given expression again. Alternatively, this response to the originating event is thought to become overlaid by subsequent emotional experience and development, such that it becomes progressively removed from – that is, temporally distant from – the present, but

40 'Jean McConville's Daughter: I'll Name Names over IRA Killing', *The Guardian* (1 May 2014) <http://www.theguardian.com/politics/2014/may/01/jean-mcconville-daughter-names-ira-killing> accessed 28 March 2016.

41 For the 'haunted' and 'liminal' position of relatives of the disappeared, see Brandon Hamber and Richard A. Wilson, 'Symbolic Closure Through Memory, Reparation and Revenge in Post-conflict Societies', *Journal of Human Rights*, 1(11) (2002), 35–53.

42 For further development of this discussion, see Graham Dawson, 'The Meaning of "Moving On": From Trauma to the History and Memory of Emotions in "Post-conflict" Northern Ireland', *Irish University Review*, 47(1) Special Issue: 'Moving Memory' (forthcoming 2017).

is capable of 'coming to the surface' (to use a spatial metaphor), or being somehow reconnected to the present. Even if located principally 'in the past', then, emotions are durational and involve complex relations between past, present and future. Their temporality may be fluid rather than fixed once and for all, and has various characteristics, including the potential for longevity but also for recurrence, re-emergence, ebb and flow, repetition, as well as a capacity for mutability (where one emotion or affective state transforms into another – commonly, for example, sadness into anger, or anger into guilt).

This complex 'afterlife' of emotion – the ways in which emotions 'live on' dynamically in time – is also shaped at the intersection of emotionality as psychic and somatic energy on one hand, and as a cultural and social phenomenon on the other. Our emotional life is not self-evident and transparent but requires work to understand and interpret what goes on in our internal world of feeling, on the basis of our interaction with 'dominant emotional codes and standards' and the ways we 'navigate between "emotional communities"'.[43] Our emotions are felt to be validated by – or to be transgressive of – cultural norms and values; and these 'structures of feeling' shape patterns and models of experience that are formed and shift historically in relation to changing conditions of life.[44] This points to the sense in which emotion and affect are not only produced by and attached to an originating event 'in the past', but are also produced and expressed, lived and interpreted in the context of circumstances and concerns of a subsequent moment located 'in the present'. In this sense there is always scope for the making of new meanings, namings and interpretations of experience, including states of feeling, that are produced in retrospect, even many years

43 Claire Langhamer, *The English in Love: The Intimate Story of an Emotional Revolution* (Oxford: Oxford University Press, 2013), 8.

44 For Raymond Williams's concept of 'structure of feeling', see his *The Long Revolution* (London: Chatto and Windus, 1961), 64–88; *Marxism and Literature* (Oxford: Oxford University Press, 1977), 128–33; *Politics and Letters: Interviews with New Left Review* (London: Verso, 1981), 156–74. For use of the term in recent histories of emotion, see, for example, Michael Roper, *The Secret Battle: Emotional Survival in the Great War* (Manchester: Manchester University Press, 2010), 159–88.

afterwards. Making new meanings of past events may transform what is felt about them now, as Nicola King argues in her take on Freud's concept of *Nachträglichkeit*, translated as 'afterwardsness'.[45] The oral historian Alistair Thomson captures the dynamics of temporality within the ordinary life course, and the way life transitions associated with the process of ageing and looking back over time may trigger a re-evaluation of feelings long attached to past events, in his phrase 'experience never ends'.[46] In a further sense, emotions are also constituted in relation to perceptions of futurity which involve uncertainties, fears and hopes concerning the potential future consequences of emotional expression in the past or in the present, and about anticipated states of feeling in the future.

Consideration of these complex temporal dynamics of emotion opens up new perspectives on the politics of time, memory and conflict transformation during and after the Northern Ireland Troubles. Analysis of the afterlife of emotion requires the making of careful distinctions between the 'past' and the 'present' moments of its production and expression, and brings into focus a number of questions about the relationship between past and present.[47] One question concerns our understanding of what has happened to 'original' wartime emotions subsequently, as it were 'in the meantime', within temporal experience over the intervening years, which now stretch to half a century since the first lives were lost to the conflict in 1966. A second question concerns the relation between this long afterlife of wartime emotions and the emotional dynamics that propel memory-based activism in the present moment of 'now'. A third considers how conflict-derived emotions that are recovered, reflected on, expressed, asserted and acted upon in the context of the peace process, are reframed according to

45 Nicola King, *Memory, Narrative, Identity: Remembering the Self* (Edinburgh: Edinburgh University Press, 2000).

46 Alistair Thomson, 'Experience Never Ends: Migrant Memories, Unsettled Identities and Historical Change', in *Crossroads of History: Experience, Memory, Orality*, Proceedings of the XIth International Oral History Conference, Istanbul (June 2000), 1081–7; 'Experience Never Ends: Connecting Life History Work and Experiential Learning', *Australian Journal of Experiential Learning*, 35 (1996), 24–9.

47 My thinking here is influenced by Marek Tamm, ed., *Afterlife of Events: Perspectives on Mnemohistory* (Basingstoke: Palgrave Macmillan, 2015).

'post-conflict' discourses of dealing with the past, reconciliation, truth, justice and closure.[48] Attention to the complex temporalities of emotion problematizes any notion of a static post-conflict 'present' as the basis for remembrance of, and engagement with, conflict 'in the past'. More than twenty years after the ceasefires, that present moment of the peace process has grown a history and is no longer as it was. 'Post-conflict' time now has its own 'past', and as this extends 'forwards' into the future, new kinds of relation to the time of conflict emerge and existing relations are transformed.

Whether before or after the ceasefires of 1994, speaking about the emotional impacts of the Troubles, 'taking them in' (in a psychoanalytical sense) and acting upon them, has involved breaking out of a 'culture of silence'.[49] The politics of time in the Northern Ireland conflict depends upon the effectiveness of strategies for silencing, or at least marginalizing, emotional expression of what is unresolved and not past, and the deployment of narratives tasked with containing or concealing its potentially disruptive affects. Powerful representations that intervene in 'post-conflict' memory by drawing a line under the past – as seen in public debate on the CGP report, Attorney General Larkin's proposal to the Haass/O'Sullivan consultation, and the arrest of Gerry Adams – work to privatize dangerous emotions that cannot be incorporated, seeking to restrict their articulation to, at best, private arenas, rendering them publicly unspeakable, and leaving those who feel and bear such emotions to do so without wider social recognition.[50] Those subjected to such a strategy

48 See, for example, Graham Dawson, 'Masculinities and "the Terrorist" in Conflict Transformation: Representation, Identity and Reconciliation in "Post-conflict" Northern Ireland', in Sue Malvern and Gabriel Koureas, eds, *Terrorist Trangressions: Gender and the Visual Culture of the Terrorist* (London: I. B. Tauris, 2014), 157–80.

49 See Dawson, *Making Peace*, 68, 237–8; Marie Breen Smyth, *Truth Recovery and Justice after Conflict: Managing Violent Pasts* (London: Routledge, 2007), 34, 48–54, 68, 86; Brandon Hamber, *Transforming Societies after Political Violence: Truth, Reconciliation, and Mental Health* (London: Springer, 2009).

50 This account rests on the theory of popular memory developed in T. G. Ashplant, Graham Dawson and Michael Roper, 'The Politics of War Memory and Commemoration: Contexts, Structures and Dynamics', in T. G. Ashplant, Graham Dawson and Michael Roper, eds, *The Politics of War Memory and Commemoration* (London: Routledge, 2000), 12–32.

also find themselves experiencing further emotional effects produced by its implementation. Contesting this process calls for some form of public performance of emotion, where it is channelled as a motivational energy in the service of political or moral agency that erupts into public visibility to demand recognition and response. The conditions for securing and reproducing a silence are contingent and susceptible to transformation. In the history of the peace process, spaces have opened up for some stories to get a hearing and some truths to be told, usually as a result of intense and time-consuming struggle, whilst others remain cloaked in silence.

The afterlife of the Kingsmills massacre of 1976 provides an illuminating case study for the investigation of these questions and concerns. On 5 January that year a minibus carrying home twelve workmen from their factory in Glenanne, South Armagh, was stopped at Kingsmills crossroads near Whitecross village by Republican gunmen who lined up and shot dead ten Protestant workmen; an eleventh man, Alan Black, survived, and a Catholic in the party was spared and ushered away. The massacre was claimed by a group calling itself the South Armagh Republican Action Force, believed to be renegade gunmen operating outside the IRA's command structure, who described it as a reprisal for the killing by loyalist gunmen of five members of the Catholic Reavey and O'Dowd families on the previous day.[51] No one has been convicted for Kingsmills, and it is a case long shrouded in silence that only began to emerge into public hearing much later and after a convoluted history. Justice campaigner Karen Armstrong, the sister of John McConville who was killed in the attack, tells how:

> There was a wall of silence for a long, long time. I used to sort of ask Mum quite often, 'Have you heard anything, Mum? Has anybody came and spoken to you?' And the answer was always the same. And then I just stopped asking Mum, because I knew in my heart that for some reason, and I don't know what that reason is, as yet, but there was never going to be any proper information or any investigation.[52]

51 David McKittrick et al., *Lost Lives: The Stories of the Men, Women and Children who Died as a Result of the Northern Ireland Troubles*, 2nd edn (Edinburgh: Mainstream, 2007), 611–14; Toby Harnden, *'Bandit Country': The IRA and South Armagh* (London: Hodder and Stoughton, 1999), 133–40.

52 Karen Armstrong interview, *Britain's Secret Terror Deals*, BBC Panorama, broadcast BBC One, 28 May 2015.

In this privatizing of emotion, recognition of the bereaved is denied along with the abnegation of responsibility for undertaking a proper investigation. Only in 2003 – twenty-six years after the original coroner's inquest produced an open verdict without taking evidence, and in what was then the 'present moment' of the politics of victimhood early in the peace process provoked by the Saville Inquiry into Bloody Sunday – did the newly established Police Service of Northern Ireland respond to strong pressure from Unionist politicians and Ulster-British victims' groups by reopening the police investigation into this and other 'forgotten atrocities' from the Troubles in South Armagh.[53] After a series of obstructive delays in which the police claimed to have 'lost the files' due variously to flood, fire, explosion and asbestos contamination[54] – delays that provoked fresh emotions of anger and frustration in the families – the PSNI's Historical Enquiries Team (HET) eventually reported in 2011 that the attack was pre-planned and carried out by members of the Provisional IRA.[55]

The HET's report overturned the hitherto broadly accepted narrative that rogue Republicans carried out a spontaneous revenge attack. It also laid the ground for the relatives' and survivors' justice campaign to push for a thorough and independent inquiry that would also consider the police's mishandling of the case, and allegations that official silence shrouding the events of 1976 was designed to protect an agent employed by the British State to operate within the IRA.[56] Karen Armstrong, explaining how she had become a leading figure in the campaign, was quoted in the *News Letter* as saying that: 'I am the oldest sister, having lost our parents, so I

53 Dawson, *Making Peace*, 252–3. See 'Police "to Reopen Murder Files"', *BBC News* (5 August 2003) <http://news.bbc.co.uk/1/hi/northern_ireland/3127289.stm> accessed 3 March 2016.

54 Alan Black interview, *Britain's Secret Terror Deals.*

55 'IRA Blamed for "Sectarian Slaughter" of 10 at Kingsmill', *Irish Times* (22 June 2011) <http://www.irishtimes.com/news/ira-blamed-for-sectarian-slaughter-of-10-at-kingsmill-1.602655> accessed 3 March 2016.

56 *Britain's Secret Terror Deals*; 'IRA Massacre Survivor Suspects State Involvement', *RTE* News (17 February 2014) <http://www.rte.ie/news/2014/0217/504876-kingsmill-massacre/> accessed 3 March 2016.

felt we could not go through the rest of my life not doing something'.[57] In 2012 the campaign secured an independent investigation by the Police Ombudsman of Northern Ireland, and in 2013 won a decision by Attorney General Larkin to reopen the coroner's inquest, which began preliminary hearings in February 2014. Further anger was provoked in the families when budget cuts caused postponements and temporarily threatened the cancellation of both enquiries.[58] Responding to this threat, Karen Armstrong pointed out that: 'so many people down through the years in positions [of authority] have turned their backs and have made excuses and they are still doing that. And we certainly will do our best to make sure that will not continue to happen'.[59] In November 2015 an investigatory team from the Police Ombudsman's office began its work, while the full coroner's inquest reopened in May 2016.

The words used by Karen Armstrong to tell the story of her family's history in the aftermath of the Kingsmills massacre point to the intertwining of emotion and temporality in their experience, and to the tangible effects of a specific politics of time, and the possibility of 'now' confronting it. That the family 'loved my brother John so much and had such respect for him'[60]

57 'Kingsmills Massacre Inquest Begins – 38 Years after Murders', *News Letter* (17 February 2014) <http://www.newsletter.co.uk/news/northern-ireland-news/kingsmills-massacre-inquest-begins-38-years-after-murders-1-5881555> accessed 3 March 2016.

58 'Kingsmills IRA Massacre Familes Consider Civil Action against Suspects', *Belfast Telegraph* (22 June 2011) <http://www.belfasttelegraph.co.uk/news/northern-ireland/kingsmills-ira-massacre-familes-consider-civil-action-against-suspects-28629671.html>; 'Ulster Police Watchdog Says Cuts Are Delaying Inquiries into Troubles Killings', *The Guardian* (30 September 2014) <http://www.theguardian.com/uk-news/2014/sep/30/ulster-police-watchdog-cuts-delaying-inquiries-troubles-killings>; 'Police Ombudsman Team to Probe Kingsmills Massacre', *News Letter* (11 November 2015) <http://www.newsletter.co.uk/news/northern-ireland-news/police-ombudsman-team-to-probe-kingsmills-massacre-1-7061258> all accessed 3 March 2016.

59 'Kingsmills Victims' Anger over Inquest Delay', *News Letter* (23 April 2015) <http://www.newsletter.co.uk/news/northern-ireland-news/kingsmills-victims-anger-over-inquest-delay-1-6705364#ixzz41rfW1S7C//> accessed 3 March 2016.

60 'Kingsmills Massacre Inquest Begins – 38 Years after Murders'.

during his lifetime gives rise at the time of his loss to the desire to understand what occurred and to see those responsible identified and held accountable. This desire prompts her own questioning of her mother, repeated 'quite often', until the realization dawns: 'I knew in my heart that [...] there was never going to be any proper information or any investigation.' Active agency is quashed, and 'I stopped asking'.[61] A 'wall of silence' then persists 'for a long, long time'. Its inhibiting effects eventually come to be broken as a result of temporal experience in an ordinary life course; the death of parents, and perhaps a heightened awareness of one's own mortality that this commonly brings, gives rise to a new sense of moral responsibility as falling upon those who are now the 'oldest', together with the feeling that we, I, 'could not go through the rest of my life not doing something'. This renewal of agency is motivated by a temporal perception; that my life in the future is limited, and with it the time left to me to 'do something' in order to fulfil the family's unmet but undissipated desire, formed half a lifetime ago. In a sense, history begins again at that new present moment when the question can be asked afresh, and what was accepted as 'never going to be' becomes something achievable once more. Now, while the 'wall of silence' is still in place and the reason for it remains unknown, this is only 'as yet', and the determination to discover why makes visible those who have practised concealment 'down through the years' and 'are still doing that', and grounds an undertaking to break up the temporal reproduction of silence so as 'to make sure that will not continue to happen' into the future.

The Kingsmills case also illuminates how the complex temporal dynamics of emotion after political violence are concerned not only with relations between past and present, but also with futurity. Perhaps the most fundamental way in which the future manifests emotionally in the aftermath of such an event is in awareness of the lost futures of the dead. Reflecting on her family's role in securing the reopening of the Kingsmills inquest, Karen Armstrong was moved to point out that her brother 'would have been fifty-eight years old now'.[62] A year later, writing about the inquest's significance

61 Karen Armstrong interview.

62 'Kingsmills Massacre Inquest Begins – 38 Years after Murders'.

for all the families, she noted that: 'John's future lay in the foreign mission field and no one will ever know the positive effect he may have had on other people'.[63] Grief, at this irreparable loss of human potential for the good, marks time in the birthdays and anniversaries of every advancing year.

In another manifestation of futurity, apprehensions of a future characterized by the unbroken reproduction of a silence may motivate the desire to 'do something' to forestall that future, as we see in the case of the Kingsmills justice campaign. Such action entails the faculty of hope, defined as the cherishing of 'a desire of good with some expectation of fulfilment'.[64] This is a desire which depends upon the cultivation of confidence and courage grounded in realistic assessment of the conditions of possibility, akin to what the Marxist thinker Antonio Gramsci, writing in another context, terms 'concrete will'.[65] Inherently, hope involves an element of vulnerability since its motivating anticipation of fulfilment necessarily risks disappointment. To hope is to let down, or to move outside of, the psychic defences constructed and maintained to guard against such exposure to the frustration of desire. Just as we must 'dare to hope' – an expression used in the opening quotation of this chapter – so, too, our hopes may be 'dashed', 'shattered' or 'extinguished', all common terms that convey the emotional violence accompanying a withdrawal of hope. The risk of hoping stems from uncertainty about what the future may bring. Uncertainty generates contradictory feelings of anticipation, ranging from the fear of what speaking out might cause to happen, experienced by the McConville family, to the complex compound of emotions suggested by

63 Karen Armstrong, 'Kingsmills 40 Years On: We Cannot Move On until the Truth is Finally Revealed', *News Letter* (5 January 2016) <http://www.newsletter.co.uk/news/northern-ireland-news/kingsmills-40-years-on-we-cannot-move-on-until-the-truth-is-finally-revealed-1-7143209> accessed 13 June 2016.

64 *Chambers Twentieth Century Dictionary*, ed. A. M. Macdonald (Edinburgh: Chambers, 1972).

65 Antonio Gramsci, 'The Modern Prince', in Quentin Hoare and Geoffrey Nowell Smith, eds, *Selections from Prison Notebooks* (London: Lawrence and Wishart, 1971), 172.

Karen Armstrong's remark, on the eve of the reopened Kingsmills inquest, that: 'Even though we have this hearing tomorrow it is still very difficult for us as a family to have to face up to listening to the hard facts that may come out. Obviously tomorrow it is a preliminary hearing and we are not sure what the outcome of that will be'.[66]

Once the time of the full hearing eventually arrives, such anticipatory emotions intersect with fresh emotional responses provoked as the hearing unfolds, whether by disturbing information that may be presented, or by further efforts to obstruct the investigative process. When the Kingsmills inquest eventually reopened at Belfast High Court on 23 May 2016, the sole survivor of the attack, Alan Black, was reportedly in distress as he listened to his own eyewitness testimony, given to police shortly after the attack from the hospital bed where he believed he was dying, and now read out in court. He was then able to tell, for the first time in a court, his harrowing memory of the shooting of his teenaged apprentice, Robert Chambers, 'a happy-go-lucky lad who was much loved by [his family]' and who died 'crying for his mummy'; and later described to the news media what he had learned to call his own 'survivor guilt', as well as the comfort gained from his belief that, through his testimony, 'I've really done something for them now'.[67]

One week into the inquest, however, the unfolding of the 'unvarnished truth'[68] about Kingsmills, so desired by Black and his fellow campaigners, appeared to be checked once again when the PSNI announced the opening of a new police investigation. This was prompted by the discovery of 'new forensic evidence' in the form of a matching record for a palm print found on a what is alleged to have been a get-away vehicle used by the

66 'Kingsmills Massacre Inquest Begins – 38 Years after Murders'.

67 'Kingsmills Inquest: Teenage Victim Cried Out for Mother before Being Shot', *BBC News* (25 May 2016) <http://www.bbc.co.uk/news/uk-northern-ireland-36380099> accessed 11 June 2016.

68 Alan Black, quoted in 'Inquest into Kingsmill Massacre in Northern Ireland Begins', *The Guardian* (23 May 2016) <https://www.theguardian.com/uk-news/2016/may/23/inquest-kingsmill-massacre-northern-ireland-ira> accessed 11 June 2016.

killers.[69] The announcement generated conflicting emotions amongst the families and campaigners. Colin Worton, whose brother Kenneth died in the attack, told the press: 'I do feel a mixture of emotions, but one of the worst ones I have is anger. Why has it taken more than forty years to come out?'[70] Alan Black described feeling 'totally baffled' because 'Tests have been run on that print seven or eight times over few years [sic] and they found absolutely nothing'; while the refusal of the police to discuss their new investigation with either the families or himself left him 'feeling very, very sore'.[71] The identity of the 'new suspect' in the case was revealed by the *Irish News* only after its successful contestation in the High Court of an injunction to prevent publication requested by the PSNI's Chief Constable, George Hamilton.[72]

The new suspect, Colm Murphy, described as 'a long-time republican extremist' by the judge passing sentence after his conviction in connection with the Omagh bombing of 1998,[73] had undergone eleven years of police investigation and legal process before his conviction was overturned on appeal in 2010, and claimed to have been arrested on more than thirty occasions and fingerprinted over forty times by police in the Republic of

69 'Northern Ireland Police Open New Investigation into Kingsmill Massacre', *The Guardian* (31 May 2016) <https://www.theguardian.com/uk-news/2016/may/31/northern-ireland-police-open-new-investigation-into-kingsmill-massacre> accessed 11 June 2016; 'Kingsmill Survivor Alan Black Asks What Dirty Secret Are the Security Forces Trying to Hide?', *Irish News* (7 June 2016) <http://www.irishnews.com/news/northernirelandnews/2016/06/07/news/kingsmill-survivor-alan-black-s-fears-of-greater-dirty-secret-behind-the-massacre-550244/> accessed 11 June 2016.

70 'Northern Ireland Police Open New Investigation into Kingsmill Massacre'.

71 'Kingsmill Survivor Alan Black Asks What Dirty Secret Are the Security Forces Trying to Hide?'.

72 '"Complete Vindication" of Irish News Stance in High Court', *Irish News* (7 June 2016) <http://www.irishnews.com/news/2016/06/07/news/-complete-vindication-of-irish-news-stance-in-high-court-550059/> accessed 11 June 2016.

73 'Omagh Bomb Plot Man Is Sentenced to 14 Years' Jail', *Independent* (26 January 2002) <http://www.independent.co.uk/news/uk/this-britain/omagh-bomb-plot-man-is-sentenced-to-14-years-jail-9224896.html> accessed 11 June 2016.

Ireland and in the North since 1976.[74] Murphy suggested that he was being used as a 'scapegoat' to 'scupper the inquest and save details emerging that could damage the peace process and embarrass Sinn Féin': 'People who are now in government with unionists [...] don't want reminded of the past. That's why you have to ask why this is being linked to me all of a sudden, much easier to deflect onto me than allow uncomfortable truth's (sic) to come spilling out'.[75] According to Murphy, one of those uncomfortable truths was an IRA plan to drive all Unionists out of South Armagh in the event of retaliation by loyalist paramilitaries after Kingsmills: 'You know that was the times we were living in, it sounds incredible now forty years on but that's what was being talked about at that time'.[76] For Black, on the other hand, the 'dirty secret' hidden behind the Kingsmills massacre concerns the security services[77] and his suspicion that 'there may have been agents of the state involved in the attack itself'.[78] For Black and other justice campaigners, the 'new investigation' by the PSNI thus appears to be consistent with the long history of police obstruction to proper inquiry into Kingsmills: having lost the battle to prevent the reopening of the coroner's inquest, tactics have been readjusted to create an indefinitely delay to its proceedings, since the police investigation would take legal precedence. This stymieing of the inquest, for which campaigners have fought for so long and which has been achieved at such effort, represents a direct attack on their capacity to sustain hopeful agency. As Black put it: 'I really do feel that they are just waiting for us to die off [...] I thought that they would stop the inquest and then just put us in a drawer and lock

74 'IRA Plot to "Ethnically Cleanse" Border after Kingsmill', *Irish News* (6 June 2016) <http://www.irishnews.com/news/northernirelandnews/2016/06/06/news/exclusive-ira-were-going-to-ethnically-cleanse-south-armagh-if-there-was-retaliation-to-kingsmill-548939/> accessed 11 June 2016.

75 Ibid.

76 Colm Murphy quoted in 'IRA Plot to "Ethnically Cleanse" Border after Kingsmill'.

77 'Kingsmill Survivor Alan Black Asks What Dirty Secret Are the Security Forces Trying to Hide?'.

78 'IRA Massacre Survivor Suspects State Involvement'.

it. And then if we ever asked when the inquest would start again, they'd just say it couldn't because there was an investigation'.[79]

Intimations of futurity are also implicit in the desire for 'closure' which is regularly expressed by those harmed by political violence and has become common currency in public discourse on the 'legacy issues' of truth and justice in the context of 'dealing with the past'. The desire for closure may be understood as a 'yearning [...] to achieve psychic and emotional composure on the basis of fully revealed information [... and] the wish for a line to be drawn under the now-resolved "past" so that a new life can begin, one no longer enmeshed in the pain and difficulty of loss compounded by uncertainty and a sense of injustice, but oriented towards new aspirations'.[80] In this sense, the pursuit of 'closure' may appear to be a necessary '[i]magining [of] an achieved conclusion' for those who have lived with the emotional pressure of unresolved bereavement.[81] It may also be read as a temporal boundary-setting practice consistent with official strategies to 'put the conflict in the past', as evoked in Alan Black's own words, to 'put us in a drawer and lock it'.[82] However, the demand for closure by those

79 'Kingsmill Survivor Alan Black Asks What Dirty Secret Are the Security Forces Trying to Hide?'. The police investigation and inquest were continuing as this book went to press. See 'Man Arrested in Kingsmill Massacre Probe', *Irish News* (5 August 2016) <http://www.irishnews.com/news/northernirelandnews/2016/08/05/news/man-held-over-kingsmill-massacre-639383> accessed 15 September 2016; 'Kingsmill Massacre: Call for Top Secret Material to be Released to Inquest', *Irish News* (12 September 2016) <http://www.irishnews.com/news/northernirelandnews/2016/09/12/news/kingsmill-massacre-call-for-top-secret-material-to-be-released-to-inquest-692290> accessed 15 September 2016. On criticism of the delays and limitations endemic to the operation of inquests in cases of 'disputed killings involving state agencies', see Mark McGovern, 'Inquiring into Collusion? Collusion, the State and the Management of Truth Recovery in Northern Ireland', *State Crime*, 2(1) (2013), 21.

80 Dawson, *Making Peace*, 315.

81 Ibid.

82 Indeed, the very term 'closure' was appropriated by the Blair Government in 2005 to justify its proposal, subsequently revealed to be the result of a secret agreement with Sinn Féin during peace negotiations, to give Republican fugitives 'on the run' indemnity from prosecution. The proposal was rejected by the House of Commons

harmed by political violence carries with it as well a riposte to the politics of temporal demarcation in its insistence that conflict-era emotions are not 'over and finished' but persist, and a challenge that laying them to rest cannot be a purely private, internal matter of individual emotion-management but is necessarily contingent on securing public redress in terms of truth and justice.

Karen Armstrong makes such a demand, riposte and challenge in her article for *News Letter*, 'Kingsmills 40 years on: We cannot move on until the truth is finally revealed'.[83] Her riposte was to those who, 'in recent years', have criticized 'victims' relatives [...] as "living in the past" and refusing to "move on"'. Her challenge is to the men responsible for the massacre, some of whom 'still walk free (some will already have met their Maker)', but also to 'those who have attempted to play with people's lives as though it were a game, and have knowingly protected the perpetrators'; all of whom are 'tethered to this event for the remainder of their lives'. Her demands are twofold. The first is for recognition from critics that the 'thousands of families affected by our troubled past' have 'had no choice but to "move on" regardless of the loved ones so cruelly taken or injured'. The second, reiterating the formal request 'for those with information to please forward it to the Coroner's Service', is an 'appeal in the hope that those with a conscience will provide such information that will help give us the peace of mind we so crave'. Posing the question, 'Will we get closure?', Karen Armstrong's own answer is that this is 'difficult because so much time has elapsed and the years without John are almost double the years we had with him'; but that 'perhaps there may be a little, depending very much on the honesty and remorse of those involved. The minimum we would want is truth, especially from those who failed to protect the eleven victims and

but implemented covertly in the form of so-called 'comfort letters' issued to the fugitives, who included two suspects connected to Kingsmills. See Dawson, *Making Peace*, 317; 'IRA Fugitives Given "Letters of Comfort" are Linked to 300 Killings', *The Guardian* (8 May 2014) <https://www.theguardian.com/uk-news/2014/may/07/ira-fugitives-linked-to-killings> accessed 11 June 2016; 'Northern Ireland Police Open New Investigation into Kingsmill Massacre'.

83 Armstrong, 'Kingsmills 40 Years On'.

instead continue to protect their assailants'. In this powerful statement, an intimate understanding of lives lived within the temporality of emotions meshes with the hope that the forthcoming and anticipated inquest 'gives us a measure of relief', but also with the recognition that any such relief is contingent on others, and an ethical call for those others – who are also 'tethered to this event' for their lifetime – to acknowledge and act upon their responsibility to those they have harmed.

Here, 'closure' in a more limited sense is conceived, not as the drawing of a line under 'the past', but as the prospect of gaining some relief from the endless reproduction and extension into the future of conflict-derived emotions. This conception of a desired, future 'peace of mind' also entails a transformation in social relationships, grounded on what Judith Butler calls 'a sense of human vulnerability' and recognition of 'our collective responsibility for the physical lives of one another'.[84] Through the opportunities they offer for the making of transformative understandings through collaborative participation, investigations into the truth of the Troubles, such as the Kingsmills Coroner's inquest, open possibilities for bringing into being a complex, new political community of the future; one that foregrounds 'relational ties' that stem from our 'fundamental dependency and ethical responsibility'.[85]

## Conclusion

I have argued in this chapter that efforts towards 'dealing with the past' in Northern Ireland after the Troubles are fraught with contradictions, ambivalences and clashes of interest involving competing discourses of temporality. At the centre of such contestation is a tension between the desire to 'move on' into a post-conflict future predicated on 'leaving the

84 Judith Butler, 'Violence, Mourning, Politics', in *Precarious Life: The Powers of Mourning and Violence* (London: Verso, 2006), 30.

85 Ibid., 22.

past behind', and the countervailing desire to address – and redress – 'unfinished business' from the time of armed conflict, that manifests in the form of emotion, feeling and affect. What I have called the temporal dynamics of 'post-conflict' emotion involves a non-linear inter-weaving of emotions attached to violent events in the past, with emotions produced in the socio-political process of remembering and forgetting those events in peace-time, and with anticipatory emotions associated with imagining the future. This complex afterlife of emotions encounters a politics of time practised in the peace process by various organizations and institutions intent on performative demarcation of the boundaries between past, present and future for their own various ends. By developing a clearer understanding of these temporal strategies for managing, silencing and privatizing conflict-related emotions, and the effects of such strategies in terms of the coercion or resistance of those subjected to them, central dynamics in the politics of memory and the handling of legacy issues in Northern Ireland may be illuminated.

Karen Armstrong from the Kingsmills justice campaign articulates a vision of the broader significance of the reopened public inquest into the killing of the ten men in 1976, when she writes:

> Our society is currently sailing a fragile sea we call the peace process. We all must look forward to a future free of violence, where the physical and psychological peace walls can finally be removed. This cannot be successful if we continually attempt to rewrite, forget or bury our history. It is often said that a people who cannot confront their past will be condemned to repeat it. We need to build a foundation based on honesty, mutual respect and properly administered justice. We need to confront that ostrich head we call the Troubles and own up to the truth, warts and all.[86]

Speaking from a position rooted in her own painful family history, Armstrong makes a case for the social necessity of confronting the Troubles in a way consistent with demands for truth and justice, an ethics of human rights, and the making of new social ties of care and responsibility. Moved by an urgent, deeply felt sense that the time to resolve this history is now, her argument points to the importance of strengthening the connection

86 Armstrong, 'Kingsmills 40 Years On'.

between 'dealing with the past' and widening the 'horizon of expectation' that constitutes the imaginable future.[87] At a time when confidence in the transformative power and potential of the peace process in Northern Ireland has eroded[88] – in common with the wider sense of living in a 'stalled present' characterized by 'loss of the former future' that has permeated European societies since the global financial crisis of 2007–2008;[89] and with the cross-border foundations of the peace agreement itself now thrown into doubt by the United Kingdom of Great Britain and Northern Ireland's 'Brexit' referendum result of June 2016 – the renewal of cultural imaginaries of hope invested in peace-building and socio-economic regeneration has become paramount. The unlocking of hopeful social agency, with the capacity to engage and transform the afterlife of emotion from the Troubles as this manifests in the course of everyday life, will be crucial to that task.

## Acknowledgements

This chapter has benefitted from discussion of its earlier iterations at the 'Irish Studies and the Dynamics of Memory' conference at Radboud University Nijmegen, 31 March–2 April 2015; the 'Ireland: Shared Futures?' conference at Université Rennes II, 10–12 September 2015; the Modern History Research Centre, University of Winchester, 14 January 2016; and the Critical Studies Research Group work-in-progress seminar, University of Brighton, 31 May 2016. I am indebted to the Complex Temporalities Reading Group at the Centre for Research in Memory, Narrative and Histories, University of Brighton, January–June 2016, with particular

87 For Reinhart Koselleck's influential thinking on the temporal horizons of experience and expectation, see Michael Pickering, 'Experience as Horizon: Koselleck, Expectation and Historical Time', *Cultural Studies*, 18(2–3) (2004), 271–89.

88 See Colin Knox, 'Northern Ireland: Where Is the Peace Dividend?', *Policy and Politics*, 44(3) (2016), 485–503.

89 Scott, *Omens of Adversity*, 6, 108.

thanks to its coordinator, Garikoitz Gómez Alfaro, who introduced me to recent work in this field and commented on the draft, as did Lucy Newby and Fearghus Roulston, to whom my thanks also go.

## Bibliography

Alexander, Steven, 'Gerry Adams: Haunted by the Shadow of a Shocking Troubles Crime', *Belfast Telegraph* (1 May 2014) <http://www.belfasttelegraph.co.uk/news/northern-ireland/gerry-adams-haunted-by-the-shadow-of-a-shocking-troubles-crime-30234297.html> accessed 28 March 2016.

Armstrong, Karen, interview, *Britain's Secret Terror Deals*, BBC Panorama, broadcast BBC One, 28 May 2015.

——, 'Kingsmills 40 Years On: We Cannot Move On until the Truth is Finally Revealed', *News Letter* (5 January 2016) <http://www.newsletter.co.uk/news/northern-ireland-news/kingsmills-40-years-on-we-cannot-move-on-until-the-truth-is-finally-revealed-1-7143209> accessed 13 June 2016.

Ashplant, T. G., Dawson, Graham, and Roper, Michael, 'The Politics of War Memory and Commemoration: Contexts, Structures and Dynamics', in T. G. Ashplant, Graham Dawson and Michael Roper, eds, *The Politics of War Memory and Commemoration* (London: Routledge, 2000), 3–85.

'Attorney General John Larkin: It's Time to Call Halt to All Troubles Cases', *Belfast Telegraph* (20 November 2013) <http://www.belfasttelegraph.co.uk/news/northern-ireland/attorney-general-john-larkin-its-time-to-call-halt-to-all-troubles-cases-29770396.html> accessed 28 March 2015.

Beesley, Arthur, 'Belfast Agreement is Fragile and Can Be Very Quickly Disrupted by Events', *Irish Times* (6 May 2014) <http://www.irishtimes.com/news/politics/analysis-belfast-agreement-is-fragile-and-can-be-very-quickly-disrupted-by-events-1.1784509> accessed 28 March 2016.

Bevernage, Berber, *History, Memory and State-sponsored Violence: Time and Justice* (Abingdon: Routledge, 2012).

——, and Lorenz, Chris, 'Breaking up Time. Negotiating the Borders between Present, Past and Future', *Storia della Storiografia*, 63(1) (2013), 31–51.

'Bitter Memories of the Troubles Threaten to Wreck £300m Ulster Healing Plan', *Times* (29 January 2009) <http://www.thetimes.co.uk/tto/news/politics/article2026830.ece> accessed 30 June 2016.

Boston College Subpoena News website <https://bostoncollegesubpoena.wordpress.com> accessed 28 March 2016.

Breen, Suzanne, 'Gerry Adams' Arrest for Murder Is a First – but It's Still Hard To Believe He Won't Survive This', *Daily Mail* (1 May 2014) <http://www.nuzhound.com/articles/breen/arts2014/may1_Gerry_Adams_arrest_for_murder_is_a_first___SBreen_Daily-Mail.php>.

——, 'Time to Take Care of Unfinished Business', *Belfast Telegraph* (3 March 2011) <http://www.belfasttelegraph.co.uk/opinion/news-analysis/time-to-take-care-of-unfinished-business-28593789.html> accessed 30 June 2016.

Breen Smyth, Marie, *Truth Recovery and Justice after Conflict: Managing Violent Pasts* (London: Routledge, 2007).

Bryant, Rebecca, 'History's Remainders: On Time and Objects after Conflict in Cyprus', *American Ethnologist*, 41(4) (2014), 681–97.

Butler, Judith, 'Violence, Mourning, Politics', in *Precarious Life: The Powers of Mourning and Violence* (London: Verso, 2006), 19–49.

*Chambers Twentieth Century Dictionary*, ed. A. M. Macdonald (Edinburgh: Chambers, 1972).

'"Complete Vindication" of Irish News Stance in High Court', *Irish News* (7 June 2016) <http://www.irishnews.com/news/2016/06/07/news/-complete-vindication-of-irish news-stance-in-high-court-550059/> accessed 11 June 2016.

Davis, David, 'John Larkin is Right – It's Time to Stop Probing the Killings of the Troubles', *The Guardian*, (20 November 2013) <https://www.theguardian.com/commentisfree/2013/nov/20/john-larkin-troubles killings-northern-ireland-david-davis> accessed 20 June 2016.

Dawson, Graham, *Making Peace with the Past? Memory, Trauma and the Irish Troubles* (Manchester: Manchester University Press, 2007).

——, 'Masculinities and "the Terrorist" in Conflict Transformation: Representation, Identity and Reconciliation in "Post-conflict" Northern Ireland', in Sue Malvern and Gabriel Koureas, eds, *Terrorist Trangressions: Gender and the Visual Culture of the Terrorist* (London: I. B. Tauris, 2014), 157–80.

——, 'The Meaning of "Moving On": From Trauma to the History and Memory of Emotions in "Post-conflict" Northern Ireland', *Irish University Review*, 47(1) Special Issue: 'Moving Memory' (forthcoming 2017).

'DwP programme' <http://www.swisspeace.ch/topics/dealing-with-the-past.html> accessed 28 June 2016.

Finlay, Fergus, 'Adams' Arrest a Grim Reminder of Just How Fragile the Peace Process Is', *Irish Examiner* (6 May 2014) <http://www.irishexaminer.com/viewpoints/columnists/fergus-finlay/fergus-finlay-adamsrsquo-arrest-a-grim-reminder-of-how-fragile-the-peace-process-truly-is-267577.html> accessed 28 March 2016.

'Freed Gerry Adams: the IRA Has Gone, the Past Is Past', *Telegraph* (4 May 2014) <http://www.telegraph.co.uk/news/uknews/northernireland/10808288/Freed-Gerry-Adams-the-IRA-has-gone-the-past-is-past.html> accessed 28 March 2016.

'Full Text of Report Launch Address' (28 January 2009) <http://cain.ulst.ac.uk/victims/docs/consultative_group/cgp_280109_address.pdf>; 'Dealing with the Legacy' accessed 25 February 2016.

Gordon, Avery, *Ghostly Matters: Haunting and the Sociological Imagination* (Minneapolis, MN: Minnesota University Press, 2008).

Gramsci, Antonio, 'The Modern Prince', in Quentin Hoare and Geoffrey Nowell Smith, eds, *Selections from Prison Notebooks* (London: Lawrence and Wishart, 1971), 123–205.

Gready, Paul, *Political Transition: Politics and Cultures* (London: Pluto, 2003).

Hamber, Brandon, *Transforming Societies after Political Violence: Truth, Reconciliation, and Mental Health* (London: Springer, 2009).

——, and Kelly, Gráinne, *A Place for Reconciliation? Conflict and Locality in Northern Ireland* (Belfast: Democratic Dialogue, 2005).

——, and Wilson, Richard A., 'Symbolic Closure Through Memory, Reparation and Revenge in Post-conflict Societies', *Journal of Human Rights*, 1(11) (2002), 35–53.

Harnden, Toby, *'Bandit Country': The IRA and South Armagh* (London: Hodder and Stoughton, 1999).

Hopkins, Stephen, 'Sinn Féin, the Past and Political Strategy: The Provisional Irish Republican Movement and the Politics of "Reconciliation"', *Irish Political Studies*, 30(1) (2015), 79–97.

Huyssen, Andreas, *Present Pasts: Urban Palimpsests and the Politics of Memory* (Stanford, CA: Stanford University Press, 2003).

'Inishowen Families Slam Report on Troubles', *Derry Journal* (2 February 2009) <http://www.derryjournal.com/news/inishowen-families-slam-report-on-troubles-12134212> accessed 30 June 2016.

'Inquest into Kingsmill Massacre in Northern Ireland Begins', *The Guardian* (23 May 2016) <https://www.theguardian.com/uk-news/2016/may/23/inquest-kingsmill-massacre northern-ireland-ira> accessed 11 June 2016.

'IRA Blamed for "Sectarian Slaughter" of 10 at Kingsmill', *Irish Times* (22 June 2011) <http://www.irishtimes.com/news/ira-blamed-for-sectarian-slaughter-of-10-at-kingsmill-1.602655> accessed 3 March 2016.

'IRA Fugitives Given "Letters of Comfort" are Linked to 300 Killings', *The Guardian* (8 May 2014) <https://www.theguardian.com/uk-news/2014/may/07/ira-fugitives-linked-to-killings> accessed 11 June 2016.

'IRA Massacre Survivor Suspects State Involvement', *RTE* News (17 February 2014) <http://www.rte.ie/news/2014/0217/504876-kingsmill-massacre/> accessed 3 March 2016.

'IRA Plot to "Ethnically Cleanse" Border after Kingsmill', *Irish News* (6 June 2016) <http://www.irishnews.com/news/northernirelandnews/2016/06/06/news/exclusive-ira-were-going-to-ethnically-cleanse-south-armagh-if-there-was-retaliation-to-kingsmill-548939/> accessed 11 June 2016.

'Jean McConville Murder: Son Says He Knows Killers', *BBC News* (1 May 2014) <http://www.bbc.co.uk/news/uk-27236414> accessed 28 March 2016.

'Jean McConville's Daughter: I'll Name Names over IRA Killing', *The Guardian* (1 May 2014) <http://www.theguardian.com/politics/2014/may/01/jean-mcconville-daughter-names-ira-killing> accessed 28 March 2016.

'Jean McConville's Daughter Tells of Fears over 40 Years after IRA Killing', *Belfast Telegraph* (6 May 2014) <http://www.belfasttelegraph.co.uk/news/northern-ireland/jean-mcconvilles-daughter-tells-of-fears-over-40-years-after-ira-killing-im-relieved-my-mother-hasnt-been-forgotten-30246116.html> accessed 28 March 2016.

Kane, Alex, 'Present and Future on Hold until We Deal with the Past', *News Letter* (5 May 2014) <http://www.newsletter.co.uk/news/present-and-future-on-hold-until-we-deal-with-the-past-1-6039969#ixzz44DCuW8eY> accessed 28 March 2016.

King, Nicola, *Memory, Narrative, Identity: Remembering the Self* (Edinburgh: Edinburgh University Press, 2000).

'Kingsmill Massacre: Call for Top Secret Material to be Released to Inquest', *Irish News* (12 September 2016) <http://www.irishnews.com/news/northernirelandnews/2016/09/12/news/kingsmill-massacre-call-for-top-secret-material-to-be-released-to-inquest-692290/> accessed 15 September 2016.

'Kingsmill Survivor Alan Black Asks What Dirty Secret Are the Security Forces Trying to Hide?', *Irish News* (7 June 2016) <http://www.irishnews.com/news/northernirelandnews/2016/06/07/news/kingsmill-survivor-alan-black-s-fears-of-greater-dirty-secret-behind-the-massacre-550244/> accessed 11 June 2016.

'Kingsmills Inquest: Teenage Victim Cried Out for Mother before Being Shot', *BBC News* (25 May 2016) <http://www.bbc.co.uk/news/uk-northern-ireland-36380099> accessed 11 June 2016.

'Kingsmills IRA Massacre Families Consider Civil Action against Suspects', *Belfast Telegraph* (22 June 2011) <http://www.belfasttelegraph.co.uk/news/northern-ireland/kingsmills-ira-massacre-familes-consider-civil-action-against-suspects-28629671.html> accessed 3 March 2016.

'Kingsmills Massacre Inquest Begins – 38 Years after Murders', *News Letter* (17 February 2014) <http://www.newsletter.co.uk/news/northern-ireland-news/kingsmills-massacre-inquest-begins-38-years-after-murders-1-5881555> accessed 3 March 2016.

'Kingsmills Victims' Anger over Inquest Delay', *News Letter* (23 April 2015) <http://www.newsletter.co.uk/news/northern-ireland-news/kingsmills-victims-anger-over-inquest-delay-1-6705364#ixzz41rfW1S7C//> accessed 3 March 2016.

Knox, Colin, 'Northern Ireland: Where Is the Peace Dividend?', *Policy and Politics*, 44(3) (2016), 485–503.

Langhamer, Claire, *The English in Love: The Intimate Story of an Emotional Revolution* (Oxford: Oxford University Press, 2013).

Lorenz, Chris, 'Blurred Lines: History, Memory and the Experience of Time', *International Journal for History, Culture and Modernity*, 2(1) (2014), 43–62.

McCartney, Jenny, 'Jean McConville: The Stubborn Ghost Who Won't Go Away', *Telegraph* (2 May 2014) <http://www.telegraph.co.uk/news/uknews/northernireland/10802227/Jean-McConville-The-stubborn-ghost-who-wont-go-away.html> accessed 28 March 2016.

McDonald, Henry, 'Jean McConville Case Continues to Haunt Gerry Adams', *The Guardian* (1 May 2014) <http://www.theguardian.com/uk-news/2014/may/01/jean-mcconville-gerry-adams-sinn-fein> accessed 28 March 2016.

McGovern, Mark, 'Inquiring into Collusion? Collusion, the State and the Management of Truth Recovery in Northern Ireland', *State Crime*, 2(1) (2013), 4–29.

'McGuinness: SF Will Not Allow These Dark Elements to Succeed', *Irish News* (3 May 2014) <http://www.irishnews.com/paywall/tsa/irishnews/irishnews/irishnews//news/2014/05/03/news/mcguinness-sf-will-not-allow-these-dark-elements-to-succeed-90652/content.html> accessed 28 March 2016.

McKittrick, David, 'Adams Events Show the Fragility of Peace and How Easily It Can Fracture', *Irish Independent* (5 May 2014) <http://www.independent.ie/opinion/analysis/adams-events-show-the-fragility-of-peace-and-how-easily-it-can-fracture-3024396> accessed 28 March 2016.

——, et al., *Lost Lives: The Stories of the Men, Women and Children who Died as a Result of the Northern Ireland Troubles*, 2nd edn (Edinburgh: Mainstream, 2007).

'Man Arrested in Kingsmill Massacre Probe', *Irish News* (5 August 2016) <http://www.irishnews.com/news/northernirelandnews/2016/08/05/news/man-held-over-kingsmill-massacre-639383/> accessed 15 September 2016.

Moriarty, Gerry, 'Boston Tapes Are Reason Why Adams Is Being Questioned over McConville Murder', *Irish Times* (2 May 2014) <http://www.irishtimes.com/news/ireland/irishnews/boston-tapes-are-reason-why-adams-is-being-questioned-over-mcconville-murder-1.1780908> accessed 28 March 2016.

'Northern Ireland Police Open New Investigation into Kingsmill Massacre', *The Guardian* (31 May 2016) <https://www.theguardian.com/uk-news/2016/may/31/northern-ireland-police-open-new-investigation-into-kingsmill-massacre> accessed 11 June 2016.

'The Nuts and Bolts of a Plan Aimed at Drawing a Line under the Troubles', *Belfast Telegraph* (20 November 2013) <http://www.belfasttelegraph.co.uk/news/northern-ireland/the-nuts-and-bolts-of-a-plan-aimed-at-drawing-a-line-under-the-troubles-the-views-of-attorney-general-john-larkin-on-29768511.html> accessed 26 June 2016.

'Omagh Bomb Plot Man Is Sentenced to 14 Years' Jail', *Independent* (26 January 2002) <http://www.independent.co.uk/news/uk/this-britain/omagh-bomb-plot-man-is-sentenced-to-14-years-jail-9224896.html> accessed 11 June 2016.

'Payment Plan for Northern Ireland Reconciliation Provokes Outrage', *New York Times* (29 January 2009) <http://www.nytimes.com/2009/01/29/world/europe/29ireland.html?_r=0> accessed 30 June 2016.

Pickering, Michael, 'Experience as Horizon: Koselleck, Expectation and Historical Time', *Cultural Studies*, 18(2–3) (2004), 271–89.

'Police Ombudsman Team to Probe Kingsmills Massacre', *News Letter* (11 November 2015) <http://www.newsletter.co.uk/news/northern-ireland-news/police-ombudsman-team-to-probe-kingsmills-massacre-1-7061258> accessed 3 March 2016.

'Police "to Reopen Murder Files"', *BBC News* (5 August 2003) <http://news.bbc.co.uk/1/hi/northern_ireland/3127289.stm> accessed 3 March 2016.

'Report of the Consultative Group on the Past' (23 January 2009) <http://cain.ulst.ac.uk/victims/docs/consultative_group/cgp_230109_report.pdf> accessed 25 February 2016.

Roper, Michael, *The Secret Battle: Emotional Survival in the Great War* (Manchester: Manchester University Press, 2010).

Scott, David, *Omens of Adversity: Tragedy, Time, Memory, Justice* (Durham, NC; London: Duke University Press, 2014).

'Sinn Féin Trying to Blackmail Police over Gerry Adams, Says First Minister', *The Guardian* (4 May 2014) <http://www.theguardian.com/politics/2014/may/04/sinn-fein-police-gerry-adams-peter-robinson> accessed 28 March 2016.

Tamm, Marek, ed., *Afterlife of Events: Perspectives on Mnemohistory* (Basingstoke: Palgrave Macmillan, 2015).

Thomson, Alistair, 'Experience Never Ends: Connecting Life History Work and Experiential Learning', *Australian Journal of Experiential Learning*, 35 (1996), 24–9.

——, 'Experience Never Ends: Migrant Memories, Unsettled Identities and Historical Change', in *Crossroads of History: Experience, Memory, Orality*, Proceedings of the XIth International Oral History Conference, Istanbul, (June 2000).

'Troubles Prosecution Call Slammed', *Belfast Telegraph* (20 November 2013) <http://www.belfasttelegraph.co.uk/news/northern-ireland/troubles-prosecution-call-slammed-29770237.html> accessed 26 June 2016.

'Ulster Police Watchdog Says Cuts Are Delaying Inquiries into Troubles Killings', *The Guardian* (30 September 2014) <http://www.theguardian.com/uk-news/2014/sep/30/ulster-police-watchdog-cuts-delaying-inquiries-troubles-killings> accessed 3 March 2016.

'Victims' Anger at "Amnesty for Terrorists"', *News Letter* (9 January 2008). <http://www.newsletter.co.uk/news//victims-anger-at-amnesty-for-terrorists-1-1862850> accessed 30 June 2016.

Walker, Gail, 'Why the Ghost of Murdered Jean McConville Won't Lie at Rest', *Belfast Telegraph* (6 May 2014) <http://www.nuzhound.com/archive/goto.php?id=244366> accessed 28 March 2016.

Watt, Nicholas, 'Arrest of Gerry Adams Raises Fears for Northern Ireland Peace Process', *The Guardian* (1 May 2014) <http://www.theguardian.com/politics/2014/may/01/arrest-gerry-adams-northern-ireland-sinn-fein-stormont> accessed 28 March 2016.

Whelan, Noel, 'Wounds Still Too Raw for an Examination of the Troubles', *Irish Times* (31 January 2009) <http://www.irishtimes.com/opinion/wounds-still-too-raw-for-an-examination-of-the-troubles-1.1238308> accessed 20 June 2016.

Williams, Raymond, *The Long Revolution* (London: Chatto and Windus, 1961).

——, *Marxism and Literature* (Oxford: Oxford University Press, 1977).

——, *Politics and Letters: Interviews with New Left Review* (London: Verso, 1981).

GUY BEINER

# 13 Irish Studies and the Dynamics of Disremembering

In the decade and a half since the turn of the millennium, Irish Studies has taken on the 'memory boom' with vigour. The significant advances that have been made in our understanding of Ireland's rich memorial cultures and social practices can be sampled in the diverse, though somewhat disjointed, assortment of essays collected in the four volumes of *Memory Ireland*, edited by Oona Frawley.[1] Further impressive achievements are evident in an ever-growing body of scholarly literature, not least in the explosion of new publications issued for the 'Decade of Centenaries' (commemorating the landmark events around the time of the Great War and the Irish Revolution), of which it is still too early to take stock. As a suggestion for future progress, I would like to propose that the study of the dynamics of remembering could benefit from a sharper awareness of the too-often overlooked dynamics of disremembering.

In introducing the concept of 'disremembering' to Irish Studies – and to Memory Studies at large – it is worth noting the seemingly trivial absence of a hyphen. This term is not the product of arcane wordsmithing, as in the fetishized hyphenation of 're-membering' in writings of postmodernists.[2] 'Disremember', a verb that means 'forget', is an Irish vernacular term, with

1 Oona Frawley, *Memory Ireland* (Syracuse, NY: Syracuse University Press, 2010–2014), volume 1: *History and Modernity*; volume 2: *Diaspora and Memory Practices*; volume 3: *The Famine and the Troubles*; volume 4 (co-edited with Katherine O'Callaghan): *James Joyce and Cultural Memory*.

2 Examples in an Irish context of a postmodern excess of hyphens, slashes, brackets, italics and inverted commas can be found in Rebecca Graff-McRae, *Remembering and Forgetting 1916: Commemoration and Conflict in Post-Peace Process Ireland* (Dublin: Irish Academic Press, 2010).

a noted northern regional provenance. Its oral use has been documented, for example, in a late nineteenth-century *Glossary of Words in Use in the Counties of Antrim and Down*.[3] This specifically Irish, if not Ulster, synonym for forgetting can serve to call attention to the imperative to focus critical attention on the study of oblivion as a neglected, I hesitate to say forgotten, form of memory. In order to trace a wider intellectual genealogy for the study of forgetting, I will first consider some classical references, elucidated through later commentary. I will then tease out specific directions for the study of disremembering, which will be demonstrated through examples from modern Irish history.

Upon advocating the need to study memory as essential training in rhetoric, Cicero recounted a tale about Simonides of Ceos, who was commissioned by the nobleman Scopas of Thessaly to write a lyric panegyric to be read out at a feast. Upon completion of the task, the tight-fisted patron refused to pay the full fee and gracelessly advised the poet to turn for recompense to the twin deities Castor and Pollux, to whom part of the poem had been devoted. In an act of true poetic justice, Simonides was indeed rewarded by the Dioscuri, when he was mysteriously summoned from the banquet hall to meet two strangers, after which the building collapsed and all those inside perished under the rubble. The mangled bodies of Scopas and his guests could not be identified by their relatives and would not have been properly buried – or memorialized – had it not been for the remarkable memory of Simonides, who was able to precisely recall where each one of them was seated. This was the founding myth of the classical *ars memoriae* – the 'art of memory' and its elaborate mnemonic method of *loci* (perhaps better known nowadays as the technique of the 'memory palace', popularized once more in the BBC/WGBH reinterpretation of Sherlock Homes). Intriguingly, Cicero prefaced his rendition of the legend with a reference to an apocryphal anecdote about Themistocles of Athens, who allegedly spurned an offer to be taught how 'to remember everything'

3 William Hugh Patterson, *A Glossary of Words in Use in the Counties of Antrim and Down* (London: Published for the English Dialect Society by Trübner & Co., 1880), 30.

and, in the words of Mark Anthony, preferred 'the science of forgetting to that of remembering'.[4]

The yearning of Themistocles to receive instruction on forgetting was derided by Cicero. In the same vein, Frances Yates, in her magisterial study on *The Art of Memory*, dismissed it as a 'frivolous remark', repeating Anthony's verdict that the study of forgetting was a diversion that 'must not cause us to neglect the training of the memory'.[5] However, on further thought, it becomes apparent that forgetting cannot be simply brushed aside. On the margins of mainstream preoccupation with memory, it is possible to trace a fragmented history of illuminating, and far from inconsequential, contemplations on forgetting. Harald Weinrich's inspirational study *Lethe* compiled references found in Homer, Ovid, Plato, Cervantes, Descartes, Goethe, and a host of other philosophers, all the way through to Nietzsche, Freud, and more recent writers, such as Sartre, Semprún and, of course, Borges.[6] One of the many fascinating cases examined by Weinrich relates to Immanuel Kant, who in his advanced age had cause to discharge his long-time manservant Martin Lampe. The elderly Kant was apparently troubled by lingering memories of his old companion and among his papers was found a note, which reads: 'the name Lampe must now be completely forgotten'. It would be too easy to dismiss this seemingly ridiculous reminder to forget as a sign of the great thinker's infirmity, as did Kant's amanuensis – Ehregott Andreas Wasianski (whose discovery of the memorandum paradoxically ensured that the otherwise obscure name of Lampe would be remembered by posterity). However, our

4 Cicero, *de Oratore* 2.74.299–300 and 2.86.351–4; reproduced in Cicero, *On the Orator*, Books 1–2, trans. E. W. Sutton, H. Rackham, Loeb Classical Library 348 (Cambridge, MA: Harvard University Press, 1942), 427; 464–7. Another version of the story of Simonides (described as 'well-known') appears in Quintilian's *Institutio Oratoria*, 11.2.12–17; see Quintilian, *The Orator's Education*, Vol. V: Books 11–12, trans. Donald A. Russell, Loeb Classical Library 494 (Cambridge, MA: Harvard University Press, 1942), 63–7.

5 Frances A. Yates, *The Art of Memory* (London; New York: Routledge, 1999; orig. edn 1966), 17.

6 Harald Weinrich, *Lethe: The Art and Critique of Forgetting* (Ithaca, NY: Cornell University Press, 2004).

intuitive understanding of writing as an *aide de mémoire*, rather than an *aide d'oubli* (to coin a phrase), can be called into question.[7]

In Plato's *Phaedrus*, Socrates tells the story of the Egyptian god Theuth who offered his invention of writing to Thamus, the king of Upper Egypt, promising that it 'will make the Egyptians wiser and will improve their memories'.[8] Thamus remained unconvinced and replied that 'this invention will produce forgetfulness in the minds of those who learn to use it, because they will not practise their memory'. Presenting an argument that seems all too familiar in our present age, in which we tend to consign the things we need to remember to digital appliances, the mythical king explained that putting 'their trust in writing, produced by external characters which are no part of themselves, will discourage the use of their own memory within them'. In a discussion of this text, Jacques Derrida highlighted the inherent ambiguity of writing, which he labelled a *pharmakon* – a drug that is both 'the remedy and the poison'. Derrida's critique is characteristically insightful:

> [I]t is this life of the memory that the *pharmakon* of writing would come to hypnotize: fascinating it, taking it out of itself by putting it to sleep in a monument. Confident of the permanence and independence of its *types* (*tupoi*), memory will fall asleep, will not keep itself up, will no longer keep to keeping itself alert, present, as close as possible to the truth of what is. Letting itself get stoned [*médusée*] by its own signs, its own guardians, by the types committed to the keeping and surveillance of knowledge, it will sink down into *lēthē*, overcome by non-knowledge and forgetfulness.[9]

Studies of cultural memory tend to assume that any representation of the past is an undeniable expression of remembrance, though this is not necessarily the case. The dynamics of social remembrance in relation to literary sources, or, for that matter, in relation to monuments and practically all other productions of cultural memory, are dependent on the reception

7 Weinrich, *Lethe*, 67–77. For an English adaptation of Wasianski's account, see Thomas De Quincey, 'The Last Days of Immanuel Kant', in *Narrative and Miscellaneous Papers*, vol. 2 (Boston, MA: Ticknor, Reed, and Fields, 1853), 267.

8 Plato, *Euthyphro. Apology. Crito. Phaedo. Phaedrus*, trans. Harold North Fowler, Loeb Classical Library 36 (Cambridge, MA: Harvard University Press, 1914), 562–3.

9 See 'Plato's Pharmacy' in Jacques Derrida, *Dissemination*, trans. Barbara Johnson (London: The Athlone Press, 1981), 105; 61–172.

and interaction of readerships and audiences. Scholars who take pride in uncovering a rare and neglected source should also be willing to contemplate that it may offer evidence of disremembering rather than of remembering. There is an evident need for an 'amnesiology', to reiterate Liedeke Plate's case for a 'study of cultural oblivion'.[10]

Umberto Eco recalled that he once indulged with friends in a humorous game of inventing and categorizing 'advertisements for university positions in nonexistent disciplines'. One of the most curious of these figments was an *Ars Oblivionis*, which would presume to elaborate techniques of deliberate forgetting. Eco, who understood mnemotechnics to be a form of semiotics, and as such makes the absent present, argued that memories could at best be obscured but could not be intentionally forgotten. Having introduced the concept of an 'Art of Forgetting', he then negated its viability and advised, tongue in cheek, to 'forget it!'.[11] Nonetheless, cognitive psychologists, unencumbered by the logical stipulations posited by Eco, have investigated 'directed forgetting' since the early experimental work of Robert Bjork in the late 1960s.[12] These studies have since progressed beyond autobiographical forgetting and have also examined intentional forgetting in situations of 'social memory, which occur in conversation or community with other people'.[13]

One is reminded of the proverbial courtroom scenario, familiar to many of us from movies or television dramas, in which a judge determines that certain evidence is inadmissible and should therefore be struck from the record and 'forgotten'. In a treatise on *The Ethics of Memory*, Avishai Margalit observed that such a ruling does not really achieve forgetting, on the contrary: 'If the judge asks the jurors to forget this evidence, this request

10 Liedeke Plate, 'Amnesiology: Towards the Study of Cultural Oblivion', *Memory Studies*, 9(2) (April 2016), 143–55.

11 Umberto Eco, 'An Ars Oblivionalis? Forget It!' *PMLA*, 103(3) (1988), 254–61.

12 Colin M. MacLeod, 'Directed Forgetting', in Colin M. MacLeod and Jonathan M. Golding, eds, *Intentional Forgetting: Interdisciplinary Approaches* (Mahwah, NJ: Lawrence Erlbaum Associates, 1998), 1–57.

13 Celia B. Harris, John Sutton and Amanda J. Barnier, 'Autobiographical Forgetting, Social Forgetting and Situated Forgetting: Forgetting in Context', in Sergio Della Sela, ed., *Forgetting* (Hove: Psychology Press, 2010), 254; 253–84.

would merely guarantee that they would remember it'.[14] The evocation of a subject in order to sanction its effacement, effectively ensuring that it will be remembered in an obscured form, deserves further attention as a peculiar cultural practice that pairs memory with forgetting.

Cutting-edge neurobiological research on memory reconsolidation suggests that it may be possible to erase painful memories and to perhaps even replace them with alternative memories.[15] It would seem that science has caught up with the science fiction of such movies as *Eternal Sunshine of the Spotless Mind* (2004). As of yet, however, these experiments test reactions to a narrow range of specific sensory stimuli, which are a far cry from the complexity of the narratives that are considered in the humanities and social sciences as expressions of traumatic memories. Attempts to adapt clinical research findings into models of collective memory can result in crude concepts of an illusory 'social hippocampus', which fail to take on board the analytical depth in the scholarly literature on history and memory.[16] Though the term 'collective memory' has not gone entirely out of use, there has been a move away from homogenous constructs of memory that are ostensibly shared by an entire society, or at least by substantial sectors within it. In line with a well-established body of more sophisticated studies on 'social memory', which has already been introduced to Irish Studies, I suggest conceptualizing and developing a corresponding term: 'social forgetting'.[17] This label, it seems to me, would be better suited than reviving

14 Avishai Margalit, *The Ethics of Memory* (Cambridge, MA, and London: Harvard University Press, 2002), 201.

15 Elizabeth A. Phelps and Daniela Schiller, 'Reconsolidation in Humans', in Cristina M. Alberini, ed., *Memory Reconsolidation* (London; New York: Elsevier Academic Press, 2013), 185–211.

16 See, for example, Thomas J. Anastasio, Kristen Ann Ehrenberger, Patrick Watson and Wenyi Zhang, *Individual and Collective Memory Consolidation: Analogous Processes on Different Levels* (Cambridge, MA: MIT Press, 2012), esp. 185-7.

17 For the move from collective memory to social memory, see Jeffrey K. Olick and Joyce Robbins, 'Social Memory Studies: From "Collective Memory" to the Historical Sociology of Mnemonic Practices', *Annual Review of Sociology*, 24 (1998), 105–40; see also James Fentress and Chris Wickham, *Social Memory* (Oxford; Cambridge, MA: Blackwell, 1992). For the application of this approach to Irish Studies, see Guy

the similar term 'social amnesia', which was originally employed by Russell Jacoby in an intellectual history of neo-Marxist neglect of psychoanalytical theory that is of less relevance to the discussion here.[18]

Significantly, social forgetting is not 'collective amnesia', which is too readily taken to mean that certain troublesome recollections have been thoroughly purged from a simplistically conceived notion of collective memory. Take, for example, the common assumption that the Great War was subject to collective amnesia in independent Ireland. This allegation was famously articulated by F. X. Martin, who noted in 1967 that 'outside of the Six Counties [of Northern Ireland], it is difficult to find men and women who will acknowledge that they are children of the men who were serving during 1916 in the British Army'. For Martin, this was 'the "Great Oblivion", an example of national amnesia'.[19] Since then, the contention has often been repeated, most recently by Kevin Myers, and is now widely accepted. Nevertheless, it is more than likely that further scrutiny would reveal numerous lesser-known memorial practices through which recollections of Irish participation in the First World War were retained. While decrying a state of 'utter amnesia', Myers conceded that 'amnesia in Ireland can sometimes be no more than a deep morning mist on an entire landscape of personal knowledge'.[20]

Contrary to prevailing assumptions, Irish remembrance of the Great War was not exclusively limited to Protestants of unionist background. An irate letter addressed to the editor of the popular magazine *History Ireland* took issue with the endorsement given by Myles Dungan to the sweeping claim put forward by Myers that 'Ireland, in particular official Ireland, had largely forgotten about one of the most significant passages

---

Beiner, *Remembering the Year of the French: Irish Folk History and Social Memory* (Madison, WI: University of Wisconsin Press, 2006).

18 Russell Jacoby, *Social Amnesia: A Critique of Contemporary Psychology* (New Brunswick, NJ: Transaction Publishers, 1997; orig. edn 1975). Somewhat similar use of amnesia in reference to neglected schools of thought can be found in Alex Law and Eric Royal Lybeck, eds, *Sociological Amnesia: Cross-Currents in Disciplinary History* (Farnham and Burlington, VT: Ashgate, 2015).

19 F. X. Martin, '1916: Myth, Fact, and Mystery', *Studia Hibernica*, 7 (1967), 68.

20 Kevin Myers, *Ireland's Great War* (Dublin: Lilliput Press, 2014), 18.

in our recent history' and that 'we had forgotten that we had forgotten'. In response, the exasperated letter writer, who had grown up in a Dublin working-class neighbourhood in the 1960s and 70s, recalled:

> In the area where I lived, most of the people I knew had some relative or other who had taken part in the First World War. For example, I can still recall an old gentleman (the grand-father of a friend of mine), who lived a few doors away from me, walking down the road with a chest full of jangling medals pinned to his overcoat as he headed for Armistice Day commemorations. We all thought that he was a hero, whatever war he had fought in.[21]

Already before the Great War's eightieth anniversary and the marked rediscovery of interest during the 1990s, the war experiences of Irishmen were recalled in private. These family traditions of veterans re-emerged in the enthusiastic popular response to the radio broadcasts of the 2008 Thomas Davis lecture series on 'Our War' and their subsequent publication. Public recognition of the memory of the war was given a further boost following the visit of Queen Elizabeth II to the Irish War Memorial Garden in Islandbridge, Dublin in 2011.[22]

Herein lies the crux of social forgetting, which pivots on tensions between public silence and private remembrance that can sustain seemingly forgotten memories, allowing them to periodically resurface, though they may subsequently sink back again into oblivion. With these dialectics in mind, the socio-historical approach to memory, inspired by Maurice Halbwachs, who focussed attention on *les cadres sociaux de la mémoire* – the social frameworks through which memory is constructed and maintained, should be revisited in order to chart in the same way *les cadres sociaux de l'oubli*, that is the social frameworks that make social forgetting possible in

21 Eoghan Ó hÁinle, in *History Ireland*, 23(3) (May/June 2015), 12–13; responding to Myles Dungan in *History Ireland*, 23(2) (March/April 2015), 61.

22 John Horne, ed., *Our War: Ireland and the Great War* (Dublin: Royal Irish Academy, 2008). For the transformative impact of the Queen's visit, see Keith Jeffery, 'Irish Varieties of Great War Commemoration', in John Horne and Edward Madigan, eds, *Towards Commemoration: Ireland in War and Revolution, 1912–1923*, (Dublin: Royal Irish Academy, 2013), 117–25.

specific historical contexts.[23] Curiously, in conceptualizing social memory, Halbwachs repressed his own personal recollections of service in the Great War. The lone reference in his writings to the war that shattered his generation (and decimated the cohort of students of Émile Durkheim, to which Halbwachs belonged) is revealing in its omission. Halbwachs merely noted in passing that 'now twelve to fifteen years separate me from the Great War' and then went on to refer to his ability to recall the period before the war, avoiding discussion of his experiences between 1914–1918.[24] This passage unwittingly offers yet another illustration of how a topic can be marked out in order for it to be forgotten. More generally, Halbwachs refrained in his writings on memory from engaging in sustained consideration of forgetting, leaving an impression that he was consciously intent to forget about it. By regarding forgetting simply as the negative of remembering, he apparently assumed that it did not need to be theorized in its own right. The striking absence of social forgetting in Halbwachs's theory of social memory begs redress.

In order to set provisional boundaries for the study of social forgetting, I propose that we first rethink the commonly assumed linear sequence of development, whereby history (as in the event itself) is followed by memory and ultimately succumbs to forgetting. We may be surprised to discover that aspects of forgetting are present from the start, even before the historical event runs its course. When the northern United Irishman William Orr – Ireland's republican protomartyr – went to his execution months before the Great Rebellion of 1798, he expressed anxieties about being forgotten. Orr's 'Dying Declaration' ended with a desperate plea: 'I trust that all my virtuous countrymen will bear me in kind Remembrance'. Almost instantly,

23 Maurice Halbwachs, *Les cadres sociaux de la mémoire* (Paris: Félix Alcan, 1925); for a partial English translation by Lewis A. Coser see Maurice Halbwachs, *On Collective Memory* (Chicago, MA; London: University of Chicago Press, 1992).

24 Maurice Halbwachs, *The Collective Memory*, trans. Francis J. Ditter and Vida Yazdi Ditter (New York: Harper & Row, 1980), 67–8; originally published posthumously as 'Mémoire et société', in *L'Année sociologique*, 3e série, 1 (1949) and reissued as *La Mémoire Collective* (Paris: Presses universitaires de France, 1968; orig. edn 1950). See also Annette Becker, 'Memory Gaps: Maurice Halbwachs, Memory and the Great War', *Journal of European Studies*, 35(1) (March 2005), 102–13.

commemorative memorabilia came into circulation, which incorporated this concern, as evident in verses that appeared on memorial cards: 'When ye forget him ... May you be debar'd that Liberty he sought, and forgotten in the Hist'ry of Nations'.[25] Schema theory, originally introduced by Frederic C. Bartlett in his landmark study *Remembering* (1932), can be adjusted, so as to recognize that apprehensions over a significant event being forgotten, already before it is duly remembered, function as 'pre-forgetting' schemata, which condition the way in which the event is subsequently remembered and initiate traditions of social forgetting.[26]

Shortly after the execution of William Orr, the United Irish poet William Drennan composed the elegy 'The Wake', which advocated a muted form of remembrance:

> Here our worthy brother lies;
> Wake not Him with women's cries;
> Mourn the way that manhood ought,
> Sit in silent trance of thought.

In this poem, which was considered 'a piece written with great power, and which, probably, had more effect on the public mind than any production of the day in prose or verse', the memory of the unnamed Orr is purposely subdued.[27] Thomas Moore, the most celebrated Irish romantic poet of the early nineteenth century, was inspired by a reference Drennan made elsewhere to 'the cold chain of silence'. In a well-known tribute to Robert Emmet, the archetypal republican martyr, Moore perfected Drennan's

25 Guy Beiner, 'Forgetting to Remember Orr: Death and Ambiguous Remembrance in Modern Ireland', in James Kelly and Mary Ann Lyons, eds, *Death and Dying in Ireland, Britain, and Europe: Historical Perspectives* (Dublin: Irish Academic Press, 2013), 171–202.

26 Frederic C. Bartlett, *Remembering: A Study in Experimental and Social Psychology* (Cambridge, MA; New York: Cambridge University Press, 1995; orig. edn 1932).

27 Originally appeared in the *Press* (13 January 1798); subsequently popularized in William Drennan, *Fugitive Pieces, in Verse and Prose* (Belfast: F. D. Finlay, 1815), 79–81. For its contemporary impact see Richard Robert Madden, *The United Irishmen: Their Lives and Times*, 1st series, volume 3 (London: J. Madden & Co., 1842), 45.

formulation of a noiseless ritual of remembrance that is permeated with forgetting:

> Oh! breathe not his name, let it sleep in the shade,
> Where cold and unhonor'd his relics are laid:
> Sad, silent and dark, Be the tears that we shed,
> As the night dew that falls on the grass o'er his head.

Here too, nonverbal remembrance of an intentionally unnamed hero is offered as a powerful alternative to standard practices of memorialization. Silent mourning in private is trusted with preserving memory, in defiance of a prohibition on public commemoration:

> And the tear that we shed, though in secret it rolls,
> Shall long keep his memory, Green in our souls.[28]

In contrast to the tendency of romantic poetry to proclaim memory out loud, in these poems the coded identification of the remembered subject did not need to be spelt out for the intended republican-nationalist readers, who were in the know.

Forgetting is typically diagnosed through omissions, which are noticed in gaps of silence, yet the immediate association of silence with forgetting proves to be misleading. In a sociological study of silence and denial in everyday circumstances, Eviatar Zerubavel discussed the 'heavy sound' of 'thundering silence' that can be heard in situations when a sensitive subject is conspicuously avoided. Such a 'conspiracy of silence' can function as a mnemonic device.[29] When silencing is imposed by the authorities, through the proscription of resistance and protest, dissident remembrance often goes underground, where it is maintained through a less audible subculture of

28 Thomas Moore, *A Selection of Irish Melodies*, 1(1) (Dublin: W. Powers, 1808), 16–18; see also Thomas Moore, *Irish Melodies* (Philadelphia: M. Carey, 1815), 147 fn.

29 Eviatar Zerubavel, *The Elephant in the Room: Silence and Denial in Everyday Life* (Oxford; New York: Oxford University Press, 2006); Eviatar Zerubavel, 'The Social Sound of Silence: Toward a Sociology of Denial', in Efrat Ben-Ze'ev, Ruth Ginio and J. M. Winter, eds, *Shadows of War: A Social History of Silence in the Twentieth Century* (Cambridge; New York: Cambridge University Press, 2010), 32–44.

secrecy. In these situations, recalcitrant memories persevere in private, on the margins of public silence.

The history of Irish secret societies offers ample examples of silenced remembrance. Apart from supressing the well-known republican revolutionary movements – namely the United Irishmen, Young Ireland and the Irish Republican Brotherhood – each of which developed visible rituals of commemoration, police repression also targeted numerous agrarian secret societies such as the Whiteboys, Oakboys, Rightboys, Steelboys, Defenders, Terry Alts, Rockites, Ribbonmen and Molly Maguires. The subversive activities of these clandestine agrarian organizations, which were not inclined to stage high-profile commemorations, surreptitiously stoked bitter recollections of dispossession that fed into disputes over ownership of land and tenant rights. These hidden traditions fuelled such pre-Famine struggles as the Tithe War of the 1830s and were subsequently harnessed to great effect in the Land War of the late nineteenth century. Traces of such silenced memory, which recall lesser-known incidents of violence that were generically labelled 'outrages', are still to be found in local oral traditions.[30]

Silencing can also be the outcome of voluntary adherence to a social taboo, as in the reluctance to openly discuss clerical and institutional abuse in present-day Irish society. The recollections of the victims were shunned from public discourse until shocking revelations began to emerge in the 1990s, in consequence of which the Taoiseach Bertie Ahern issued in May 1999 a public apology to the children that had been abused while in state care. In 2009, the findings of official investigations were made public in the report of the Commission to Inquire into Child Abuse (also known as the Ryan Report) and the report of the Commission of Investigation into the Catholic Archdiocese of Dublin (the Murphy Report). In addition, as shown by Emilie Pine, various genres of cultural productions, namely memoir, prose fiction, film and in particular theatre, contributed to

30 Breandán Mac Suibhne, *The End of the Outrage* (Oxford: Oxford University Press, forthcoming); see also Fergus Campbell, '"Killing Time" in Rural Ireland, c.1881–2013', in *Irish Studies Review*, 21(3) (2013), 1–19.

undermining the 'deference to a culture of silence and amnesia'.[31] Whether imposed though official prohibition or social disapproval, maintaining silence can make the 'elephant in the room' all the more apparent. Staged silence is a mechanism of social forgetting to cultivate troubling recollections that unsettle the hegemonic narratives of collective memory. Disremembering bides its time and seizes upon opportunities in which repressed memories can, often momentarily, re-emerge into the open.

Although silencing is undoubtedly a universal phenomenon, by tuning in to the 'sounds of silence' (to borrow from Simon and Garfunkel), we can try and discern national and regional distinctions. Ireland may have developed its own patterns of 'vernacular silence', which in turn generated distinctly Irish practices of social forgetting. The mapping of silence, it seems to me, stands to confirm that Northern Ireland is particularly fertile ground for such an exploration, which returns us to the northern provenance of the colloquial term 'disremember'. The inclination in Ulster to refrain from speaking one's mind in public was captured in Seamus Heaney's pithy reference to 'northern reticence, the tight gag of place' in his poem 'Whatever You Say, Say Nothing'. The poem's evocative title plays on a paramilitary slogan, which is also shared with a shrewdly humorous song by the Co. Down balladeer Colum Sands.[32] For centuries, the politicized sectarian tensions entrenched in Ulster society discouraged free speech and muffled memories that did not conform to the essentialist dualism of the 'Two Traditions' paradigm, which positions the memory of Catholic nationalists in binary opposition to the memory of Protestant loyalists. Under the surface, however, the landscape of memory is considerably more complicated. An in-depth historical examination of the social forgetting

31 Emilie Pine, *The Politics of Irish Memory: Performing Remembrance in Contemporary Irish Culture* (Houndmills; New York: Palgrave Macmillan, 2011), 18–51; Emilie Pine, 'The Abuse of History/A History of Abuse: Theatre as Memory and the Abbey's "Darkest Corner"', in Christopher Collins and Mary P. Caulfield, eds, *Ireland, Memory and Performing the Historical Imagination* (Houndmills; New York: Palgrave Macmillan, 2014), 207–22. The Northern Ireland Historical Institutional Abuse Inquiry, est. 2013, has yet to publish its findings.

32 Seamus Heaney, *North* (London: Faber and Faber, 2001; orig. edn 1975), 52–5; Colum Sands, *Unapproved Road* (Spring Records, 1981).

of the 1798 rebellion in counties Antrim and Down – popularly known as 'The Turn-Out' – reveals a vernacular historiography, which gave voice to the many ways in which Presbyterians continued to recall their United Irish ancestry, while maintaining a silent public façade of oblivion.[33]

During 'The Troubles' – a characteristic understatement for the thirty-year-long upsurge in violence in Northern Ireland – paramilitary intimidation and terrorism, alongside military surveillance and repression, reinforced the local culture of silence. An IRA poster from the 1970s, illustrated with a daunting photo of a masked gunman, imitated wartime propaganda posters in cautioning that 'loose-talk costs lives' and issuing strict instructions for self-censorship:

> In taxis
> On the phone
> In clubs and bars
> At football matches
> At home with friends
> Anywhere!
> Whatever you say – say nothing[34]

At the same time, the British authorities were determined to censor and silence republicans, going so far as to impose in the United Kingdom a television and radio ban on broadcasting the voices of Sinn Féin representatives.[35]

In certain cases, the barriers of silencing have been broken. The conclusions of the tribunal appointed in February 1972 to investigate the killings of civilian protesters by paratroopers in Derry on Bloody Sunday (the Widgery Report) dismissively ignored the recollections of aggrieved

33 Guy Beiner, 'Disremembering 1798?: An Archaeology of Social Forgetting and Remembrance in Ulster', *History & Memory*, 25(1) (2013), 9–50.

34 Marc Mulholland, *The Longest War: Northern Ireland's Troubled History* (Oxford; New York: Oxford University Press, 2002), 100.

35 Ed Moloney, 'Closing Down the Airwaves: The Story of the Broadcasting Ban', in Bill Rolston, ed., *The Media and Northern Ireland: Covering the Troubles* (Houndmills; London: Macmillan, 1991), 8–50. For contemporaneous censorship in the Republic of Ireland see in the same volume: Betty Purcell, 'The Silence in Irish Broadcasting', 51–68.

eyewitnesses. For over three decades, the demand to lend an ear to these silenced memories was tirelessly repeated by the Bloody Sunday Justice Campaign, until it was ultimately vindicated with the full disclosure in 2010 of the findings of the Bloody Sunday Inquiry (the Saville Report).[36] However, countless other traumatic memories remain unaddressed, even though post-conflict Northern Ireland is trying hard to face its demons.

Initiatives to tackle the legacy of the Northern Irish conflict through setting up in 2005 a Historical Enquiries Team (HET) within the Police Service of Northern Ireland (PSNI), or through the establishment in 2008 of a Commission for Victims and Survivors (CVS) and a Victims and Survivors Service (VSS), have stumbled when attempting to address the thorny issues of memory. The multiparty talks chaired in 2013 by the American diplomats Dr Richard Haass and Prof Meghan O'Sullivan failed to reach agreements on 'dealing with the past'. Remembrance of the victims of the Troubles continues to be a contentious issue, which the Stormont House Agreement (2014) has tried to resolve by establishing an Oral History Archive, in what has been labelled 'a new politics of storytelling'. Yet it is not at all clear whether an official archive can become a repository for recollections that have long remained outside of the public domain.[37] Meanwhile, opinions on public commemoration of the victims of the Troubles seem to be irreconcilable.[38]

The *raison d'être* of disremembering is the removal of memory from the public sphere, rather than its total obliteration. Social forgetting encourages

36 Charlotte Barcat, '"A Truth for the World": From Widgery to Saville, the Campaign for Truth and Justice About Bloody Sunday', in Lesley Lelourec and Grainne O'Keeffe-Vigneron, eds, *Ireland and Victims: Confronting the Past, Forging the Future* (Oxford; New York: Peter Lang, 2012), 59–73.

37 Cillian McGrattan, 'The Stormont House Agreement and the New Politics of Storytelling in Northern Ireland', *Parliamentary Affairs* (2015), 1–19; Brandon Hamber and Gráinne Kelly, 'Practice, Power and Inertia: Personal Narrative, Archives and Dealing with the Past in Northern Ireland', *Journal of Human Rights Practice* (2016), 1–20.

38 John D. Brewer and Bernadette C. Hayes, 'Victimhood and Attitudes towards Dealing with the Legacy of a Violent Past: Northern Ireland as a Case Study', *British Journal of Politics & International Relations*, 17(3) (2015), 512–30.

active opposition to commemoration, often leading to action that can be labelled 'decommemorating'. This can take the form of a range of hostile reactions that result in the enforcement of prohibitions by vigilantes or in outbursts of vandalism and iconoclasm. There is a cultural history waiting to be written of the destruction of monuments in Ireland, which would cover, among other episodes, the purge of loyalist and imperial statues from the streets of Dublin and other sites in independent Ireland.[39] Assaults on memorials do not result in the elimination of memory. As shown in studies of *damnatio memoriae* in ancient Rome, the defacement of monuments enacts rituals of 'sham oblivion' that actually preserve memory in mutilated forms.[40]

An archaeological study of contemporary material culture commented on the inherent interdependence of commemoration and 'decommemoration':

> It is not just that memorials commemorate and iconoclasm causes forgetfulness; the relation between remembrance and forgetfulness is not a linear process but a struggle, a tension – in every memorial, something has been left out or forgotten, in every removal, something is left behind, remembered. In both cases, it is what is *not there*, what is *absent* that causes this tension.[41]

39 See Yvonne Whelan, 'The Construction and Destruction of a Colonial Landscape: Monuments to British Monarchs in Dublin before and after Independence', *Journal of Historical Geography*, 28(4) (2002), 508–33.

40 Charles W. Hedrick, *History and Silence: Purge and Rehabilitation of Memory in Late Antiquity* (Austin, TX: University of Texas Press, 2000), 88–130; Peter Stewart, *Statues in Roman Society: Representation and Response* (Oxford: Oxford University Press, 2003), 267–99. See also Jaś Elsner, 'Iconoclasm and the Preservation of Memory', in Robert S. Nelson and Margaret Olin, eds, *Monuments and Memory, Made and Unmade* (Chicago: University of Chicago Press, 2003), 209–32; Eric R. Varner, *Mutilation and Transformation: Damnatio Memoriae and Roman Imperial Portraiture* (Leiden; Boston, MA: Brill, 2004); Harriet I. Flower, *The Art of Forgetting: Disgrace and Oblivion in Roman Political Culture* (Chapel Hill, NC: University of North Carolina Press, 2006).

41 Victor Buchli and Gavin Lucas, 'Between Remembering and Forgetting', in Victor Buchli and Gavin Lucas, eds, *Archaeologies of the Contemporary Past* (London and New York: Routledge, 2001), 80 (italics as in the original).

There are numerous examples throughout late-modern history of how a gaping void left by the obliteration of a cultural site can acquire symbolic meaning.[42] The location of the 1987 Remembrance Day bombing in Enniskillen, marked since 2002 by a Clinton Centre dedicated to peace and reconciliation, has become a compound site of remembrance that evokes memories of the Great War and of the Troubles.[43] In this way, decommemorating, though intended as an act of social forgetting, can actually serve to recharge memory.

In light of the realization that memory can be revitalized, accepted notions of transience – defined by Daniel Shacter as one of the 'sins of memory', whereby remembrance ultimately succumbs to forgetting – need to be rethought. In place of the forgetting curve first measured by Hermann Ebbinghaus in 1885, an alternative hypothesis might come up with a model more akin to a sine graph, which would look beyond decline and follow the dynamics of regeneration of social memory.[44] Likewise, practices of social forgetting can also be regenerated, insofar as they can be transmitted to a second generation (and beyond) in a fashion similar to what has been labelled postmemory, thereby creating trans-generational traditions that facilitate a lasting culture of social forgetting.[45]

It follows that what is needed is a lethnohistory, which would identify and map sites of social forgetting – the Irish *lieux d'oubli* – and then chart the cycles of decline and renewal, evident in moments of silencing of memories: moments in which these memories come out into the limelight, and moments when they are forced back into retreat. Periods

42 See Robert Bevan, *The Destruction of Memory: Architecture at War* (London: Reaktion, 2006).

43 Graham Dawson, *Making Peace with the Past? Memory, Trauma and the Irish Troubles* (Manchester: Manchester University Press, 2007), 288–305.

44 For forgetting as a function of the passing of time, see Hermann Ebbinghaus, *Memory: A Contribution to Experimental Psychology* (New York: Teachers College, Columbia University, 1913), 62–80; Daniel L. Schacter, *The Seven Sins of Memory: How the Mind Forgets and Remembers* (Boston, MA: Houghton Mifflin, 2001), 12–40.

45 Guy Beiner, 'Probing the Boundaries of Irish Memory: From Postmemory to Prememory and Back', *Irish Historical Studies*, 39(154) (2014), 296–307 (esp. 298–304).

of intense commemoration – such as the excessively fêted 'Decade of Commemorations' – give an illusory impression that social forgetting has come to an end, though this does not necessarily prove to be the case in the long term. The Bicentenary of the 1798 Rebellion (which was part of a previous decade of commemorations) was applauded as a time in which northern unionist Presbyterians finally overcame their inhibitions about remembering in public their United Irish republican ancestors.[46] Since then, there have been various indications that this honeymoon with commemoration was short-lived and that the willingness to remember in public soon diminished amongst Presbyterians in Antrim and Down. For example, a proposal in 2004 to name a new bridge in Toome, Co. Antrim, after the local '98 folk hero Roddy McCorley was thwarted by unionists on the local council; in 2008 an initiative to promote tourism to a site associated with the '98 folk heroine Betsy Gray was shot down by the unionist-dominated Ards Borough Council in Co. Down; and in 2014 a blue plaque commemorating the '98 working-man hero James (Jemmie) Hope was vandalized shortly after it was put up in Mallusk Cemetery, Co. Antrim. These relatively minor incidents of local opposition to commemorating 1798 are part of a renewed decommemorating drive, which reveals that at least some Presbyterian unionists are no longer interested in sharing in public their private traditions of the Turn-Out with republicans. It would seem that deep-rooted traditions of social forgetting in Ulster have made a comeback.[47]

It has become commonplace in Memory Studies to associate our times with forgetting. Pierre Nora famously declared that 'we speak so much of memory because there is so little of it left', Andreas Huyssen wrote about a 'culture of amnesia', and Paul Connerton authored a study on 'how

46 Peter Collins, *Who Fears to Speak of '98? Commemoration and the Continuing Impact of the United Irishmen* (Belfast: Ulster Historical Foundation, 2004), 84–121.

47 For the Roddy McCorley bridge controversy, see *News Letter* (16 March and 15 April 2004); *Irish News* (15 April 2004); *An Phoblacht* (15 April 2004). For the Betsy Gray heritage controversy, see *Irish News* (11 September 2008); *An Phoblacht* (2 October 2008). For the vandalism of the James Hope plaque see *News Letter* (25 April 2014); *Belfast Telegraph* (25 April 2014).

modernity forgets'.[48] This commentary fails to take on board the cyber revolution. Viktor Mayer-Schönberger argued in a monograph on 'The Virtue of Forgetting in the Digital Age' that the balance between remembering and forgetting has undergone a paradigm change:

> Since the beginning of time, for us humans, forgetting has been the norm and remembering the exception. Because of digital technology and global networks, however, this balance has shifted. Today, with the help of widespread technology, forgetting has become the exception, and remembering the default.[49]

Far from succumbing to oblivion, it seems that we may actually be moving towards a digital age of total memory, which promises to eradicate forgetting altogether. Anxiety over losing the ability to forget is an old trope, which was masterfully depicted in Jorge Louis Borges's short story on '*Funes el memorioso*' ['Funes the Memorious'], who 'remembered not only every leaf of every tree of every wood, but also every one of the times he had perceived or imagined it'.[50] A much older variant appeared already in an ancient Irish legend, set in Ulster, about the warrior scholar Cenn Faeladh Mac Aillela, who was believed to have sustained an injury at the battle of Magh Rath [Moire, Co. Down] in 636 and to have 'lost his brain of forgetting' [*inchind dermaid*].[51] Forgetting is not a malfunction of memory, but is in fact indispensable for everyday functioning: our ability to remember relies on our ability to forget.

In 2012, the European Commission disclosed a draft European Data Protection Regulation, which includes a clause on 'the right to be forgotten'

48 Pierre Nora, 'Between Memory and History: *Les Lieux de Mémoire*', *Representations*, 26 (1989), 7–24; Andreas Huyssen, *Twilight Memories: Marking Time in a Culture of Amnesia* (New York: Routledge, 1995); Paul Connerton, *How Modernity Forgets* (Cambridge; New York: Cambridge University Press, 2009).

49 Viktor Mayer-Schönberger, *Delete: The Virtue of Forgetting in the Digital Age* (Princeton, NJ: Princeton University Press, 2009), 2.

50 Jorge Luis Borges, *Labyrinths: Selected Stories & Other Writings* (New York: New Directions, 1964), 69–75.

51 David Georgi, 'A Stunning Blow on the Head: Literacy and the Anxiety of Memory in the Legend of Cenn Faelad's Brain of Forgetting', *Proceedings of the Harvard Celtic Colloquium*, 16(17) (1996), 195–205.

(Article 17), and officials within the European Union have subsequently been pushing for legislation that would enable citizens to demand the removal of personal information from Google search engine results in order that it could be forgotten.[52] Putting aside technical questions over whether it is at all feasible to completely erase information once it has appeared on the World Wide Web, the debate can be reconfigured not as a juristic conflict between rights of privacy and freedom of speech, but as a conflict between rights of memory and rights of oblivion, which is reflective of the interplay between rites of social memory and rites of social forgetting. Milan Kundera's famous quip that 'the struggle of man against power is the struggle of memory against forgetting' (which appeared in a critique of the airbrushing of historical memory under totalitarian communism) positions memory and forgetting as separate antithetical entities. Paul Ricoeur, however, has persuasively argued for the 'imbrication of forgetting in memory'.[53] The validity of this claim has been strengthened by research in cognitive psychology, which has similarly concluded that 'the study of forgetting cannot be separated from the study of memory'.[54] Somewhat like the Dioscuri, mentioned earlier in the story of Simonides, memory and forgetting are inseparably intertwined and are more like Siamese twins than stand-alone siblings. We can therefore benefit from examining forgetting, not as the Other of memory, but as part and parcel of memory. Accordingly, redirecting critical attention to explorations of social forgetting, as a form of remembering, shows considerable potential for Irish Studies, as well as for Memory Studies at large.

52 Christiana Markou, 'The "Right to Be Forgotten": Ten Reasons Why It Should Be Forgotten', in Serge Gutwirth, Ronald Leenes and Paul de Hert, eds, *Reforming European Data Protection Law* (Dordrecht, Heidelberg, New York and London: Springer, 2015), 203–26.

53 Paul Ricoeur, *Memory, History, Forgetting* (Chicago, IL: University of Chicago Press, 2004), 426.

54 Roberto Cubelli, 'A New Taxonomy of Memory and Forgetting', in Sergio Della Sela, ed., *Forgetting* (Hove; New York: Psychology Press, 2010), 42.

# Bibliography


*An Phoblacht* (15 April 2004); (2 October 2008).

Anastasio, Thomas J., Ehrenberger, Kristen Ann, Watson, Patrick and Zhang, Wenyi, *Individual and Collective Memory Consolidation: Analogous Processes on Different Levels* (Cambridge, MA: MIT Press, 2012).

Barcat, Charlotte, '"A Truth for the World": From Widgery to Saville, the Campaign for Truth and Justice About Bloody Sunday', in Lesley Lelourec and Grainne O'Keeffe Vigneron, eds, *Ireland and Victims: Confronting the Past, Forging the Future* (Oxford; New York: Peter Lang, 2012), 59–73.

Bartlett, Frederic C., *Remembering: A Study in Experimental and Social Psychology* (Cambridge; New York: Cambridge University Press, 1995; orig. edn 1932).

Becker, Annette, 'Memory Gaps: Maurice Halbwachs, Memory and the Great War', *Journal of European Studies*, 35(1) (2005), 102–13.

Beiner, Guy, 'Disremembering 1798?: An Archaeology of Social Forgetting and Remembrance in Ulster', *History & Memory*, 25(1) (2013), 9–50.

——, 'Forgetting to Remember Orr: Death and Ambiguous Remembrance in Modern Ireland', in James Kelly and Mary Ann Lyons, eds, *Death and Dying in Ireland, Britain, and Europe: Historical Perspectives* (Dublin: Irish Academic Press, 2013), 171–202.

——, 'Probing the Boundaries of Irish Memory: From Postmemory to Prememory and Back', *Irish Historical Studies*, 39(154) (2014), 296–307.

——, *Remembering the Year of the French: Irish Folk History and Social Memory* (Madison, WI: University of Wisconsin Press, 2006).

*Belfast News Letter* (15 and 16 April 2004); (25 April 2014).

*Belfast Telegraph* (25 April 2014).

Bevan, Robert, *The Destruction of Memory: Architecture at War* (London: Reaktion, 2006).

Borges, Jorge Luis, *Labyrinths: Selected Stories & Other Writings* (New York: New Directions, 1964).

Brewer, John D., and Hayes, Bernadette C., 'Victimhood and Attitudes towards Dealing with the Legacy of a Violent Past: Northern Ireland as a Case Study', *British Journal of Politics & International Relations*, 17(3) (2015), 512–30.

Buchli, Victor, and Lucas, Gavin, 'Between Remembering and Forgetting', in Victor Buchli and Gavin Lucas, eds, *Archaeologies of the Contemporary Past* (London and New York: Routledge, 2001).

Campbell, Fergus, '"Killing Time" in Rural Ireland, c.1881–2013', in *Irish Studies Review*, 21(3) (2013), 1–19.

Cicero, *On the Orator*, Books 1–2, trans. E. W. Sutton, H. Rackham, Loeb Classical Library 348 (Cambridge, MA: Harvard University Press, 1942).
Collins, Peter, *Who Fears to Speak of '98? Commemoration and the Continuing Impact of the United Irishmen* (Belfast: Ulster Historical Foundation, 2004).
Colum Sands, *Unapproved Road* (Spring Records, 1981).
Connerton, Paul, *How Modernity Forgets* (Cambridge; New York: Cambridge University Press, 2009).
Cubelli, Roberto, 'A New Taxonomy of Memory and Forgetting', in Sergio Della Sela, ed., *Forgetting* (Hove; New York: Psychology Press, 2010), 35–47.
Dawson, Graham, *Making Peace with the Past? Memory, Trauma and the Irish Troubles* (Manchester: Manchester University Press, 2007).
Derrida, Jacques, *Dissemination*, trans. Barbara Johnson (London: The Athlone Press, 1981).
Drennan, William, *Fugitive Pieces, in Verse and Prose* (Belfast: F. D. Finlay, 1815).
Ebbinghaus, Hermann, *Memory: A Contribution to Experimental Psychology* (New York: Teachers College, Columbia University, 1913).
Eco, Umberto, 'An Ars Oblivionalis? Forget It!', *PMLA*, 103(3) (1988), 254–61.
Elsner, Jaś, 'Iconoclasm and the Preservation of Memory', in Robert S. Nelson and Margaret Olin, eds, *Monuments and Memory, Made and Unmade* (Chicago, IL: University of Chicago Press, 2003), 209–32.
Fentress, James, and Wickham, Chris, *Social Memory* (Oxford; Cambridge, MA: Blackwell, 1992).
Flower, Harriet I., *The Art of Forgetting: Disgrace and Oblivion in Roman Political Culture* (Chapel Hill, NC: University of North Carolina Press, 2006).
Frawley, Oona, *Memory Ireland*, volume 1: *History and Modernity* (Syracuse, NY: Syracuse University Press, 2010).
——, *Memory Ireland*, volume 2: *Diaspora and Memory Practices* (Syracuse, NY: Syracuse University Press, 2012).
——, *Memory Ireland*, volume 3: *The Famine and the Troubles* (Syracuse, NY: Syracuse University Press, 2014).
——, *Memory Ireland*, volume 4: *James Joyce and Cultural Memory* (Syracuse, NY: Syracuse University Press, 2014).
Georgi, David, 'A Stunning Blow on the Head: Literacy and the Anxiety of Memory in the Legend of Cenn Faelad's Brain of Forgetting', *Proceedings of the Harvard Celtic Colloquium*, 16(17) (1996), 195–205.
Halbwachs, Maurice, *Les cadres sociaux de la mémoire* (Paris: Félix Alcan, 1925).
——, *The Collective Memory*, trans. Francis J. Ditter and Vida Yazdi Ditter (New York: Harper & Row, 1980).

Hamber, Brandon, and Kelly, Gráinne, 'Practice, Power and Inertia: Personal Narrative, Archives and Dealing with the Past in Northern Ireland', *Journal of Human Rights Practice* (2016), 1–20.
Harris, Celia B., Sutton, John, and Barnier, Amanda J., 'Autobiographical Forgetting, Social Forgetting and Situated Forgetting: Forgetting in Context', in Sergio Della Sela, ed., *Forgetting* (Hove: Psychology Press, 2010), 253–84.
Heaney, Seamus, *North* (London: Faber and Faber, 2001; orig. edn 1975).
Hedrick, Charles W., *History and Silence: Purge and Rehabilitation of Memory in Late Antiquity* (Austin, TX: University of Texas Press, 2000).
Horne, John, ed., *Our War: Ireland and the Great War* (Dublin: Royal Irish Academy, 2008).
Huyssen, Andreas, *Twilight Memories: Marking Time in a Culture of Amnesia* (New York: Routledge, 1995).
*Irish News* (15 April 2004); (11 September 2008).
Jacoby, Russell, *Social Amnesia: A Critique of Contemporary Psychology* (New Brunswick, NJ: Transaction Publishers, 1997; orig. edn 1975).
Jeffery, Keith, 'Irish Varieties of Great War Commemoration', in John Horne and Edward Madigan, eds, *Towards Commemoration: Ireland in War and Revolution, 1912–1923*, (Dublin: Royal Irish Academy, 2013), 117–25.
Mac Suibhne, Breandán, *The End of the Outrage* (Oxford: Oxford University Press, forthcoming).
McGrattan, Cillian, 'The Stormont House Agreement and the New Politics of Storytelling in Northern Ireland', *Parliamentary Affairs* (2015), 1–19.
MacLeod, Colin M., 'Directed Forgetting', in Colin M. MacLeod and Jonathan M. Golding, eds, *Intentional Forgetting: Interdisciplinary Approaches* (Mahwah, NJ: Lawrence Erlbaum Associates, 1998).
Madden, Richard Robert, *The United Irishmen: Their Lives and Times*, 1st series, volume 3 (London: J. Madden & Co., 1842).
Margalit, Avishai, *The Ethics of Memory* (Cambridge, MA; London: Harvard University Press, 2002).
Markou, Christina, 'The "Right to Be Forgotten": Ten Reasons Why It Should Be Forgotten', in Serge Gutwirth, Ronald Leenes and Paul de Hert, eds, *Reforming European Data Protection Law* (Dordrecht, Heidelberg, New York and London: Springer, 2015), 203–26.
Martin, F. X., '1916: Myth, Fact, and Mystery', *Studia Hibernica*, 7 (1967), 7–126.
Mayer-Schönberger, Viktor, *Delete: The Virtue of Forgetting in the Digital Age* (Princeton, NJ: Princeton University Press, 2009).
Moloney, Ed, 'Closing Down the Airwaves: The Story of the Broadcasting Ban', in Bill Rolston, ed., *The Media and Northern Ireland: Covering the Troubles* (Houndmills; London: Macmillan, 1991), 8–50.

Moore, Thomas, *A Selection of Irish Melodies*, 1(1) (Dublin: W. Powers, 1808).
——, *Irish Melodies* (Philadelphia: M. Carey, 1815).
Mulholland, Marc, *The Longest War: Northern Ireland's Troubled History* (Oxford; New York: Oxford University Press, 2002).
Myers, Kevin, *Ireland's Great War* (Dublin: Lilliput Press, 2014).
Nora, Pierre, 'Between Memory and History: *Les Lieux de Mémoire*', *Representations*, 26 (1989), 7–24.
Olick, Jeffrey K., and Robbins, Joyce, 'Social Memory Studies: From "Collective Memory" to the Historical Sociology of Mnemonic Practices', *Annual Review of Sociology*, 24 (1998), 105–40.
Patterson, William Hugh, *A Glossary of Words in Use in the Counties of Antrim and Down* (London: Published for the English Dialect Society by Trübner & Co., 1880).
Phelps, Elizabeth A., and Schiller, Daniela, 'Reconsolidation in Humans', in Cristina M. Alberini, ed., *Memory Reconsolidation* (London; New York: Elsevier Academic Press, 2013), 185–211.
Pine, Emilie, 'The Abuse of History/A History of Abuse: Theatre as Memory and the Abbey's "Darkest Corner"', in Christopher Collins and Mary P. Caulfield, eds, *Ireland, Memory and Performing the Historical Imagination* (Houndmills; New York: Palgrave Macmillan, 2014), 207–22.
——, *The Politics of Irish Memory: Performing Remembrance in Contemporary Irish Culture* (Houndmills; New York: Palgrave Macmillan, 2011).
Plate, Liedeke, 'Amnesiology: Towards the Study of Cultural Oblivion', *Memory Studies*, 9(2) (2016), 143–55.
Plato, *Euthyphro. Apology. Crito. Phaedo. Phaedrus*, trans. Harold North Fowler, Loeb Classical Library 36 (Cambridge, MA: Harvard University Press, 1914).
Ricoeur, Paul, *Memory, History, Forgetting* (Chicago: University of Chicago Press, 2004).
Schacter, Daniel L., *The Seven Sins of Memory: How the Mind Forgets and Remembers* (Boston, MA: Houghton Mifflin, 2001).
Stewart, Peter, *Statues in Roman Society: Representation and Response* (Oxford: Oxford University Press, 2003).
Varner, Eric R., *Mutilation and Transformation: Damnatio Memoriae and Roman Imperial Portraiture* (Leiden; Boston, MA: Brill, 2004).
Weinrich, Heinrich, *Lethe: The Art and Critique of Forgetting* (Ithaca, NY: Cornell University Press, 2004).
Whelan, Yvonne, 'The Construction and Destruction of a Colonial Landscape: Monuments to British Monarchs in Dublin before and after Independence', *Journal of Historical Geography*, 28(4) (2002), 508–33.

Yates, Frances A., *The Art of Memory* (London; New York: Routledge, 1999; orig. edn 1966).
Zerubavel, Eviatar, *The Elephant in the Room: Silence and Denial in Everyday Life* (Oxford; New York: Oxford University Press, 2006).
——, 'The Social Sound of Silence: Toward a Sociology of Denial', in Efrat Ben Ze'ev, Ruth Ginio and J. M. Winter, eds, *Shadows of War: A Social History of Silence in the Twentieth Century* (Cambridge; New York: Cambridge University Press, 2010), 32–44.

# Notes on Contributors

NELSON BARRE is Visiting Assistant Professor of Theatre Arts at the University of Oregon. He received his PhD from the National University of Ireland, Galway, where he was awarded a Hardiman Research Scholarship for his doctoral work on the plays of Enda Walsh. His research focuses on memory, ritual and performance in contemporary theatre and performance. His writing has appeared in *New Hibernia Review*, *Comparative Drama*, *Theatre Journal* and essay collections on Enda Walsh, Mark O'Rowe and American Theatre.

GAIL BAYLIS lectures at the University of Ulster where she teaches the history and theory of photography, visual culture and gender studies. Her latest publications include 'Boy culture and Ireland 1916', *Early Popular Visual Culture* 11(3) (August 2015), 192–208; 'Remembering to Forget: Marginalised Visual Narratives in the Irish Nation Narrative', *Kynmpa/Culture* 7 (2014), 123–135; 'Gender in the Frame: photography and the performance of the nation narrative in early twentieth-century Ireland', *Irish Studies Review* 22(2) (May 2014), 184–206 and 'A Few Too Many Photographs? Indexing Digital Histories', *History of Photography* 38(1) (February 2014), 3–20.

GUY BEINER is Senior Lecturer of Modern History at Ben-Gurion University of the Negev in Israel and has held research fellowships at University College Dublin, Trinity College Dublin, the University of Notre Dame, the Central European University and the University of Oxford. He specializes in the study of memory, with a particular interest in forgetting, and is the author of the prize-winning book *Remembering the Year of the French: Irish Folk History and Social Memory* (University of Wisconsin Press, 2007).

RUUD VAN DEN BEUKEN is Lecturer at the University of Groningen and Radboud University Nijmegen. In 2015, he was awarded the Irish Society for Theatre Research's (ISTR) New Scholars' Prize. He is one of the editors of *Global Legacies of the Great Irish Famine: Transnational and Interdisciplinary Perspectives* (Oxford: Peter Lang, 2014) and has published an article on commemorations of the Easter Rising at the Gate Theatre in *Irish Studies Review* (2015). He is currently finishing a PhD thesis on the Gate Theatre's original playwrights titled *Memory, Modernity and (Inter) nationalist Identities at the Dublin Gate Theatre, 1928–1940*.

MARGUÉRITE CORPORAAL is Associate Professor in English Literature at Radboud University, Nijmegen. She was the principal investigator and coordinator of the research project *Relocated Remembrance: The Great Famine in Irish (Diaspora) Fiction, 1847–1921*, for which she obtained a Starting Grant from the European Research Council (2010–2015). Furthermore, she is the director of the International Network of Irish Famine Studies, which is funded by the Dutch research council, NWO (2014–2017). Among her recent and forthcoming international publications are: *Recollecting Hunger: An Anthology. Cultural Memories of the Great Famine in Irish Fiction, 1847–1920* (co-authored, Irish Academic Press, 2012); *Global Legacies of the Great Irish Famine: Transnational and Interdisciplinary Perspectives* (co-edited, Peter Lang, 2014); *Travelling Irishness in the Long Nineteenth Century* (co-edited, forthcoming with Palgrave, 2017); and *Relocated Memories of the Great Famine in Irish and Diaspora Fiction, 1846–1870* (forthcoming with Syracuse UP, 2017).

CHRISTOPHER CUSACK is completing his PhD at Radboud University, the Netherlands, and is a lecturer at HAN University of Applied Sciences. He has published articles and chapters on representations of the Great Famine in Irish and Irish diaspora fiction, as well as on the historiography of the Famine and the diaspora. He is a co-editor of *Recollecting Hunger* (2012) and *Global Legacies of the Great Irish Famine* (2014).

GRAHAM DAWSON is Professor of Historical Cultural Studies and Director of the Centre for Research in Memory, Narrative and Histories at the University of Brighton, England. He is author of *Soldier Heroes: British*

*Adventure, Empire and the Imagining of Masculinities* (1994) and *Making Peace with the Past? Memory, Trauma and the Irish Troubles* (2007). He is also co-editor of *Trauma and Life Stories* (1999), *The Politics of War Memory and Commemoration* (2000) and *Contested Spaces: Sites, Representations and Histories of Conflict* (2007), and is currently co-editing a collection on *The Northern Ireland Troubles in Britain: Impacts, Engagements, Legacies and Memories*, to be published by Manchester University Press in 2016. Alongside continuing work on the Irish peace process and legacies of the Troubles in Ireland and Britain, his current interests lie in the cultural dimensions of dealing with the past within conflict transformation processes, involving questions of memory and subjectivity, representation, imaginative geography, historical justice and human rights.

TRACY FAHEY is Head of the Department of Fine Art and Head of the Centre of Postgraduate Studies in Limerick School of Art and Design (LSAD). In 2013, she established the LSAD research centre ACADEMY, where she also acts as principal investigator. Her primary research area is the Gothic with special reference to the visual art with chapters on this subject in collections published by Routledge, Palgrave, Manchester University Press, McParland, Cork University Press and Rowman & Littlefield. She has also published on medical Gothic, contemporary art, transgressive art and a/r/tography. She is a founder member of the Gothic Association of New Zealand and Australia (2013) and the Irish Network for Gothic Scholars (2013). In 2010 she founded the collaborative fine art practice, Gothicise, who work on site-specific projects related to ideas of site, traces and narrative and who have received Grants Under The Arts funding for their projects *Waking St. Munchin* (2014) and *Death Café Limerick* (2015). Since 2015 she has worked with thanatologist Jennifer Moran Stritch as the research collective, Femmes Fatales, on several papers, symposia and a Creative Europe project. She also works as a creative writer with her short fiction published in thirteen anthologies, and her first short story collection *The Unheimlich Manoeuvre* was published in July 2016.

STEPHEN HOPKINS is Lecturer in Politics in the Department of Politics and International Relations at the University of Leicester. His book, *The Politics of Memoir and the Northern Ireland Conflict*, was published in

2013 by Liverpool University Press. He is also the author of several recent articles and book chapters, including 'The Chronicles of Long Kesh: Irish Republican Memoirs and the Contested Memory of the Hunger Strikes', in *Memory Studies* (2014); and 'Victims and Memoir-Writing in Northern Ireland: Leaving the Troubles Behind?' in Lesley Lelourec and Grainne O'Keeffe-Vigneron, eds, *Ireland and Victims: Confronting the Past, Forging the Future* (Peter Lang, 2012). He is co-editor (with Graham Dawson and Jo Dover) of *The Northern Ireland Troubles in Britain: Impacts, Engagements, Legacies and Memories* (Manchester University Press, forthcoming).

LINDSAY JANSSEN was awarded her PhD in 2016 from Radboud University, the Netherlands, where she has also worked as an instructor. She has published essays on Irish and Irish diaspora Famine fiction. She is co-editor of *Recollecting Hunger* (2012), *Holodomor and Gorta Mór* (2012) and *Global Legacies of the Great Irish Famine* (2014). She currently works as an educational consultant.

SARA DYBRIS MCQUAID is Associate Professor in British and Irish History, Society and Culture and Director of the Center for Irish Studies at Arhus University. She is a member of the steering committee and core researcher in the Centre for Resolution of International Conflicts (CRIC) at Copenhagen University, funded by the Danish Strategic Research Council. Her research focus is on the role of memory in protracted conflict; conflict resolution and peace-building (particularly in Northern Ireland); legacy issues in political and societal transitions; transnational memory and diasporas in Britain; and the role of public policy in governing memory discourses.

EAMON MAHER is Director of the National Centre for Franco-Irish Studies in IT Tallaght. He has published two monographs on John McGahern and is General Editor of the *Reimagining Ireland* book series with Peter Lang, Oxford.

EVE MORRISON studied history at Trinity College, Dublin, receiving her BA in 2003. She continued her studies in modern Irish history as an

Irish Research Council of the Humanities and Social Sciences (now IRC) postgraduate scholar at TCD, and was awarded her PhD in 2011. Her particular area of expertise is legacy interviews with veterans of the Irish independence struggle and civil war. The subject of her doctoral studies was the Bureau of Military History, and her postdoctoral research focused on the Ernie O'Malley notebook interviews. She was an IRC postdoctoral fellow at University College Dublin from 2013–2015, working on the O'Malley notebooks. She is also writing a book based on her doctoral work for Liverpool University Press.

NIAMH NICGHABHANN is Course Director for the MA Festive Arts Programme at the Irish World Academy of Music and Dance at the University of Limerick. Her research focuses on Irish studies, with an emphasis on the art and architecture of Ireland, concepts of public space, memory, performance and the body. She is particularly interested in exploring these concepts through visual images, built landscapes, poetry and prose fiction. Her current research also includes work on arts management, cultural entrepreneurship and innovation, and ideas of measurement and value in the arts and cultural sector. She is a member of the editorial board of *Artefact*, the journal of the Irish Association of Art Historians, and her monograph, *Medieval Ecclesiastical Buildings in Ireland, 1789–1915: Building on the Past*, was published by Four Courts Press in 2015.

EMILIE PINE lectures in Modern Drama and Irish Studies at University College Dublin. Her main research interests are in the interdisciplinary study of modern Irish culture, with a specific focus on memory and trauma studies, theatre and film. Her book *The Politics of Irish Memory: Performing Remembrance in Contemporary Irish Culture* was published by Palgrave Macmillan in 2011. She is the editor of the *Irish University Review* and is a Fulbright Ireland alumnus. She continues to work on Irish culture, memory and trauma and is founding Director of the Irish Memory Studies Research Network, funded by the Irish Research Council and supported by the UCD Humanities Institute. She is currently leading a major IRC project *Industrial Memories* (2015–2018) on the history of child abuse in Ireland.

# Index

# Reimagining Ireland

Series Editor: Dr Eamon Maher, Institute of Technology, Tallaght

The concepts of Ireland and 'Irishness' are in constant flux in the wake of an ever-increasing reappraisal of the notion of cultural and national specificity in a world assailed from all angles by the forces of globalisation and uniformity. Reimagining Ireland interrogates Ireland's past and present and suggests possibilities for the future by looking at Ireland's literature, culture and history and subjecting them to the most up-to-date critical appraisals associated with sociology, literary theory, historiography, political science and theology.

Some of the pertinent issues include, but are not confined to, Irish writing in English and Irish, Nationalism, Unionism, the Northern 'Troubles', the Peace Process, economic development in Ireland, the impact and decline of the Celtic Tiger, Irish spirituality, the rise and fall of organised religion, the visual arts, popular cultures, sport, Irish music and dance, emigration and the Irish diaspora, immigration and multiculturalism, marginalisation, globalisation, modernity/postmodernity and postcolonialism. The series publishes monographs, comparative studies, interdisciplinary projects, conference proceedings and edited books.

Proposals should be sent either to Dr Eamon Maher at eamon.maher@ittdublin.ie or to ireland@peterlang.com.

Vol. 1 Eugene O'Brien: 'Kicking Bishop Brennan up the Arse': Negotiating Texts and Contexts in Contemporary Irish Studies
ISBN 978-3-03911-539-6. 219 pages. 2009.

Vol. 2 James P. Byrne, Padraig Kirwan and Michael O'Sullivan (eds): Affecting Irishness: Negotiating Cultural Identity Within and Beyond the Nation
ISBN 978-3-03911-830-4. 334 pages. 2009.

Vol. 3 Irene Lucchitti: The Islandman: The Hidden Life of Tomás O'Crohan
ISBN 978-3-03911-837-3. 232 pages. 2009.

Vol. 4 Paddy Lyons and Alison O'Malley-Younger (eds): No Country for Old Men: Fresh Perspectives on Irish Literature
ISBN 978-3-03911-841-0. 289 pages. 2009.

Vol. 5 Eamon Maher (ed.): Cultural Perspectives on Globalisation and Ireland
ISBN 978-3-03911-851-9. 256 pages. 2009.

Vol. 6 Lynn Brunet: 'A Course of Severe and Arduous Trials': Bacon, Beckett and Spurious Freemasonry in Early Twentieth-Century Ireland
ISBN 978-3-03911-854-0. 218 pages. 2009.

Vol. 7 Claire Lynch: Irish Autobiography: Stories of Self in the Narrative of a Nation
ISBN 978-3-03911-856-4. 234 pages. 2009.

Vol. 8 Victoria O'Brien: A History of Irish Ballet from 1927 to 1963
ISBN 978-3-03911-873-1. 208 pages. 2011.

Vol. 9 Irene Gilsenan Nordin and Elin Holmsten (eds): Liminal Borderlands in Irish Literature and Culture
ISBN 978-3-03911-859-5. 208 pages. 2009.

Vol. 10 Claire Nally: Envisioning Ireland: W.B. Yeats's Occult Nationalism
ISBN 978-3-03911-882-3. 320 pages. 2010.

Vol. 11 Raita Merivirta: The Gun and Irish Politics: Examining National History in Neil Jordan's *Michael Collins*
ISBN 978-3-03911-888-5. 202 pages. 2009.

Vol. 12 John Strachan and Alison O'Malley-Younger (eds): Ireland: Revolution and Evolution
ISBN 978-3-03911-881-6. 248 pages. 2010.

Vol. 13 Barbara Hughes: Between Literature and History: The Diaries and Memoirs of Mary Leadbeater and Dorothea Herbert
ISBN 978-3-03911-889-2. 255 pages. 2010.

Vol. 14 Edwina Keown and Carol Taaffe (eds): Irish Modernism: Origins, Contexts, Publics
ISBN 978-3-03911-894-6. 256 pages. 2010.

Vol. 15 John Walsh: Contests and Contexts: The Irish Language and Ireland's Socio-Economic Development
ISBN 978-3-03911-914-1. 492 pages. 2011.

Vol. 16 Zélie Asava: The Black Irish Onscreen: Representing Black and Mixed-Race Identities on Irish Film and Television
ISBN 978-3-0343-0839-7. 213 pages. 2013.

Vol. 17 Susan Cahill and Eóin Flannery (eds): This Side of Brightness: Essays on the Fiction of Colum McCann
ISBN 978-3-03911-935-6. 189 pages. 2012.

Vol. 18 Brian Arkins: The Thought of W.B. Yeats
ISBN 978-3-03911-939-4. 204 pages. 2010.

Vol. 19 Maureen O'Connor: The Female and the Species: The Animal in Irish Women's Writing
ISBN 978-3-03911-959-2. 203 pages. 2010.

Vol. 20 Rhona Trench: Bloody Living: The Loss of Selfhood in the Plays of Marina Carr
ISBN 978-3-03911-964-6. 327 pages. 2010.

Vol. 21 Jeannine Woods: Visions of Empire and Other Imaginings: Cinema, Ireland and India, 1910–1962
ISBN 978-3-03911-974-5. 230 pages. 2011.

Vol. 22 Neil O'Boyle: New Vocabularies, Old Ideas: Culture, Irishness and the Advertising Industry
ISBN 978-3-03911-978-3. 233 pages. 2011.

Vol. 23 Dermot McCarthy: John McGahern and the Art of Memory
ISBN 978-3-0343-0100-8. 344 pages. 2010.

Vol. 24 Francesca Benatti, Sean Ryder and Justin Tonra (eds): Thomas Moore: Texts, Contexts, Hypertexts
ISBN 978-3-0343-0900-4. 220 pages. 2013.

Vol. 25 Sarah O'Connor: No Man's Land: Irish Women and the Cultural Present
ISBN 978-3-0343-0111-4. 230 pages. 2011.

Vol. 26 Caroline Magennis: Sons of Ulster: Masculinities in the Contemporary Northern Irish Novel
ISBN 978-3-0343-0110-7. 192 pages. 2010.

Vol. 27 Dawn Duncan: Irish Myth, Lore and Legend on Film
ISBN 978-3-0343-0140-4. 181 pages. 2013.

Vol. 28 Eamon Maher and Catherine Maignant (eds): Franco-Irish Connections in Space and Time: Peregrinations and Ruminations
ISBN 978-3-0343-0870-0. 295 pages. 2012.

Vol. 29 Holly Maples: Culture War: Conflict, Commemoration and the Contemporary Abbey Theatre
ISBN 978-3-0343-0137-4. 294 pages. 2011.

Vol. 30 Maureen O'Connor (ed.): Back to the Future of Irish Studies: Festschrift for Tadhg Foley
ISBN 978-3-0343-0141-1. 359 pages. 2010.

Vol. 31 Eva Urban: Community Politics and the Peace Process in Contemporary Northern Irish Drama
ISBN 978-3-0343-0143-5. 303 pages. 2011.

Vol. 32 Mairéad Conneely: Between Two Shores / *Idir Dhá Chladach*: Writing the Aran Islands, 1890–1980
ISBN 978-3-0343-0144-2. 299 pages. 2011.

Vol. 33 Gerald Morgan and Gavin Hughes (eds): Southern Ireland and the Liberation of France: New Perspectives
ISBN 978-3-0343-0190-9. 250 pages. 2011.

Vol. 34 Anne MacCarthy: Definitions of Irishness in the 'Library of Ireland' Literary Anthologies
ISBN 978-3-0343-0194-7. 271 pages. 2012.

Vol. 35 Irene Lucchitti: Peig Sayers: In Her Own Write
ISBN 978-3-0343-0253-1. Forthcoming.

Vol. 36 Eamon Maher and Eugene O'Brien (eds): Breaking the Mould: Literary Representations of Irish Catholicism
ISBN 978-3-0343-0232-6. 249 pages. 2011.

Vol. 37 Mícheál Ó hAodha and John O'Callaghan (eds): Narratives of the Occluded Irish Diaspora: Subversive Voices
ISBN 978-3-0343-0248-7. 227 pages. 2012.

Vol. 38 Willy Maley and Alison O'Malley-Younger (eds): Celtic Connections: Irish–Scottish Relations and the Politics of Culture
ISBN 978-3-0343-0214-2. 247 pages. 2013.

Vol. 39 Sabine Egger and John McDonagh (eds): Polish–Irish Encounters in the Old and New Europe
ISBN 978-3-0343-0253-1. 322 pages. 2011.

Vol. 40 Elke D'hoker, Raphaël Ingelbien and Hedwig Schwall (eds): Irish Women Writers: New Critical Perspectives
ISBN 978-3-0343-0249-4. 318 pages. 2011.

Vol. 41 Peter James Harris: From Stage to Page: Critical Reception of Irish Plays in the London Theatre, 1925–1996
ISBN 978-3-0343-0266-1. 311 pages. 2011.

Vol. 42 Hedda Friberg-Harnesk, Gerald Porter and Joakim Wrethed (eds): Beyond Ireland: Encounters Across Cultures
ISBN 978-3-0343-0270-8. 342 pages. 2011.

Vol. 43 Irene Gilsenan Nordin and Carmen Zamorano Llena (eds): Urban and Rural Landscapes in Modern Ireland: Language, Literature and Culture
ISBN 978-3-0343-0279-1. 238 pages. 2012.

Vol. 44 Kathleen Costello-Sullivan: Mother/Country: Politics of the Personal in the Fiction of Colm Tóibín
ISBN 978-3-0343-0753-6. 247 pages. 2012.

Vol. 45 Lesley Lelourec and Gráinne O'Keeffe-Vigneron (eds): Ireland and Victims: Confronting the Past, Forging the Future
ISBN 978-3-0343-0792-5. 331 pages. 2012.

Vol. 46 Gerald Dawe, Darryl Jones and Nora Pelizzari (eds): Beautiful Strangers: Ireland and the World of the 1950s
ISBN 978-3-0343-0801-4. 207 pages. 2013.

Vol. 47 Yvonne O'Keeffe and Claudia Reese (eds): New Voices, Inherited Lines: Literary and Cultural Representations of the Irish Family
ISBN 978-3-0343-0799-4. 238 pages. 2013.

Vol. 48 Justin Carville (ed.): Visualizing Dublin: Visual Culture, Modernity and the Representation of Urban Space
ISBN 978-3-0343-0802-1. 326 pages. 2014.

Vol. 49 Gerald Power and Ondřej Pilný (eds): Ireland and the Czech Lands: Contacts and Comparisons in History and Culture
ISBN 978-3-0343-1701-6. 243 pages. 2014.

Vol. 50 Eoghan Smith: John Banville: Art and Authenticity
ISBN 978-3-0343-0852-6. 199 pages. 2014.

Vol. 51 María Elena Jaime de Pablos and Mary Pierse (eds): George Moore and the Quirks of Human Nature
ISBN 978-3-0343-1752-8. 283 pages. 2014.

Vol. 52 Aidan O'Malley and Eve Patten (eds): Ireland, West to East: Irish Cultural Connections with Central and Eastern Europe
ISBN 978-3-0343-0913-4. 307 pages. 2014.

Vol. 53 Ruben Moi, Brynhildur Boyce and Charles I. Armstrong (eds): The Crossings of Art in Ireland
ISBN 978-3-0343-0983-7. 319 pages. 2014.

Vol. 54 Sylvie Mikowski (ed.): Ireland and Popular Culture
ISBN 978-3-0343-1717-7. 257 pages. 2014.

Vol. 55 Benjamin Keatinge and Mary Pierse (eds): France and Ireland in the Public Imagination
ISBN 978-3-0343-1747-4. 279 pages. 2014.

Vol. 56 Raymond Mullen, Adam Bargroff and Jennifer Mullen (eds): John McGahern: Critical Essays
ISBN 978-3-0343-1755-9. 253 pages. 2014.

Vol. 57 Máirtín Mac Con Iomaire and Eamon Maher (eds): 'Tickling the Palate': Gastronomy in Irish Literature and Culture
ISBN 978-3-0343-1769-6. 253 pages. 2014.

Vol. 58 Heidi Hansson and James H. Murphy (eds): Fictions of the Irish Land War
ISBN 978-3-0343-0999-8. 237 pages. 2014.

Vol. 59 Fiona McCann: A Poetics of Dissensus: Confronting Violence in Contemporary Prose Writing from the North of Ireland
ISBN 978-3-0343-0979-0. 238 pages. 2014.

Vol. 60 Marguérite Corporaal, Christopher Cusack, Lindsay Janssen and Ruud van den Beuken (eds): Global Legacies of the Great Irish Famine: Transnational and Interdisciplinary Perspectives
ISBN 978-3-0343-0903-5. 357 pages. 2014.

Vol. 61 Katarzyna Ojrzyńska: 'Dancing As If Language No Longer Existed': Dance in Contemporary Irish Drama
ISBN 978-3-0343-1813-6. 318 pages. 2015.

Vol. 62 Whitney Standlee: 'Power to Observe': Irish Women Novelists in Britain, 1890–1916
ISBN 978-3-0343-1837-2. 288 pages. 2015.

Vol. 63 Elke D'hoker and Stephanie Eggermont (eds): The Irish Short Story: Traditions and Trends
ISBN 978-3-0343-1753-5. 330 pages. 2015.

Vol. 64 Radvan Markus: Echoes of the Rebellion: The Year 1798 in Twentieth-Century Irish Fiction and Drama
ISBN 978-3-0343-1832-7. 248 pages. 2015.

Vol. 65 B. Mairéad Pratschke: Visions of Ireland: Gael Linn's *Amharc Éireann* Film Series, 1956–1964
ISBN 978-3-0343-1872-3. 301 pages. 2015.

Vol. 66 Una Hunt and Mary Pierse (eds): France and Ireland: Notes and Narratives
ISBN 978-3-0343-1914-0. 272 pages. 2015.

Vol. 67 John Lynch and Katherina Dodou (eds): The Leaving of Ireland: Migration and Belonging in Irish Literature and Film
ISBN 978-3-0343-1896-9. 313 pages. 2015.

Vol. 68 Anne Goarzin (ed.): New Critical Perspectives on Franco-Irish Relations
ISBN 978-3-0343-1781-8. 271 pages. 2015.

Vol. 69 Michel Brunet, Fabienne Gaspari and Mary Pierse (eds): George Moore's Paris and His Ongoing French Connections
ISBN 978-3-0343-1973-7. 279 pages. 2015.

Vol. 70 Carine Berbéri and Martine Pelletier (eds): Ireland: Authority and Crisis
ISBN 978-3-0343-1939-3. 296 pages. 2015.

Vol. 71 David Doolin: Transnational Revolutionaries: The Fenian Invasion of Canada, 1866
ISBN 978-3-0343-1922-5. 348 pages. 2016.

Vol. 72 Terry Phillips: Irish Literature and the First World War: Culture, Identity and Memory
ISBN 978-3-0343-1969-0. 297 pages. 2015.

Vol. 73 Carmen Zamorano Llena and Billy Gray (eds): Authority and Wisdom in the New Ireland: Studies in Literature and Culture
ISBN 978-3-0343-1833-4. 263 pages. 2016.

Vol. 74 Flore Coulouma (ed.): New Perspectives on Irish TV Series: Identity and Nostalgia on the Small Screen
ISBN 978-3-0343-1977-5. 222 pages. 2016.

Vol. 75 Fergal Lenehan: Stereotypes, Ideology and Foreign Correspondents: German Media Representations of Ireland, 1946–2010
ISBN 978-3-0343-2222-5. 306 pages. 2016.

Vol. 76 Jarlath Killeen and Valeria Cavalli (eds): 'Inspiring a Mysterious Terror': 200 Years of Joseph Sheridan Le Fanu
ISBN 978-3-0343-2223-2. 260 pages. 2016.

Vol. 77 Anne Karhio: 'Slight Return': Paul Muldoon's Poetics of Place
ISBN 978-3-0343-1986-7. 272 pages. 2017.

Vol. 78 Margaret Eaton: Frank Confessions: Performance in the Life-Writings of Frank McCourt
ISBN 978-1-906165-61-1. 294 pages. 2017.

Vol. 79 Marguérite Corporaal, Christopher Cusack and Ruud van den Beuken (eds): Irish Studies and the Dynamics of Memory: Transitions and Transformations
ISBN 978-3-0343-2236-2. 360 pages. 2017